CAMBRIDGE TRACTS IN MATHEMATICS

---

79. *Chain conditions in topology*

W. W. COMFORT
Professor of Mathematics, Wesleyan University

S. NEGREPONTIS
Professor of Mathematics, Athens University

Chain conditions in topology

CAMBRIDGE UNIVERSITY PRESS

CAMBRIDGE
LONDON NEW YORK NEW ROCHELLE
MELBOURNE SYDNEY

CAMBRIDGE UNIVERSITY PRESS
Cambridge, New York, Melbourne, Madrid, Cape Town, Singapore, São Paulo, Delhi

Cambridge University Press
The Edinburgh Building, Cambridge CB2 8RU, UK

Published in the United States of America by Cambridge University Press, New York

www.cambridge.org
Information on this title: www.cambridge.org/9780521234870

First published 1982
This digitally printed version (with corrections) 2008

A catalogue record for this publication is available from the British Library

Library of Congress Catalogue Card Number 81-6092

ISBN 978-0-521-23487-0 hardback
ISBN 978-0-521-09062-9 paperback

Contents

Introduction

Our monograph passed through several stages before acquiring its present form. It began, several years ago, as material gathered for our earlier work on ultrafilters but eventually discarded as too peripheral to the principal subject. Lying inert for some time, and slowly gaining some unity, it appeared later in our minds as a systematization of existing applications of the Erdős–Rado principle on quasi-disjoint sets to topological situations (mainly, in product spaces). Finally, though, through the contributions of researchers such as Gaifman, Laver, Galvin, Hajnal, Kunen, Argyros, Tsarpalias, and Shelah, it became something more fascinating and delightful: a study of the fine structure of the (countable) chain condition; and, a study of topological spaces (and also of partially ordered sets, and of Boolean algebras, and even of Banach spaces) as a function of their Souslin number.

The tools for the most part are the classical, yet constantly developing and inexhaustibly fertile, principles of infinitary combinatorics (given in Chapter 1). Early in the development of topology, especially in the Moscow School of Alexandroff and Urysohn, in the work of Lusin and Souslin, in the Polish School, and in the work of Hausdorff, informal set-theoretic and infinitary combinatorial considerations were prominent. The subsequent systematic development of infinitary combinatorics by the Hungarian School, led by Erdős, based on Dedekind's box (pigeonhole) principle and inspired by Ramsey's theorem, provided concrete techniques through which topological questions could be examined. Combinatorial tools returned to a central position in the work of Shanin, who studied fundamental questions on the intersection properties of families of open sets (the chain conditions, defined in detail in Chapter 2) in product spaces, using quasi-disjoint sets. More recently in the same spirit Arhangel'skiĭ, Hajnal, Juhász, Šapirovskiĭ and others have produced significant results on cardinal invariants associated with topological spaces.

A simple but quite useful extension to singular cardinals of the Erdős–Rado theorem on quasi-disjoint sets, noted independently by Shelah and Argyros, allows for positive statements concerning the conservation of chain conditions in cartesian products or powers (in fact, in various stronger box topologies) in Shanin's spirit. However, methods involving quasi-disjoint sets, for both regular and singular cardinals, have their limitations; their usefulness lies with spaces that are products, or have a product-like structure. This is the case with the results of Chapter 3, where we study some classes of chain conditions (calibres, compact-calibres, and pseudo-compactness numbers); with Shelah's result in Chapter 4, which systematically exploits calibres of Σ-products of dyadic powers to define (non-compact) spaces whose calibre gaps are created more or less at will; and with the results in Chapter 10, where the pseudo-compactness properties given in Chapter 3 are applied to determine the dependence of continuous functions defined on 'large' subsets of products (with the cartesian or various stronger box topologies) on a 'small' set of coordinates. Furthermore some of the results in Chapters 6 and 7, concerning which we say more below, where a limited use of quasi-disjoint sets can be observed, also concern the dyadic powers $\{0, 1\}^I$ with topologies quite different from the cartesian product topology.

Results to the effect that a class of cardinals satisfying certain obvious restrictions is realized as the set of non-compact-calibres of a space, analogous to the results of Shelah for calibres, are obtained in Chapter 8; here we use spaces of non-uniform ultrafilters rather than Σ-products of dyadic powers. The corresponding statements for pseudo-compactness numbers are not yet available and indeed it is not clear in this case whether there are conditions analogous to the 'obvious restrictions' dealing with calibre and compact-calibre. The difficulty, as described in Chapter 9 using (permutation) types of ultrafilters, derives from the fact that properties of pseudo-compactness type are not finitely productive.

In Chapter 5 we study (arbitrary) topological spaces as a function of their Souslin number, enlarging greatly Shanin's original program.

Combinatorial concepts more powerful than quasi-disjoint sets are needed to deal with general spaces, where there is no explicit or implicit product structure. Such principles have been formed by Argyros and Tsarpalias, and used to determine a large class of regular and singular calibres of compact spaces. In fact, as is mentioned below, assuming the generalized continuum hypothesis, most calibres of compact spaces are determined by these methods (see the second chart in section 7.18). It is worth noting that the proof of the regular cardinal case uses a combinatorial kernel sufficiently strong that it yields in Chapter 1 several related and classical results (quasi-disjoint families and the early 'arrow relations' of Erdős and Rado); we indicate in Chapter 1 that these results can be proved also by use of a simple form of the pressing-down lemma.

These techniques were in fact inspired in part by questions in functional analysis not directly concerned with chain conditions and described only informally in this monograph. These questions concern the existence of large independent families, and the resulting isomorphic embedding of the Banach space l^1_α into α-dimensional subspaces of the space $C(X)$ with X a compact, Hausdorff space whose Souslin number is 'small' relative to α.

We noted above that Shelah's result in Chapter 4 allows the creation of completely regular, Hausdorff spaces whose classes of calibres are assigned in advance. In contrast, the class of non-calibres of a compact space is more restricted. In fact, assuming the generalized continuum hypothesis, a regular cardinal α either is not a calibre of a compact space for the trivial reason that α is smaller than the Souslin number of the space, or α is indeed a calibre of the space (with some possible 'boundary' exceptions of the form $\alpha = \beta^+$ with the cofinality of β smaller than the Souslin number of the space, in which case the space still has calibre (α, β)).

The deeper results describing the fine structure of the countable chain condition, especially the examples of Chapters 6 and 7, rely on infinitary combinatorial techniques that in addition contain dialectical (mostly diagonal) arguments. The success of a large part of this undertaking depends on the continuum hypothesis: although some remarkable fragments hold without any special hypotheses (essentially, the statements concerning the existence of spaces with no strictly positive measure), other parts definitely collapse.

These examples lie, in the countable case, between the very strong property of separability and the very general countable chain condition (here abbreviated c.c.c.). This spectrum of chain conditions – including calibre ω^+, the existence of a strictly positive measure and the related properties (*) and (**), Knaster's property (K) and the related properties K_n for natural numbers $n \geq 2$, and the productive countable chain condition – grew up around the problem posed in 1920 by Souslin. The examples given in Chapter 7 by Laver, Hajnal, Galvin and Kunen serve to differentiate some of these properties. Thus, assuming the continuum hypothesis, there is a c.c.c. space whose square is not a c.c.c. space, and there is a productively c.c.c. space that does not have Knaster's property. The celebrated example in Chapter 6 of Gaifman, a c.c.c. space with no strictly positive measure (and, assuming CH, with no calibre), was for many years the only known example of its kind. Then came the Galvin–Hajnal example of a compact space, obtained without any special set-theoretic assumptions (and solving as well a problem of Horn and Tarski), with no strictly positive measure and for which every regular uncountable cardinal is a calibre. Recently for every natural number $n \geq 2$ Argyros found a space with no strictly positive measure, with property K_n and, assuming CH, without property K_{n+1}; motivated by a problem in model theory, Rubin and Shelah found, assuming CH, another example of a space with K_n and without K_{n+1}. Argyros found also, given an infinite cardinal α, a c.c.c. space X such that for every set $\{\mu_i : i < \alpha\}$ of regular, Borel measures on X there is a non-empty, open subset U of X with $\mu_i(U) = 0$ for all $i < \alpha$; subsequently Galvin observed that an appropriate modification of the Galvin–Hajnal example mentioned above produces the same phenomenon. In the opposite direction it is shown, usually assuming appropriate segments of the generalized continuum hypothesis, that the Stone spaces of the homogeneous algebras are – indeed, they are the only examples known so far – compact spaces with strictly positive measures but without various (arbitrarily large) calibres. This class of examples, due to Erdős for the separable homogeneous measure algebra, together with a further example of Argyros (given in Chapter 5) related to his examples on strictly positive measure, serves as well to delineate the class of allowed calibres of compact spaces.

In the Notes for Chapter 7, we describe the study of the chain conditions on certain classes of spaces (for example, the Eberlein-compact and the Corson-compact spaces) arising in the theory of Banach spaces. As M. Wage has remarked, 'the fields of analysis, general topology and set theory have another happy reunion in the study of weakly compact subsets of Banch spaces'.

After all is said, not everything is settled and complete. We are at a point where fascinating problems still abound while beyond, far-reaching connections can only be imagined. Our work we hope will find its rest in changing: *μεταβάλλον ἀναπαύεται.*

W. Wistar Comfort and Stylianos A. Negrepontis
Middletown, Connecticut, USA and Athens, Greece
August 20, 1981

E R R A T A

p. 4, l. 1: $|\mathcal{A}| \to |\mathcal{A}_\eta|$
p. 4, l. -11 to -6: $\overline{\xi} \to \overline{\zeta}$ (9 times)
p. 10, l. -3: $A^i_\xi(\xi') \to A^i_\xi(\xi)$
p. 26, l. 13: $N_{\eta(\sigma)} \to V_{\eta(\sigma)}$
p. 31, l. 18: $= \emptyset \to \neq \emptyset$
p. 34, l. -3: $U_{-k} \to U_k$
p. 38, l. 1: B.14 → B.21
p. 40, ll. 6, 9: 2.18(b) → 2.18(c)
p. 44, l.14: $\pi_j \to \pi_J$
p. 47, l. -6: $\bigcap_{\xi \in S} \to \bigcap_{\xi \in C}$
p. 47, l. -5: $x_J \to X_I$
p. 55, l. -10: 1.11 → 1.10
p. 58, l. 6: 1.11 → 1.10
p. 59, l. 9: $X_{-J_\sigma} \to X_{\widetilde{J}_\sigma}$
p. 65, l. 13: $X \to x$
p. 65, l. -13: $I \to A$
p. 70, l. 19: $^\kappa \to \overset{\kappa}{\smile}$ (2 times)
p. 70, l. 22: $^\kappa \to \overset{\kappa}{\smile}$ (7 times)
p. 76, l. -1: $^\kappa \to \overset{\kappa}{\smile}$
p. 148, l. 5: $T_{n,k}{}' \to T_{n,k'}$.
p. 190, l. -13: $\alpha^\eta \to \alpha$
p. 192, l. 2: $\eta \to \omega$
p. 194, l. 3: $F^i = \emptyset \to F^{\bar{i}} = \emptyset$
p. 194, l. 11: $\{\eta\} \to \{\eta_1\}$
p. 195, l. -2: $) \to \}$
p. 198, l. 3: production → product
p. 198: $(**) \Rightarrow K_n$ is 6.17
p. 204, l. -3: $N_{n \in N(x)} \to \bigcap_{n \in N(x)}$
p. 207, l. 4: c.c.c. → c.c.c.)
p. 220, l. 17: B.7 → B.8
p. 222, l. -1: B.11 → B.12
p. 228, l. 2: $U(y) \to V(y)$
p. 234, l. 11: 5.1 → 10.1
p. 234, l. 16: $\in I \to i \in I$
p. 234, l. -12: $\mathcal{P}^* \to \mathcal{P}^*_\alpha$
p. 235, l. -10: 3.5 → 3.6
p. 235, l. -9: 3.13 → 3.12
p. 236, l. 15: B.2 → B.5
p. 237, ll. -6, -2: B.4 → B.5
p. 238, l. -8: $\acute{\cup} \to \cup$
p. 239, l.-10: the → of
p. 249, l. 14: B.4 → B.5(b)
p. 249, l. -10: (a)↔(b)
p. 253, l. 14. the → then
p. 254, l. -10: $\beta_\lambda \to \beta_\xi$
p. 274, l. 14: $\mathcal{S} \to S$ (3 times)
p. 285, l. -15: $\beta X\ X \to \beta X \backslash X$
p. 288, l. -14: **52** → **102**
p. 295, l. 4: branch, 255 → branch, 256
p. 296, l. -21: Gleason space, 272 → Gleason space, 271
p. 297, l. 1: locally finite → finite cellular
p. 298, l. -14: Pèłczynski → Pełczynski
p. 298, l. -13: Phèlps → Phelps

[Note. The proof of Lemma 10.1 is flawed. A valid proof and consequences are given in Comfort, Recoder-Núñez and Gotchev, Topology and Its Applications **155** (2008).]

Acknowledgements

We acknowledge with thanks research facilities and financial support received from the following institutions: Athens University, National Research Foundation (Greece), National Science Foundation (U.S.A.), Wesleyan University.

1

Some Infinitary Combinatorics

This introductory chapter describes the basic combinatorial tools used in the proofs of most of the results contained in the present monograph.

Our treatment of this classical material is based on two (alternative) principles for regular cardinals: Argyros' ramification lemma (1.1) and the pressing-down lemma (1.3). The main combinatorial results established for regular cardinals – the Erdős–Rado theorem for quasi-disjoint families (1.4) and the Erdős–Rado arrow relations (1.5, 1.7) – are consequences of each of these principles. We need also a simple extension to singular cardinals of the Erdős–Rado theorem for quasi-disjoint families.

We note that sometimes, especially in Chapter 5, we obtain information on calibres of spaces directly from Argyros' ramification lemma (and a technique for singular cardinals due to Tsarpalias), rather than from some derived combinatorial result.

1.1 *Lemma.* (Argyros' ramification lemma.) Let $\omega \leq \beta \leq \alpha$ with β and α regular, let κ be a cardinal such that $0 < \kappa \ll \beta$, and for every $A \subset \alpha$ with $|A| = \alpha$ let $\mathscr{P}_A$ be a partition of A such that $|\mathscr{P}_A| < \beta$. Then there is a family $\{A_\eta : \eta < \kappa\}$ of subsets of α such that

$$|A_\eta| = \alpha \text{ for } \eta < \kappa,$$
$$A_{\eta+1} \in \mathscr{P}_{A_\eta} \text{ for } \eta < \kappa,$$
$$A_{\eta'} \subset A_\eta \text{ for } \eta < \eta' < \kappa, \text{ and}$$
$$\cap_{\eta<\kappa} A_\eta \neq \varnothing.$$

Proof. We define families $\{\mathscr{A}_\eta : \eta < \kappa\}$ such that

(i) $\mathscr{A}_0 = \{\alpha\}$;

(ii) $0 < |\mathscr{A}_\eta| < \beta$ for $\eta < \kappa$;

(iii) if $A, B \in \mathscr{A}_\eta$ and $A \neq B$, then $A \cap B = \varnothing$ for $\eta < \kappa$;

(iv) if $A \in \mathscr{A}_\eta$ then $|A| = \alpha$ for $\eta < \kappa$;

(v) $\mathscr{A}_{\eta+1} \subset \cup\{\mathscr{P}_A : A \in \mathscr{A}_\eta\}$ for $\eta < \kappa$; and
(vi) $|\alpha \backslash \cup \mathscr{A}_\eta| < \alpha$ for $\eta < \kappa$.

We proceed by recursion.

We define $\mathscr{A}_0$ by (i).

Next for $\eta < \kappa$ we define $\mathscr{A}_{\eta+1}$. We set

$$\mathscr{A}'_{\eta+1} = \cup\{\mathscr{P}_A : A \in \mathscr{A}_\eta\}, \text{ and}$$
$$\mathscr{A}_{\eta+1} = \{A \in \mathscr{A}'_{\eta+1} : |A| = \alpha\}.$$

We verify conditions (ii), (iii), (iv), (v), and (vi) for $\mathscr{A}_{\eta+1}$.

(ii) Let $A \in \mathscr{A}_\eta$. Since $|A| = \alpha$ and $|\mathscr{P}_A| < \beta \leq \alpha$ and $A = \cup \mathscr{P}_A$ and α is regular, there is $B \in \mathscr{P}_A$ such that $|B| = \alpha$; we have $B \in \mathscr{A}_{\eta+1}$ and hence $\mathscr{A}_{\eta+1} \neq \varnothing$. Further, since $|\mathscr{A}_\eta| < \beta$ and $|\mathscr{P}_A| < \beta$ for $A \in \mathscr{A}_\eta$ and β is regular, we have $|\mathscr{A}'_{\eta+1}| < \beta$ and hence $|\mathscr{A}_{\eta+1}| < \beta$.

(iii), (iv) and (v) for $\mathscr{A}_{\eta+1}$ are clear from the definitions.

(vi) For $A \in \mathscr{A}_\eta$ we set $S_A = \cup\{B \in \mathscr{P}_A : |B| < \alpha\}$, and we set $S = \cup\{S_A : A \in \mathscr{A}_\eta\}$. Since α is regular and $|\mathscr{P}_A| < \alpha$ we have $|S_A| < \alpha$ for $A \in \mathscr{A}_\eta$; hence $|S| < \alpha$. Since

$$\alpha \backslash \cup \mathscr{A}_{\eta+1} = (\alpha \backslash \cup \mathscr{A}_\eta) \cup S,$$

we have $|\alpha \backslash \cup \mathscr{A}_{\eta+1}| < \alpha$, as required.

Now we assume that η is a limit ordinal such that $0 < \eta < \kappa$, and that $\mathscr{A}_\xi$ has been defined for $\xi < \eta$, and we define $\mathscr{A}_\eta$. We set

$$\mathscr{A}'_\eta = \{\bigcap_{\xi<\eta} A_\xi : A_\xi \in \mathscr{A}_\xi \text{ and } \bigcap_{\xi<\eta} A_\xi \neq \varnothing\}, \text{ and}$$
$$\mathscr{A}_\eta = \{A \in \mathscr{A}'_\eta : |A| = \alpha\}.$$

We verify conditions (ii), (iii), (iv), (v) and (vi) for $\mathscr{A}_\eta$.

(ii) We define $\varphi : \mathscr{A}'_\eta \to \prod_{\xi<\eta} \mathscr{A}_\xi$ by the rule

$$\varphi(\bigcap_{\xi<\eta} A_\xi) = \langle A_\xi : \xi < \eta \rangle.$$

It is clear that φ is a one-to-one function. Since $|\mathscr{A}_\xi| < \beta$ for $\xi < \eta$ and $|\eta| < \kappa$ and $\kappa \ll \beta$, we have $|\prod_{\xi<\eta} \mathscr{A}_\xi| < \beta$ by A.5(a); it follows that

$$|\mathscr{A}_\eta| \leq |\mathscr{A}'_\eta| < \beta.$$

We set $S = \cup_{\xi<\eta}(\alpha \backslash \cup \mathscr{A}_\xi)$. Since $|\alpha \backslash \cup \mathscr{A}_\xi| < \alpha$ for $\xi < \eta$ and $|\eta| < \kappa < \alpha$ and α is regular, we have $|S| < \alpha$. Since $\alpha \backslash S = \cup \mathscr{A}'_\eta$ and $|\mathscr{A}'_\eta| < \alpha$, there is $A \in \mathscr{A}'_\eta$ such that $|A| = \alpha$; we have $A \in \mathscr{A}_\eta$ and hence $\mathscr{A}_\eta \neq \varnothing$.

(iii) and (iv) are clear for $\mathscr{A}_\eta$.

(vi) We have

$$\alpha\backslash\cup\mathscr{A}_\eta=\cup\{A\in\mathscr{A}'_\eta:|A|<\alpha\}\cup S.$$

Since $|\mathscr{A}'_\eta|<\beta\leq\alpha$ and $|S|<\alpha$, we have $|\alpha\backslash\cup\mathscr{A}_\eta|<\alpha$, as required.

The definition of the family $\{\mathscr{A}_\eta:\eta<\kappa\}$ is complete.

Since $|\alpha\backslash\cup\mathscr{A}_\eta|<\alpha$ for $\eta<\kappa$ and $\kappa<\alpha$ and α is regular, we have $|\cup_{\eta<\kappa}(\alpha\backslash\cup\mathscr{A}_\eta)|<\alpha$, and hence there is

$$\zeta\in\alpha\backslash\bigcup_{\eta<\kappa}(\alpha\backslash\cup\mathscr{A}_\eta).$$

For $\eta<\kappa$ there is $A_\eta\in\mathscr{A}_\eta$ such that $\zeta\in A_\eta$. It is clear that the family $\{A_\eta:\eta<\kappa\}$ satisfies the required conditions.

We apply Lemma 1.1 most frequently with $\beta=\alpha$. In Chapter 5 we will need the case

$$\beta=(\kappa^\kappa)^+\leq\alpha \text{ with } \kappa \text{ and } \alpha \text{ regular.}$$

The definition of a weakly compact cardinal is given in Appendix A.

1.2 *Lemma.* Let α be a weakly compact cardinal, and for every $A\subset\alpha$ with $|A|=\alpha$ let $\mathscr{P}_A$ be a partition of A such that $|\mathscr{P}_A|<\alpha$. Then there is a family $\{A_\eta:\eta<\alpha\}$ of subsets of α such that

$$|A_\eta|=\alpha \text{ for } \eta<\alpha,$$
$$A_{\eta+1}\in\mathscr{P}_{A_\eta} \text{ for } \eta<\alpha, \text{ and}$$
$$A_{\eta'}\subset A_\eta \text{ for } \eta<\eta'<\alpha.$$

Proof. We define families $\{\mathscr{A}_\eta:\eta<\alpha\}$ such that

(i) $\mathscr{A}_0=\{\alpha\}$;

(ii) $0<|\mathscr{A}_\eta|<\alpha$ for $\eta<\alpha$;

(iii) if $A,B\in\mathscr{A}_\eta$ and $A\neq B$, then $A\cap B=\varnothing$ for $\eta<\alpha$;

(iv) if $A\in\mathscr{A}_\eta$ then $|A|=\alpha$ for $\eta<\alpha$; and

(v) $\mathscr{A}_{\eta+1}\subset\cup\{\mathscr{P}_A:A\in\mathscr{A}_\eta\}$ for $\eta<\alpha$.

(The argument is essentially that of Lemma 1.1. To verify (ii) for limit ordinals η such that $0<\eta<\alpha$ we set $\gamma=\sup\{|\mathscr{A}_\xi|:\xi<\eta\}$ and we note that since $\gamma<\alpha$ we have

$$|\mathscr{A}_\eta|\leq|\mathscr{A}'_\eta|\leq\prod_{\xi<\eta}|\mathscr{A}_\xi|\leq\gamma^{|\eta|}\leq 2^\gamma\cdot 2^{|\eta|}<\alpha.)$$

Now we set $\mathscr{A}=\cup_{\eta<\alpha}\mathscr{A}_\eta$ and we define a partial order $\preccurlyeq$ on $\mathscr{A}$

by $A \preccurlyeq B$ if $A \supset B$. Then $\langle \mathscr{A}, \preccurlyeq \rangle$ is a tree of height α with $|\mathscr{A}| < \alpha$ for $\eta < \alpha$; hence there is a branch

$$\Sigma = \{A_\eta : \eta < \alpha\}$$

of $\mathscr{A}$ with $A_\eta \in \mathscr{A}_\eta$ for $\eta < \alpha$. It is clear that the family $\{A_\eta : \eta < \alpha\}$ is as required.

1.3 *Lemma.* (The pressing-down lemma.) Let $\omega \leq \kappa < \alpha$ with α and κ regular, let

$$S = \{\xi < \alpha : \mathrm{cf}(\xi) \geq \kappa\},$$

and let f be a function from S to α such that $f(\xi) < \xi$ for $\xi \in S$. Then there are $T \subset S$ with $|T| = \alpha$ and $\bar{\zeta} < \alpha$ such that $f(\xi) < \bar{\zeta}$ for all $\xi \in T$.

Proof. We suppose the lemma fails. Then for $\zeta < \alpha$ we have $|f^{-1}(\zeta)| < \alpha$ and hence there is $g(\zeta)$ such that

$$\zeta < g(\zeta) < \alpha, \text{ and}$$

$$f(\xi) \geq \zeta \quad \text{for } \xi \in S, \xi \geq g(\zeta).$$

We define $\{\zeta(\eta) : \eta < \kappa\}$ by the rule

$\zeta(0) = 0,$

$\zeta(\eta) = \sup\{\zeta(\eta') : \eta' < \eta\}$ for non-zero limit ordinals $\eta < \kappa$, and

$\zeta(\eta + 1) = g(\zeta(\eta)) \quad$ for $\eta < \kappa$;

and we set $\bar{\xi} = \sup_{\eta < \kappa} \zeta(\eta)$. The function $\eta \to \zeta(\eta)$ is an ordered-set isomorphism of κ into $\bar{\xi}$, and since κ is a regular cardinal we have $\mathrm{cf}(\bar{\xi}) = \kappa$ and hence $\bar{\xi} \in S$. For $\eta < \kappa$ we have

$$g(\zeta(\eta)) = \zeta(\eta + 1) < \bar{\xi},$$

and hence $f(\bar{\xi}) \geq \zeta(\eta)$. From $\bar{\xi} = \sup_{\eta < \kappa} \zeta(\eta)$ it then follows that $f(\bar{\xi}) \geq \bar{\xi}$, a contradiction.

The proof is complete.

We remark, retaining the notation of Lemma 1.3, that since $T = \cup_{\zeta < \bar{\zeta}} (f^{-1}(\{\zeta\}) \cap T)$ and α is regular, there are $T' \subset T$ with $|T'| = \alpha$ and $\zeta < \bar{\zeta}$ such that $f(\xi) = \zeta$ for all $\xi \in T'$.

Definition. An indexed family $\{S_i : i \in I\}$ of sets is a *quasi-disjoint* family if

$$\bigcap_{i\in I} S_i = S_j \cap S_{j'} \text{ whenever } j, j' \in I, j \neq j'.$$

It is clear that a family $\{S_i : i \in I\}$ is quasi-disjoint if and only if there is a set S such that

$$S = S_j \cap S_{j'} \text{ whenever } j, j' \in I, j \neq j'.$$

1.4 *Theorem.* Let $\omega \leq \kappa \ll \alpha$ with α regular and let $\{S_\xi : \xi < \alpha\}$ be a family of sets such that $|S_\xi| < \kappa$ for $\xi < \alpha$. Then there are $A \subset \alpha$ with $|A| = \alpha$ and a set J such that

$$S_\xi \cap S_{\xi'} = J \text{ for } \xi, \xi' \in A, \xi \neq \xi'.$$

Proof. (In the terminology of the definition above we are to show that there is $A \in [\alpha]^\alpha$ such that $\{S_\xi : \xi \in A\}$ is a quasi-disjoint family. We give two proofs.)

First Proof (using Lemma 1.1). We suppose that if $A \subset \alpha$ and $\{S_\xi : \xi \in A\}$ is quasi-disjoint then $|A| < \alpha$; and for every $A \subset \alpha$ with $|A| = \alpha$ we set

$$J_A = \bigcap_{\xi \in A} S_\xi, \text{ and}$$

$$\mathscr{B}_A = \{B \subset A : \text{if } \xi, \xi' \in B, \xi \neq \xi', \text{ then } S_\xi \cap S_{\xi'} = J_A\}.$$

Since the set $\mathscr{B}_A$ partially ordered by inclusion is inductive, there is a maximal element $B_A \in \mathscr{B}_A$. We have $|B_A| < \alpha$, and since $\{\xi\} \in \mathscr{B}_A$ for all $\xi \in A$ we have $B_A \neq \emptyset$.

For $\xi \in A \backslash B_A$ it follows from the maximality of B_A that there is $\zeta(\xi) \in B_A$ such that

$$S_\xi \cap S_{\zeta(\xi)} \supsetneqq J_A.$$

We define

$$\varphi_A : A \backslash B_A \to \cup \{\mathscr{P}(S_\zeta) : \zeta \in B_A\}$$

by the rule

$$\varphi_A(\xi) = S_\xi \cap S_{\zeta(\xi)},$$

and we set

$$\mathscr{P}_A = \{B_A\} \cup \{\varphi_A^{-1}(\{S\}) : S \in \cup \{\mathscr{P}(S_\zeta) : \zeta \in B_A\}\}.$$

Then $\mathscr{P}_A$ is a partition of A, and since $|S_\zeta| < \kappa$ for $\zeta \in B_A$ and $\kappa \ll \alpha$, we have $|\mathscr{P}(S_\zeta)| < \alpha$ and hence (since $|B_A| < \alpha$ and α is regular) we have $|\mathscr{P}_A| < \alpha$. It follows from (the case $\beta = \alpha$ of) Lemma 1.1 that there is a family $\{A_\eta : \eta < \kappa\}$ of subsets of α such that

$$\begin{aligned} &|A_\eta| = \alpha \quad \text{for } \eta < \kappa, \\ &A_{\eta+1} \in \mathscr{P}_{A_\eta} \quad \text{for } \eta < \kappa, \\ &A_{\eta'} \subset A_\eta \quad \text{for } \eta < \eta' < \kappa, \text{ and} \\ &\bigcap_{\eta < \kappa} A_\eta \neq \varnothing. \end{aligned}$$

For $\eta < \kappa$ there is $S(\eta) \in \cup\{\mathscr{P}(S_\zeta) : \zeta \in B_{A_\eta}\}$ such that

$$\begin{aligned} &A_{\eta+1} = \varphi_{A_\eta}^{-1}(\{S(\eta)\}) \in \mathscr{P}_{A_\eta} \quad \text{and} \\ &\varphi_{A_\eta}(\xi) = S_\xi \cap S_{\zeta(\xi)} = S(\eta) \supsetneqq J_{A_\eta} \quad \text{for } \xi \in A_{\eta+1}. \end{aligned}$$

Since $S(\eta) = \varphi_{A_\eta}(\xi) \subsetneqq S_\xi$ for $\xi \in A_{\eta+1}$, we have

$$J_{A_\eta} \subsetneqq S(\eta) \subset \bigcap_{\xi \in A_{\eta+1}} S_\xi = J_{A_{\eta+1}}$$

and hence $J_{A_{\eta+1}} \backslash J_{A_\eta} \neq \varnothing$ for $\eta < \kappa$; it follows that

$$\left| \bigcup_{\eta < \kappa} J_{A_\eta} \right| \geq \sum_{\eta < \kappa} |J_{A_{\eta+1}} \backslash J_{A_\eta}| \geq \kappa.$$

Now let $\xi \in \bigcap_{\eta < \kappa} A_\eta$. Then $\xi \in A_{\eta+1}$ for $\eta < \kappa$, and from $S_\xi \supset J_{A_\eta}$ we have

$$S_\xi \supset \bigcup_{\eta < \kappa} J_{A_\eta}$$

and hence $|S_\xi| \geq \kappa$, a contradiction.

Second Proof (using Lemma 1.3). We assume without loss of generality that $S_\xi \subset \alpha$ for $\xi < \alpha$. We set

$$S = \{\xi < \alpha : \text{cf}(\xi) \geq \kappa\}$$

and we define $f : S \to \alpha$ by the rule

$$f(\xi) = \sup(S_\xi \cap \xi) \quad \text{for } \xi \in S.$$

For $\xi \in S$ we have $|S_\xi| < \kappa \leq \text{cf}(\xi)$ and hence $f(\xi) < \xi$. It follows from Lemma 1.3 that there are $T \subset S$ with $|T| = \alpha$ and $\bar{\zeta} < \alpha$ such that $f[T] \subset \bar{\zeta}$.

For $\xi \in T$ we have $S_\xi \cap \bar{\zeta} \in \mathscr{P}_\kappa(\bar{\zeta})$, and since $\kappa \ll \alpha$ and α is regular we have $|\mathscr{P}_\kappa(\bar{\zeta})| < \alpha$; it follows that there are $T' \subset T$ with $|T'| = \alpha$ and $J \in \mathscr{P}_\kappa(\bar{\zeta})$ such that $S_\xi \cap \bar{\zeta} = J$ for $\xi \in T'$.

We define a function $\varphi : \alpha \to T'$ as follows. We set $\varphi(0) = \min T'$, and if $\xi < \alpha$ and $\varphi(\xi')$ has been defined for all $\xi' < \xi$ we choose $\varphi(\xi) \in T'$ such that

$$\sup\{\varphi(\xi') : \xi' < \xi\} < \varphi(\xi), \text{ and}$$
$$\sup\Big(\bigcup_{\xi' < \xi} S_{\varphi(\xi')}\Big) < \varphi(\xi);$$

such a choice is possible because

$$\Big|\bigcup_{\xi' < \xi} S_{\varphi(\xi')}\Big| \le \sum_{\xi' < \xi} |S_{\varphi(\xi')}| \le |\xi| \cdot \kappa < \alpha$$

and T' is cofinal in α.

We set $A = \{\varphi(\xi) : \xi < \alpha\}$ and we claim that if $\xi' < \xi < \alpha$ then

$$S_{\varphi(\xi')} \cap S_{\varphi(\xi)} = J.$$

Indeed since $\varphi(\xi'), \varphi(\xi) \in T'$ we have

$$S_{\varphi(\xi')} \cap S_{\varphi(\xi)} \cap \bar{\zeta} = J \cap J = J;$$

and if $\eta \ge \bar{\zeta}$ and $\eta \in S_{\varphi(\xi)}$, then since

$$\sup(S_{\varphi(\xi)} \cap \varphi(\xi)) = f(\varphi(\xi)) < \bar{\zeta},$$

we have $\eta \ge \varphi(\xi) > \sup S_{\varphi(\xi')}$ and hence $\eta \notin S_{\varphi(\xi')}$.

The extension to singular cardinals of the Erdős–Rado theorem for quasi-disjoint families is deferred to Theorem 1.9 below.

Though it is not needed later in this work, we note a strong converse to Theorem 1.4.

Theorem. Let κ and α be cardinals with $\kappa < \alpha$ and $\alpha \ge \omega$. If for every family $\{S_\xi : \xi < \alpha\}$ of sets with $|S_\xi| < \kappa$ for $\xi < \alpha$ there is $A \subset \alpha$ such that $|A| = \alpha$ and $\{S_\xi : \xi \in A\}$ is quasi-disjoint, then $\kappa \ll \alpha$.

Proof. Suppose there are $\beta < \alpha$, $\lambda < \kappa$ such that $\beta^\lambda \ge \alpha$, and let $\{f_\xi : \xi < \alpha\}$ be a subset of β^λ with $f_\xi \ne f_{\xi'}$ for $\xi' < \xi < \alpha$. Then f_ξ is a function from λ to β, and we set

$$S_\xi = \operatorname{graph} f_\xi = \{\langle \eta, f_\xi(\eta)\rangle : \eta < \lambda\} \quad \text{for } \xi < \alpha.$$

Since $|S_\xi| = \lambda < \kappa$, there is $A \subset \alpha$ such that $|A| = \alpha$ and $\{S_\xi : \xi \in A\}$ is quasi-disjoint. Since $|A| = \alpha > \beta$, for $\eta < \lambda$ the function from A to β defined by $\xi \to f_\xi(\eta)$ is not one-to-one and hence there are distinct elements ξ, ξ' of A and $\varphi(\eta) < \beta$ such that

$$f_\xi(\eta) = f_{\xi'}(\eta) = \varphi(\eta);$$

it follows that $f_\zeta(\eta)=\varphi(\eta)$ for all $\zeta\in A, \eta<\lambda$ and hence $\{f_\zeta:\zeta\in A\}=\{\varphi\}$. Since $f_\zeta\neq f_{\zeta'}$ for $\zeta,\zeta'\in A$ with $\zeta\neq\zeta'$ we have $|A|=1$, a contradiction.

Notation. Let α, κ and λ be cardinals and $n<\omega$. The *arrow notation*

$$\alpha\to(\kappa)^n_\lambda$$

denotes the following partition relation: if

$$[\alpha]^n=\bigcup_{i<\lambda}P_i,$$

then there are $A\subset\alpha$ and $i<\lambda$ such that

$$|A|=\kappa, \text{ and}$$

$$[A]^n\subset P_i.$$

If $A\subset\alpha$ and $[A]^n\subset P_i$, then A is said to be *P_i-homogeneous.*

1.5 *Theorem.* Let $\omega\leq\kappa\ll\alpha$ with α regular.

(a) If κ is regular then $\alpha\to(\kappa)^2_\lambda$ for all $\lambda<\kappa$; and

(b) if κ is singular then $\alpha\to(\kappa^+)^2_\kappa$.

Proof. (a) Let $[\alpha]^2=\bigcup_{i<\lambda}P_i$. We are to show that there are $i<\lambda$ and a P_i-homogeneous subset A of α such that $|A|=\kappa$. In fact we show a bit more: for each $i_0<\lambda$, either there are $i<\lambda$ with $i\neq i_0$ and a P_i-homogeneous subset A of α such that $|A|=\kappa$, or there is a P_{i_0}-homogeneous subset A of α such that $|A|=\alpha$. Without loss of generality we assume in what follows that $i_0=0$, and we give two proofs.

First Proof (using Lemma 1.1). We suppose that if $A\subset\alpha$ and A is P_0-homogeneous then $|A|<\alpha$; and for every $A\subset\alpha$ with $|A|=\alpha$ we set

$$\mathscr{B}_A=\{B\subset A:[B]^2\subset P_0\}.$$

Since the set $\mathscr{B}_A$ partially ordered by inclusion is inductive, there is a maximal element B_A of $\mathscr{B}_A$. We have $|B_A|<\alpha$, and since $\{\xi\}\in\mathscr{B}_A$ for all $\xi\in A$ we have $B_A\neq\emptyset$.

For $\xi\in A\backslash B_A$ it follows from the maximality of B_A that there is $\zeta(\xi)\in B_A$ such that $\{\xi,\zeta(\xi)\}\notin P_0$. We choose $i(\xi)\in\lambda\backslash\{0\}$ such that

$\{\xi, \zeta(\xi)\} \in P_{i(\xi)}$, and we define

$$\varphi_A : A \backslash B_A \to B_A \times (\lambda \backslash \{0\})$$

by the rule

$$\varphi_A(\xi) = (\zeta(\xi), i(\xi)),$$

and we set

$$\mathscr{P}_A = \{B_A\} \cup \{\varphi_A^{-1}(\{\zeta, i\}) : \zeta \in B_A, i \in \lambda \backslash \{0\}\}.$$

Since $\mathscr{P}_A$ is a partition of A with $|\mathscr{P}_A| < \alpha$, it follows from (the case $\beta = \alpha$ of) Lemma 1.1 that there is a family $\{A_\eta : \eta < \kappa\}$ of subsets of α such that

$$\begin{aligned} &|A_\eta| = \alpha \quad \text{for } \eta < \kappa, \\ &A_{\eta+1} \in \mathscr{P}_{A_\eta} \quad \text{for } \eta < \kappa, \text{ and} \\ &A_{\eta'} \subset A_\eta \quad \text{for } \eta < \eta' < \kappa. \end{aligned}$$

For $\eta < \kappa$ there are $\bar{\zeta}(\eta) \in B_{A_\eta}$ and $i(\eta) \varepsilon \lambda \backslash \{0\}$ such that

$$A_{\eta+1} = \varphi_{A_\eta}^{-1}(\{\bar{\zeta}(\eta), \bar{i}(\eta)\}).$$

Since $\lambda < \kappa$ and κ is regular, there are $I \subset \kappa$ with $|I| = \kappa$ and $\bar{i} \in \lambda \backslash \{0\}$ such that $\bar{i}(\eta) = \bar{i}$ for all $\eta \in I$. We set

$$H = \{\bar{\zeta}(\eta) : \eta \in I\}$$

and we show that

$$\begin{aligned} &|H| = \kappa, \text{ and} \\ &[H]^2 \subset P_{\bar{i}}. \end{aligned}$$

If $\eta, \eta' \in I$ and $\eta < \eta'$, then $\bar{\zeta}(\eta) \in B_{A_\eta}$ and

$$\bar{\zeta}(\eta') \in A_{\eta+1} \subset A_\eta \backslash B_{A_\eta};$$

it follows that $\bar{\zeta}(\eta) \neq \bar{\zeta}(\eta')$. Hence $|H| = |I| = \kappa$. Further, for $\eta, \eta' \in I$ with $\eta < \eta'$ we have

$$\bar{\zeta}(\eta') \in A_{\eta+1} = \varphi_{A_\eta}^{-1}(\{\bar{\zeta}(\eta), \bar{i}(\eta)\})$$

and hence

$$\{\bar{\zeta}(\eta), \bar{\zeta}(\eta')\} \in P_{\bar{i}(\eta)} = P_{\bar{i}};$$

it follows that $[H]^2 \subset P_{\bar{i}}$, as required.

Second Proof (using Lemma 1.3). We set

$$S = \{\xi < \alpha : \mathrm{cf}(\xi) \geq \kappa\}.$$

For $0 < i < \lambda$ and $\xi \in S$ we define a family $\{A^i_\xi(\zeta) : \zeta < \xi\}$ as follows.

$$A^i_\xi(0) = \varnothing \quad \text{if } \{0, \xi\} \notin P_i$$
$$= \{0\} \quad \text{if } \{0, \xi\} \in P_i;$$

and if $\zeta < \xi$ and $A^i_\xi(\zeta')$ has been defined for $\zeta' < \zeta$ we set

$$\tilde{A}^i_\xi(\zeta) = \bigcup_{\zeta' < \zeta} A^i_\xi(\zeta')$$

and then

$$A^i_\xi(\zeta) = \tilde{A}^i_\xi(\zeta) \qquad \text{if } \tilde{A}^i_\xi(\zeta) \cup \{\zeta, \xi\} \text{ is not } P_i\text{-homogeneous}$$
$$= \tilde{A}^i_\xi(\zeta) \cup \{\zeta\} \quad \text{if } \tilde{A}^i_\xi(\zeta) \cup \{\zeta, \xi\} \text{ is } P_i\text{-homogeneous.}$$

We set $A^i_\xi = \bigcup_{\zeta < \xi} A^i_\xi(\zeta)$ and we note that $\xi \notin A^i_\xi$ (since $A^i_\xi \subset \xi$) and $A^i_\xi \cup \{\xi\}$ is P_i-homogeneous.

If there are i and ξ such that $|A^i_\xi| = \kappa$ we set $A = A^i_\xi$ and the proof is complete. We assume in what follows that $|A^i_\xi| < \kappa$ for $0 < i < \lambda$ and $\xi \in S$.

We define $g : S \to \alpha$ by

$$g(\xi) = \sup\Big(\bigcup_{0 < i < \lambda} A^i_\xi\Big) \quad \text{for } \xi \in S.$$

Since $\lambda < \kappa \leq \mathrm{cf}(\xi)$ we have $g(\xi) < \xi$ for $\xi \in S$. It follows from Lemma 1.3 that there are $T \subset S$ with $|T| = \alpha$ and $\bar{\zeta} < \alpha$ such that $g[T] \subset \bar{\zeta}$.

For $\xi \in T$ we set $\Phi(\xi) = \langle A^i_\xi : 0 < i < \lambda \rangle$. Writing as usual $\mathscr{P}_\kappa(\bar{\zeta}) = \{B \subset \bar{\zeta} : |B| < \kappa\}$ we have $|\mathscr{P}_\kappa(\bar{\zeta})| < \alpha$ (since $\kappa \ll \alpha$ and α is regular) and hence $|\prod_{0 < i < \lambda} \mathscr{P}_\kappa(\bar{\zeta})| < \alpha$. Since $\Phi(\xi) \in \prod_{0 < i < \lambda} \mathscr{P}_\kappa(\bar{\zeta})$ for $\xi \in T$ there is $T' \subset T$ with $|T'| = \alpha$ such that if $\xi, \xi' \in T'$ then $\Phi(\xi) = \Phi(\xi')$; we show $[T']^2 \subset P_0$.

Let $\xi', \xi \in T'$ with $\xi' < \xi$ and suppose there is i such that $0 < i < \lambda$ and $\{\xi', \xi\} \in P_i$. Since $\Phi(\xi') = \Phi(\xi)$ we have $A^i_\xi = A^i_{\xi'}$; from the fact that the sets

$$A^i_\xi \cup \{\xi\} \text{ and } A^i_{\xi'} \cup \{\xi'\}$$

are P_i-homogeneous it then follows that

$$A^i_\xi \cup \{\xi', \xi\}$$

is P_i-homogeneous. We have

$$\xi' \in A^i_\xi(\xi') \subset A^i_\xi = A^i_{\xi'},$$

a contradiction.

(b) If κ is singular we have $\kappa^+ \ll \alpha$ from Theorem A.5(c), and from part (a) (with κ and λ replaced by κ^+ and κ, respectively) we have $\alpha \to (\kappa^+)^2_\kappa$, as required.

1.6 *Corollary.* $(2^\alpha)^+ \to (\alpha^+)^2_\alpha$ for all infinite cardinals α.

Proof. We have $(2^\alpha)^\alpha = 2^\alpha$; thus the (regular) cardinals $\alpha^+, (2^\alpha)^+$ satisfy the relation $\alpha^+ \ll (2^\alpha)^+$ and the result follows from Theorem 1.5(a).

Notation. Let α be a cardinal and ξ an ordinal. The *arrow notation*

$$\alpha \to (\alpha, \xi)^2$$

denotes the following partition relation: if

$$[\alpha]^2 = P_0 \cup P_1,$$

then either there is $A \subset \alpha$ with $|A| = \alpha$ such that $[A]^2 \subset P_0$, or there is $A \subset \alpha$ with type $A = \xi$ such that $[A]^2 \subset P_1$.

The negation of the relation $\alpha \to (\alpha, \xi)^2$ is denoted $\alpha \not\to (\alpha, \xi)^2$.

1.7 *Theorem.* If α and κ are cardinals with $\omega \leq \kappa \ll \alpha$ and with α regular, then $\alpha \to (\alpha, \kappa + 1)^2$.

Proof. Let $[\alpha]^2 = P_0 \cup P_1$. We suppose that if $A \subset \alpha$ and A is P_0-homogeneous then $|A| < \alpha$; and for every $A \subset \alpha$ with $|A| = \alpha$ we set

$$\mathscr{B}_A = \{B \subset A : [B]^2 \subset P_0\}.$$

Since the set $\mathscr{B}_A$ partially ordered by inclusion is inductive, there is a maximal element B_A of $\mathscr{B}_A$. We have $|B_A| < \alpha$, and since $\{\xi\} \in \mathscr{B}_A$ for all $\xi \in A$ we have $B_A \neq \varnothing$. Since α is regular, there is $\xi(A) < \alpha$ such that $B_A \subset \xi(A)$.

For $\xi \in A \backslash B_A$ it follows from the maximality of B_A that there is $\zeta(\xi) \in B_A$ such that $\{\xi, \zeta(\xi)\} \notin P_0$. We define

$$\varphi_A : A \backslash \xi(A) \to B_A$$

by the rule

$$\varphi_A(\xi) = \zeta(\xi),$$

and we set

$$\mathscr{P}_A = \{\xi(A) \cap A\} \cup \{\varphi_A^{-1}(\{\zeta\}) : \zeta \in B_A\}.$$

Since $\mathscr{P}_A$ is a partition of A with $|\mathscr{P}_A| < \alpha$, it follows from (the case $\beta = \alpha$ of) Lemma 1.1 that there is a family $\{A_\eta : \eta < \kappa\}$ of subsets of α such that

$$|A_\eta| = \alpha \text{ for } \eta < \kappa,$$
$$A_{\eta+1} \in \mathscr{P}_{A_\eta} \text{ for } \eta < \kappa,$$
$$A_{\eta'} \subset A_\eta \text{ for } \eta < \eta' < \kappa, \text{ and}$$
$$\bigcap_{\eta<\kappa} A_\eta \neq \emptyset.$$

For $\eta < \kappa$ there is $\overline{\zeta}(\eta) \in B_{A_\eta}$ such that

$$A_{\eta+1} = \varphi_{A_\eta}^{-1}(\{\overline{\zeta}(\eta)\}).$$

We choose $\overline{\zeta}(\kappa) \in \bigcap_{\eta<\kappa} A_\eta$, we set

$$H = \{\overline{\zeta}(\eta) : \eta \leq \kappa\},$$

and we show that

$$\text{type } H = \kappa + 1, \text{ and}$$
$$[H]^2 \subset P_1.$$

If $\eta < \eta' \leq \kappa$ then $\overline{\zeta}(\eta) \in B_{A_\eta} \subset \xi(A_\eta)$ and

$$\overline{\zeta}(\eta') \in A_{\eta+1} \subset A_\eta \backslash \xi(A_\eta),$$

so that $\overline{\zeta}(\eta) < \overline{\zeta}(\eta')$. Thus type $H = \kappa + 1$.

For $\eta < \eta' \leq \kappa$ we have

$$\overline{\zeta}(\eta') \in A_{\eta+1} = \varphi_{A_\eta}^{-1}(\{\overline{\zeta}(\eta)\}),$$

so that $\{\overline{\zeta}(\eta'), \overline{\zeta}(\eta)\} \notin P_0$ and hence $\{\overline{\zeta}(\eta), \overline{\zeta}(\eta')\} \in P_1$; it follows $[H]^2 \subset P_1$, as required.

We indicate in Corollary 7.8 below that, at least if the contin hypothesis is assumed, the statement of Theorem 1.7 canno improved for $\alpha = \omega^+$.

1.8 *Lemma.* Let α be an (infinite) singular cardinal, $\kappa \ll \alpha$ and $\xi < \alpha\}$ a set of subsets of α such that

$$|S_\xi| < \kappa \text{ for } \xi < \alpha.$$

If $\{\alpha_\sigma : \sigma < \text{cf}(\alpha)\}$ is a set such that

$$\alpha_\sigma \text{ is a regular cardinal for } \sigma < \text{cf}(\alpha),$$
$$\kappa \ll \alpha_\sigma \text{ for } \sigma < \text{cf}(\alpha),$$
$$\text{cf}(\alpha) < \alpha_{\sigma'} < \alpha_\sigma < \alpha \text{ for } \sigma' < \sigma < \text{cf}(\alpha), \text{ and}$$
$$\sum_{\sigma<\text{cf}(\alpha)} \alpha_\sigma = \alpha,$$

then there are $\{A_\sigma : \sigma < \mathrm{cf}(\alpha)\}$ and $\{J_\sigma : \sigma < \mathrm{cf}(\alpha)\}$ such that

(i) $A_\sigma \in [\alpha]^{\alpha_\sigma}$ for $\sigma < \mathrm{cf}(\alpha)$,

(ii) $A_{\sigma'} \cap A_\sigma = \varnothing$ for $\sigma' < \sigma < \mathrm{cf}(\alpha)$, and

(iii) $S_{\xi'} \cap S_\xi = J_\sigma$ for $\xi', \xi \in A_\sigma, \xi' \neq \xi$,
$\subset J_\sigma$ for $\xi' \in A_{\sigma'}, \xi \in A_\sigma, \sigma' < \sigma < \mathrm{cf}(\alpha)$.

Proof. There is $\{D_\sigma : \sigma < \mathrm{cf}(\alpha)\} \subset \mathscr{P}(\alpha)$ such that

$$|D_\sigma| = \alpha_\sigma \text{ for } \sigma < \mathrm{cf}(\alpha), \text{ and}$$
$$D_{\sigma'} \cap D_\sigma = \varnothing \text{ for } \sigma' < \sigma < \mathrm{cf}(\alpha).$$

By the Erdős–Rado theorem for quasi-disjoint families there are $C_\sigma \in [D_\sigma]^{\alpha_\sigma}$ and J_σ such that

$$S_{\xi'} \cap S_\xi = J_\sigma \quad \text{for } \xi', \xi \in C_\sigma, \xi' \neq \xi.$$

We note that if $\sigma' \neq \sigma, \xi' \in C_{\sigma'}$ and $\eta \in S_{\xi'}$, then

$$|\{\xi \in C_\sigma : \eta \in S_\xi, \eta \notin J_\sigma\}| \leq 1.$$

It then follows, setting

$$B_\sigma(\sigma') = \{\xi \in C_\sigma : \text{there is } \xi' \in C_{\sigma'} \text{ such that } S_{\xi'} \cap S_\xi \not\subset J_\sigma\},$$
$$B_\sigma = \cup \{B_\sigma(\sigma') : \sigma' < \sigma\}, \text{ and}$$
$$A_\sigma = C_\sigma \backslash B_\sigma,$$

that

$$|B_\sigma(\sigma')| \leq |\cup \{S_{\xi'} : \xi' \in C_{\sigma'}\}| \leq \alpha_{\sigma'} \cdot \kappa = \alpha_{\sigma'},$$
$$|B_\sigma| \leq \sum_{\sigma' < \sigma} \alpha_{\sigma'} < \alpha_\sigma, \text{ and}$$
$$|A_\sigma| = \alpha_\sigma$$

for $\sigma < \mathrm{cf}(\alpha)$. It is clear that $\{A_\sigma : \sigma < \mathrm{cf}(\alpha)\}$ and $\{J_\sigma : \sigma < \mathrm{cf}(\alpha)\}$ are as required.

Conditions sufficient to ensure that there is a set $\{\alpha_\sigma : \sigma < \mathrm{cf}(\alpha)\}$ as in the statement of Lemma 1.8 are given in Theorem A.7 of the Appendix. One such condition is that $\kappa = \omega$; we use Lemma 1.8 with $\kappa = \omega$ only, in Theorem 8.10 below.

The next result is the extension to singular cardinals of the Erdős–Rado theorem.

1.9 *Theorem.* Let α be singular with $\mathrm{cf}(\alpha) > \omega$, let $\kappa \ll \alpha$ and

$\kappa \ll \mathrm{cf}(\alpha)$, and let $\{S_\xi : \xi < \alpha\}$ be a set of subsets of α such that

$$|S_\xi| < \kappa \quad \text{for } \xi < \alpha.$$

If $\{\alpha_\sigma : \sigma < \mathrm{cf}(\alpha)\}$ is a set such that

$$\begin{aligned}
&\alpha_\sigma \text{ is a regular cardinal for } \sigma < \mathrm{cf}(\alpha),\\
&\kappa \ll \alpha_\sigma \quad \text{for } \sigma < \mathrm{cf}(\alpha),\\
&\mathrm{cf}(\alpha) < \alpha_{\sigma'} < \alpha_\sigma < \alpha \quad \text{for } \sigma' < \sigma < \mathrm{cf}(\alpha), \text{ and}\\
&\sum_{\sigma < \mathrm{cf}(\alpha)} \alpha_\sigma = \alpha,
\end{aligned}$$

then there are $\Gamma, J, \{A_\sigma : \sigma \in \Gamma\}$ and $\{J_\sigma : \sigma \in \Gamma\}$ such that

(i) $\Gamma \in [\mathrm{cf}(\alpha)]^{\mathrm{cf}(\alpha)}$,

(ii) $A_\sigma \in [\alpha]^{\alpha_\sigma}$ for $\sigma \in \Gamma$,

(iii) $A_\sigma \cap A_{\sigma'} = \varnothing$ for $\sigma', \sigma \in \Gamma, \sigma' \neq \sigma$, and

(iv) $S_{\xi'} \cap S_\xi = J_\sigma$ for $\xi', \xi \in A_\sigma, \xi' \neq \xi$
$= J$ for $\xi' \in A_{\sigma'}, \xi \in A_\sigma, \sigma, \sigma' \in \Gamma, \sigma \neq \sigma'$.

Proof. From Lemma 1.8 there are $\{C_\sigma : \sigma < \mathrm{cf}(\alpha)\}$ and $\{J_\sigma : \sigma < \mathrm{cf}(\alpha)\}$ with properties (analogous to) (ii) and (iii) and such that

$$\begin{aligned}
S_{\xi'} \cap S_\xi &= J_\sigma \quad \text{for } \xi', \xi \in C_\sigma, \xi' \neq \xi\\
&\subset J_\sigma \quad \text{for } \xi' \in C_{\sigma'}, \xi \in C_\sigma, \sigma' < \sigma < \mathrm{cf}(\alpha).
\end{aligned}$$

Since $\kappa \ll \mathrm{cf}(\alpha)$ and $|J_\sigma| < \kappa$ for $\sigma < \mathrm{cf}(\alpha)$ there are (by the Erdős–Rado theorem for quasi-disjoint families) Γ and J such that

$$\begin{aligned}
&\Gamma \in [\mathrm{cf}(\alpha)]^{\mathrm{cf}(\alpha)} \text{ and}\\
&J_{\sigma'} \cap J_\sigma = J \quad \text{for } \sigma', \sigma \in \Gamma, \sigma' \neq \sigma.
\end{aligned}$$

We consider two cases.

Case 1. $J = \varnothing$. We note that if $\sigma', \sigma \in \Gamma, \sigma' \neq \sigma$ and $\eta \in J_{\sigma'}$, then

$$|\{\xi \in C_\sigma : \eta \in S_\xi\}| \leq 1$$

(since $\eta \notin J_\sigma \cap J_{\sigma'} = \varnothing$). It then follows, setting

$$\begin{aligned}
B_\sigma(\sigma') &= \{\xi \in C_\sigma : S_\xi \cap J_{\sigma'} \neq \varnothing\},\\
B_\sigma &= \cup \{B_\sigma(\sigma') : \sigma' \in \Gamma, \sigma' \neq \sigma\}, \text{ and}\\
A_\sigma &= C_\sigma \backslash B_\sigma,
\end{aligned}$$

that

$$|B_\sigma(\sigma')| \le |J_{\sigma'}| < \kappa,$$
$$|B_\sigma| \le |\Gamma|\cdot\kappa < \alpha_0, \text{ and}$$
$$|A_\sigma| = \alpha_0$$

for $\sigma\in\Gamma$. It is clear that

$$S_\xi \cap J_{\sigma'} = \varnothing \quad \text{for } \xi\in A_\sigma, \sigma, \sigma'\in\Gamma, \sigma \ne \sigma'.$$

Now let $\xi'\in A_{\sigma'}, \xi\in A_\sigma, \sigma'\ \sigma\in\Gamma$, and $\sigma' < \sigma$. Then $S_{\xi'} \cap S_\xi \subset J_{\sigma'}$, and from $S_{\xi'} \cap J_\sigma = \varnothing$ we have

$$S_{\xi'} \cap S_\xi = \varnothing,$$

as required.

Case 2. $J \ne \varnothing$. We set $\tilde{J} = \varnothing$, $\tilde{J}_\sigma = J_\sigma \backslash J$ for $\sigma\in\Gamma$ and $\tilde{S}_\xi = S_\xi \backslash J$ for $\xi\in C_\sigma, \sigma\in\Gamma$, and we note from Case 1 that there is $\{A_\sigma : \sigma\in\Gamma\}$ satisfying (ii) and (iii), and

$$\begin{aligned} \tilde{S}_{\xi'} \cap \tilde{S}_\xi &= \tilde{J}_\sigma \quad \text{for } \xi', \xi\in A_\sigma, \xi' \ne \xi \\ &= \tilde{J} \quad \text{for } \xi'\in A_{\sigma'}, \xi\in A_\sigma, \sigma, \sigma'\in\Gamma, \sigma \ne \sigma'. \end{aligned}$$

It is clear that condition (iv) also is satisfied.

The proof is complete.

We note that, according to the remark following A.7 of the Appendix the hypothesis $\kappa \ll \alpha$ in Lemma 1.8 and Theorem 1.9 is redundant; that is, the condition $\kappa \ll \alpha$ follows from the existence of the given family $\{\alpha_\sigma : \sigma < \mathrm{cf}(\alpha)\}$.

1.10 *Corollary.* Let $\omega \le \kappa \ll \alpha$ and $\kappa \ll \mathrm{cf}(\alpha)$, let $\{S_\xi : \xi < \alpha\}$ be a set of sets such that

$$|S_\xi| = \lambda < \kappa \text{ for } \xi < \alpha,$$

and for $\xi < \alpha$ let φ_ξ be a one-to-one function from S_ξ onto λ. Then there is $A \subset \alpha$ with $|A| = \alpha$ such that if $\xi, \xi'\in A$ and $\eta\in S_\xi \cap S_{\xi'}$ then $\varphi_\xi(\eta) = \varphi_{\xi'}(\eta)$.

Proof. Case 1. α is regular. From the Erdős–Rado theorem on quasi-disjoint families (Theorem 1.4) we may assume without loss of generality that there is a set J such that $S_\xi \cap S_{\xi'} = J$ for $\xi' < \xi < \alpha$. Then

$$\{\varphi_\xi | J : \xi < \alpha\} \subset \lambda^J,$$

and since $|J| \leq \lambda < \kappa < \alpha$ and $\kappa \ll \alpha$ we have $|\lambda^J| < \alpha$. Since α is regular, there is $A \in [\alpha]^\alpha$ such that $\varphi_\xi | J = \varphi_{\xi'} | J$ for $\xi, \xi' \in A$, as required.

Case 2. α is singular. From Theorem A7 there is a set $\{\alpha_\sigma : \sigma < \text{cf}(\alpha)\}$ of regular cardinals such that

$$\kappa \ll \alpha_\sigma < \alpha \quad \text{for } \sigma < \text{cf}(\alpha) \text{ and}$$
$$\sum_{\sigma < \text{cf}(\alpha)} \alpha_\sigma = \alpha,$$

and by the extension to singular cardinals of the Erdős–Rado theorem there are $\Gamma, J, \{A_\sigma : \sigma \in \Gamma\}$ and $\{J_\sigma : \sigma \in \Gamma\}$ such that

$$\begin{aligned} &\Gamma \in [\text{cf}(\alpha)]^{\text{cf}(\alpha)}, \\ &A_\sigma \in [\alpha_\sigma]^{\alpha_\sigma} \quad \text{for } \sigma \in \Gamma, \\ &A_\sigma \cap A_{\sigma'} = \varnothing \quad \text{for } \sigma', \sigma \in \Gamma, \sigma' \neq \sigma, \text{ and} \\ &S_\xi \cap S_{\xi'} = J_\sigma \quad \text{for } \xi, \xi' \in A_\sigma, \xi \neq \xi' \\ &\qquad\quad = J \quad \text{for } \xi \in A_\sigma, \xi' \in A_{\sigma'}, \sigma, \sigma' \in \Gamma, \sigma \neq \sigma'. \end{aligned}$$

It follows from Case 1 (with α_σ replacing α) that for $\sigma \in \Gamma$ there is $B_\sigma \in [A_\sigma]^{\alpha_\sigma}$ such that if $\xi, \xi' \in B_\sigma$ and $\eta \in S_\xi \cap S_{\xi'}$, then $\varphi_\xi(\eta) = \varphi_{\xi'}(\eta)$. For $\sigma \in \Gamma$ we define a function ψ_σ from J_σ to λ by the rule

$$\psi_\sigma = \varphi_\xi | J_\sigma \quad \text{for } \xi \in B_\sigma.$$

Since $\psi_\sigma[J_\sigma] \subset \lambda$ and $|\psi_\sigma[J_\sigma]| \leq \lambda < \kappa \ll \text{cf}(\alpha)$, there are $\Delta \in [\Gamma]^{\text{cf}(\alpha)}$ and a set $M \subset \lambda$ such that $\psi_\sigma[J_\sigma] = M$ for $\sigma \in \Delta$. Since ψ_σ is a one-to-one function from J_σ onto M, there is by Case 1 (with cf(α) replacing α) a set $E \subset \Delta$ with $|E| = \text{cf}(\alpha)$ such that if $\sigma, \sigma' \in E$ and $\eta \in J_\sigma \cap J_{\sigma'}$, then $\psi_\sigma(\eta) = \psi_{\sigma'}(\eta)$. We set $A = \cup_{\sigma \in E} B_\sigma$ and we claim that if $\xi, \xi' \in A$ and $\eta \in S_\xi \cap S_{\xi'}$, then $\varphi_\xi(\eta) = \varphi_{\xi'}(\eta)$. If there is $\sigma \in E$ such that $\xi, \xi' \in B_\sigma$ this is clear; and if $\xi \in B_\sigma, \xi' \in B_{\sigma'}$ with $\sigma, \sigma' \in E$ and $\sigma \neq \sigma'$, then $\eta \in J_\sigma \cap J_{\sigma'} = J$ and hence

$$\varphi_\xi(\eta) = \psi_\sigma(\eta) = \psi_{\sigma'}(\eta) = \varphi_{\xi'}(\eta),$$

as required.

1.11 *Theorem.* Let $\alpha \geq \omega$ with α regular, ρ an ordinal such that $\rho < \alpha$, and $f : \alpha \to \mathscr{P}(\alpha)$ such that

$$\text{type } f(\xi) \leq \rho \quad \text{for } \xi < \alpha, \text{ and}$$
$$\xi \notin f(\xi) \quad \text{for } \xi < \alpha.$$

Then there is $A \subset \alpha$ such that

$$|A| = \alpha, \text{ and}$$
$$\xi \notin f(\zeta) \quad \text{for } \xi, \zeta \in A$$

(i.e., A is *free for* f).

Proof. Suppose there is no set A satisfying the conclusion of the theorem. We define families $\{\zeta_\eta : \eta \leq \rho\}$, $\{A_\eta : \eta \leq \rho\}$, $\{B_\eta : \eta \leq \rho\}$, and $\{\bar{\zeta}_\eta : \eta \leq \rho\}$ recursively as follows. If $\eta \leq \rho$ and if $\zeta_{\eta'}, A_{\eta'}, B_{\eta'}$, and $\bar{\zeta}_{\eta'}$ have been defined for $\eta' < \eta$, we set

$$\zeta_\eta = \sup_{\eta' < \eta} \bar{\zeta}_{\eta'},$$

we choose for A_η a subset of $\alpha \backslash \zeta_\eta$ which is maximal (among all subsets of $\alpha \backslash \zeta_\eta$) with respect to the property that A_η is free for f, we set

$$B_\eta = A_\eta \cup (\cup \{f(\xi) : \xi \in A_\eta\}),$$

and we choose $\bar{\zeta}_\eta$ so that

$$B_\eta \subset \bar{\zeta}_\eta < \alpha;$$

such an element $\bar{\zeta}_\eta$ of α exists because $|A_\eta| < \alpha$ and $|f(\xi)| \leq |\rho| < \alpha$ for $\xi \in A_\eta$ and α is a regular cardinal.

We set $\bar{\zeta} = \sup_{\eta \leq \rho} \bar{\zeta}_\eta$ and we claim that $f(\bar{\zeta}) \cap A_\eta \neq \emptyset$ for $\eta \leq \rho$. Indeed if $\eta \leq \rho$ and $f(\bar{\zeta}) \cap A_\eta = \emptyset$, then for $\xi \in A_\eta$ we have $\xi \notin f(\bar{\zeta})$ and also $\bar{\zeta} \notin f(\xi)$ (since

$$f(\xi) \subset B_\eta \subset \bar{\zeta}_\eta \leq \bar{\zeta});$$

thus if $\eta \leq \rho$ and $f(\bar{\zeta}) \cap A_\eta = \emptyset$, the set $A_\eta \cup \{\bar{\zeta}\}$ is free for f and a subset of $\alpha \backslash \zeta_\eta$, contrary to the maximality of A_η. The claim is proved.

For $\eta \leq \rho$ we choose $\xi(\eta) \in f(\bar{\zeta}) \cap A_\eta$. For $\eta' < \eta \leq \rho$ we have

$$\xi(\eta') \in A_{\eta'} \subset B_{\eta'} \subset \bar{\zeta}_{\eta'} \leq \zeta_\eta < \min A_\eta \leq \xi(\eta);$$

thus the function from $\rho + 1$ into $f(\bar{\zeta})$ given by $\eta \to \xi(\eta)$ is an ordered-set isomorphism. Since $f(\bar{\zeta})$ contains a set of type $\rho + 1$, we have

$$\text{type } f(\bar{\zeta}) \geq \rho + 1.$$

This contradiction completes the proof.

Notes for Chapter 1

Infinitary combinatorics and the partition calculus, which is a generalization of the Dedekind–Dirichlet pigeon-hole principle and whose first important result is Ramsey's theorem, $\omega \to (\omega)^n_m$ for $0 < m, n < \omega$ (Ramsey, 1930, Theorem A), was introduced and developed into a whole theory by Erdős, Rado, Hajnal, and others. Comprehensive treatments are given by Erdős & Rado (1956), Erdős, Hajnal & Rado (1965), Williams (1977), and in the treatise of Erdős, Hajnal, Máté & Rado (1982).

Argyros' ramification lemma (Lemma 1.1) is from the doctoral dissertation of Argyros (1977b).

The statement that if $f \in (\omega^+)^{\omega^+}$ and $f(\zeta) < \zeta$ for $0 < \zeta < \omega^+$ then f is constant on a cofinal subset of ω^+, is due to Alexandroff & Urysohn (1929). Generalizations were given by Dushnik (1931), Erdős (1950) (final paragraph), Novák (1950, Theorem 1) and Bachmann (1955, section 9). Neumer (1951) proved that if ξ is an uncountable, regular cardinal, if S is a subset of ξ which meets every unbounded subset of ξ closed in the order topology of ξ, and if $f : S \to \xi$ satisfies $f(\zeta) < \zeta$ for $\zeta \neq 0$, then there are cofinal $T \subset S$ and $\bar{\zeta} < \xi$ such that $f[T] \subset \bar{\zeta}$. (Such sets S are the *stationary sets* – les ensembles stationnaires – of Bloch (1953), who showed that every stationary subset of ω^+ is the union of two disjoint stationary sets.) Fodor (1955) showed that in the statement of Neumer's theorem it is enough to assume that ξ is an ordinal number with cf $(\xi) > \omega$; and (Fodor, 1956) the set T may itself be chosen stationary.

Pairs of cardinals $\langle \kappa, \alpha \rangle$ with α strongly κ-inaccessible were introduced by Fuhrken (1965) (with different terminology) in connection with a generalization of Łoś' ultraproduct theorem (cf. Comfort & Negrepontis, 1974, section 11).

The remarkable theorem of Shanin (1946a) implies the case $\kappa = \omega, \alpha$ arbitrary (regular) of Theorem 1.4 (on 'Δ-systems', or quasi-disjoint families); see also Marczewski (1947) and Bockstein (1948) for implicit statements, and Mazur (1952, vii, p. 235) for an explicit statement and proof, of the case $\kappa = \omega, \alpha = \omega^+$ of Theorem 1.4 and for applications to product spaces.

Erdős & Rado (1960) prove Theorem 1.4 for non-limit cardinals α. Short proofs of this case were given also by Michael (1962) and Davies (1967). Theorem 1.4 in its present general form is from Erdős & Rado (1969). Alternative proofs include an unpublished proof of Davies, referred to by Erdős & Rado (1969), and the proof of Comfort & Negrepontis (1972a, Theorem 2.1); this latter is related to arguments given by Całczyńska-Karłowicz (1964), Marek (1964), and Mostowski (1969, Theorem 13.3.1). These are given in detail, together with the converse of the theorem, due also to Erdős & Rado (1969), by Comfort & Negrepontis (1974, section 3).

The first proof given here of Theorem 1.4 is from the doctoral dissertation of Argyros (1977b); the second proof is derived from an argument of Hajnal & Máté (1975, Theorem 2.3).

Preliminary formulations of the arrow relation of Theorem 1.5 appear (with different hypotheses) in Erdős & Rado (1956, Theorem 39 (ii)) and Kurepa (1959a). A different proof of Theorem 1.5, using the saturated structures of model theory, has been announced by Simpson (1970) and appears in Keisler (1971, section 14). We note that the case $(2^{\alpha})^+ \to (\alpha^+)^2_{\alpha}$ appears in Erdős (1942, Theorem I). Related results on partition calculus were also proved independently by Kurepa (1939, 1959a,b) (cf. the review by Erdős, 1962).

The first proof of Theorem 1.5 given here is due to the authors, while the second was shown to us by Hajnal.

Theorem 1.7 is due to Erdős & Rado (1956, Corollary 1). The present proof, based on Argyros' lemma (1.1), is due to the authors; clearly another proof is available, along the lines of the second proof of Theorem 1.5, using Lemma 1.3 (the pressing-down lemma).

The argument used in Lemma 1.8 is essentially the same as the one given by Noble & Ulmer (1972, Lemma 1.1 (ii)).

The extension of the Erdős–Rado theorem on quasi-disjoint families to singular cardinals (Theorem 1.9), already noted in preliminary form in the original paper of Shanin (1946a), is stated in Shelah (1977, Claim 2.4(B)) for $\kappa = \omega$, and was proved independently in its present general form by Argyros in his doctoral dissertation (1977b, Theorem 1.3); see also Argyros & Negrepontis (1977, 1980). Hagler and Haydon, independently, have also noted the result for $\kappa = \omega$ (letter to the authors of October, 1977).

Argyros (1977b, Theorem 1.3) also proves a converse to Theorem 1.9.

For some versions of the Erdős–Rado theorem on quasi-disjoint families which use roots of cardinals, consult Verbeek

(1970), Juhász (1971), and O'Callaghan (1975). Argyros (1980) gives a strong analogue of Theorem 1.10 with (weak) roots.

Results analogous to Theorem 1.11 have been obtained as follows: for ρ a cardinal and α regular, by Lázár (1936) and Piccard (1937); for ρ a cardinal and α a (regular or singular) cardinal, assuming the generalized continuum hypothesis, by Erdős (1950); for ρ and α cardinals, by Hajnal (1961); for ρ an ordinal and α a cardinal, by Erdős & Specker (1961). Related results and generalizations are given by Erdős & Fodor (1957) and by Erdős, Galvin & Hajnal (1975).

We note that for limit cardinals α (regular or singular), the conclusion of Theorem 1.11 is immediate from the theorem of Hajnal (1961).

For a generalization of Hajnal's theorem involving finitely additive measures, see Tsarpalias (1977, 1978a, 1980).

The combinatorial methods and theorems given in this section have extensive applications outside the scope of this monograph. In topology, for example, there are applications by Hajnal & Juhász (1967, 1969), Juhász (1971, 1976), Ginsburg & Woods (1976), Fleissner (1978) and Gerlits (1980a, b). In functional analysis there are: an example of an Eberlein compact space that cannot be embedded in a Hilbert space in its weak topology, given by Benyamini & Starbird (1976); results of Talagrand (1978, 1980) on the nonexistence of strong lifting; results (described in the Notes for Chapter 5) on the embedding of $l^1(\Gamma)$ into Banach spaces $C(X)$, and in the isomorphic theory of Banach spaces; results on the conjecture of Pełczynski (1968) by Haydon (1977, 1978), Argyros & Negrepontis (1980), and Argyros (1981b); and many more.

2

Introducing the Chain Conditions

Here we define in their most general form the concepts studied in this monograph, and we record numerous basic and elementary relationships and facts concerning them. The properties are, in increasing order of generality: calibre, compact-calibre, and pseudo-compactness.

On the basis of this work it becomes natural to associate with each space X, for $\lambda > 0$, a cardinal number we call the λ-Souslin number of X (denoted $S_\lambda(X)$). We show (Theorem 2.14, a generalization of a classical result of Erdős & Tarski) that for infinite Hausdorff spaces X and regular $\lambda \geq \omega$, the number $S_\lambda(X)$ is a regular cardinal.

For the most part we deal with arbitrary topological spaces (not subjected to specialized separation conditions), but it is convenient to investigate certain questions about chain conditions in the context of compact Hausdorff spaces (see Chapter 5). The device used for effecting this reduction (with no loss of generality) is the Gleason space of an (arbitrary) space; our treatment of the Gleason space and its definition follows recent work of Galvin.

Chain conditions

Definition. Let X be a space and let α, β and γ be cardinals with $\alpha \geq \beta \geq \gamma$. Then

(a) X has *calibre* (α, β, γ) if for every set $\{U_\xi : \xi < \alpha\}$ of non-empty, open subsets of X there is $B \subset \alpha$ with $|B| = \beta$ such that if $C \subset B$ and $|C| = \gamma$ then $\cap_{\xi \in C} U_\xi \neq \varnothing$; and

(b) X has *compact-calibre* (α, β, γ) if for every set $\{U_\xi : \xi < \alpha\}$ of non-empty, open subsets of X there is $B \subset \alpha$ with $|B| = \beta$ such that if $C \subset B$ and $|C| = \gamma$ there is a compact subset K of X such that $U_\xi \cap K \neq \varnothing$ whenever $\xi \in C$.

We consider in practice almost exclusively the case $\alpha \geq \omega$, with $\beta = \alpha$ or $\beta = \gamma$. If X has calibre (α, β, β) we say that X *has calibre* (α, β); and if X has calibre (α, α) we say that X *has calibre* α (or, α *is a calibre for* X). The terms *compact-calibre* (α, β) and *compact-calibre* α are defined analogously.

Definition. Let X be a space and let α and β be cardinals with $\alpha \geq \beta$. Then X is *pseudo-*(α, β)*-compact* if for every set $\{U_\xi : \xi < \alpha\}$ of non-empty, open subsets of X there is $x \in X$ such that

$$|\{\xi < \alpha : V \cap U_\xi \neq \varnothing\}| \geq \beta$$

for every neighborhood V of x.

An indexed family $\{A_i : i \in I\}$ of subsets of X is said to be *locally finite* (in X) if for every $x \in X$ there is a neighborhood V of x such that

$$|\{i \in I : V \cap A_i : \neq \varnothing\}| < \omega.$$

Hence X is pseudo-(α, ω)-compact if and only if no locally finite family of non-empty open subsets of X is indexed by α. A pseudo-(α, ω)-compact space is said to be *pseudo-*α*-compact*, and a pseudo-ω-compact space is said to be *pseudocompact*.

For later use we record in 2.1–2.5 several elementary facts of marginal intrinsic interest about the properties defined above.

2.1 *Theorem.* Let $\alpha \geq \beta \geq \gamma, \alpha \geq \omega$, and X a space.

(a) If $\alpha' \geq \beta' \geq \gamma'$ with $\alpha' \geq \alpha, \beta' \leq \beta$ and $\gamma' \leq \gamma$, and if X has calibre (α, β, γ), then X has calibre $(\alpha', \beta', \gamma')$;

(b) if $\alpha' \geq \beta' \geq \gamma'$ with $\alpha' \geq \alpha, \beta' \leq \beta$ and $\gamma' \leq \gamma$, and if X has compact-calibre (α, β, γ), then X has compact-calibre $(\alpha', \beta', \gamma')$;

(c) if $\alpha' \geq \alpha \geq \beta \geq \beta'$ and X is pseudo-(α, β)-compact, then X is pseudo-(α', β')-compact;

(d) if X has calibre (α, β, γ), then X has compact-calibre (α, β, γ);

(e) if X has calibre (α, β, γ), then X is pseudo-(α, γ)-compact;

(f) if $\gamma \geq \omega$ and X has compact-calibre (α, β, γ), then X is pseudo-(α, γ)-compact;

(g) X has compact-calibre (α, α, n) for all $n < \omega$;

(h) if X has calibre $(\alpha, 2)$, then X has calibre (α, n) for all $n < \omega$;

(i) if X is a regular, Hausdorff space and X is pseudo-$(\alpha, 2)$-compact, then X is pseudo-α-compact.

Proof. (a), (b), and (c) are immediate from the definitions.

(d) follows from the fact that if $x \in \cap_{\xi \in C} U_\xi$, then $\{x\}$ is a compact set K such that $U_\xi \cap K \neq \varnothing$ whenever $\xi \in C$.

(e) Let $\{U_\xi : \xi < \alpha\} \subset \mathscr{T}^*(X)$. Since X has calibre (α, γ, γ) there are $C \in [\alpha]^\gamma$ and $x \in X$ such that $x \in \cap_{\xi \in C} U_\xi$. It is clear that

$$|\{\xi < \alpha : V \cap U_\xi \neq \varnothing\}| \geq |\{\xi \in C : V \cap U_\xi \neq \varnothing\}| = \gamma$$

for every neighborhood V of x.

(f) If the statement fails there are $\{U_\xi : \xi < \alpha\} \subset \mathscr{T}^*(X)$ and $\{V_x : x \in X\} \subset \mathscr{T}^*(X)$ such that

(i) $x \in V_x$ for $x \in X$, and

(ii) if $A(x) = \{\xi < \alpha : V_x \cap U_\xi \neq \varnothing\}$, then $|A(x)| < \gamma$ for $x \in X$. Since X has compact-calibre (α, γ, γ), there are $C \in [\alpha]^\gamma$ and compact $K \subset X$ such that $U_\xi \cap K \neq \varnothing$ for all $\xi \in C$. There is $F \in \mathscr{P}_\omega(K)$ such that $K \subset \cup_{x \in F} V_x$, so that

$$\gamma = |C| \leq \sum_{x \in F} |C \cap A(x)| \leq \sum_{x \in F} |A(x)| < \gamma,$$

a contradiction.

(g) Let $\{U_\xi : \xi < \alpha\} \subset \mathscr{T}^*(X)$ and $A \in [\alpha]^n$. For $\xi \in A$ let $x_\xi \in U_\xi$ and set $K = \{x_\xi : \xi \in A\}$. Then K is a compact subspace of X such that $U_\xi \cap K \neq \varnothing$ whenever $\xi \in A$.

(h) It follows from part (a) that it is enough to show that if $k < \omega$ and X has calibre (α, k) then X has calibre (α, k^2). Let $\{U_\xi : \xi < \alpha\} \subset \mathscr{T}^*(X)$ and choose $\{A_\eta : \eta < \alpha\} \subset [\alpha]^\alpha$ so that

$$\alpha = \bigcup_{\eta < \alpha} A_\eta, \text{ and}$$
$$A_\eta \cap A_{\eta'} = \varnothing \quad \text{for } \eta' < \eta < \alpha.$$

Since X has calibre (α, k), for $\eta < \alpha$ there is $B_\eta \in [A_\eta]^k$ such that $\cap \{U_\xi : \xi \in B_\eta\} \neq \varnothing$; we set $V_\eta = \cap \{U_\xi : \xi \in B_\eta\}$ for $\eta < \alpha$. Again since X has calibre (α, k), there is $C \in [\alpha]^k$ such that $\cap_{\eta \in C} V_\eta \neq \varnothing$. We set $D = \cup_{\eta \in C} B_\eta$. Then $D \in [\alpha]^{k^2}$, and

$$\bigcap_{\xi \in D} U_\xi = \bigcap_{\eta \in C} V_\eta \neq \varnothing,$$

as required.

(i) If the statement fails there is locally finite $\{U_\xi : \xi < \alpha\} \subset \mathscr{T}^*(X)$. We choose $x_\xi \in U_\xi$ and we note that since

$$|\{\zeta < \alpha : x_\xi \in U_\zeta\}| < \omega \text{ for } \xi < \alpha$$

there is $A\in[\alpha]^\alpha$ such that

$$x_\xi \neq x_\zeta \text{ for } \xi,\zeta\in A, \xi\neq\zeta.$$

For $\xi\in A$ there is a neighborhood V_ξ of x_ξ such that

$$|\{\zeta\in A: V_\xi\cap U_\zeta\neq\emptyset\}|<\omega;$$

we assume without loss of generality, shrinking V_ξ if necessary, that

$$\text{cl}\, V_\xi\subset U_\xi \text{ and } x_\zeta\notin \text{cl}\, V_\xi \text{ for } \xi,\zeta\in A, \xi\neq\zeta.$$

We set

$$W_\xi = V_\xi\backslash\cup\{\text{cl}\, V_\zeta:\zeta\in A, \zeta\neq\xi\} \text{ for } \xi\in A.$$

It is easy to verify, using the fact that $\{V_\zeta:\zeta\in A\}$ is a locally finite family in X, that $\cup\{\text{cl}\, V_\zeta:\zeta\in A,\zeta\neq\xi\}$ is closed in X; hence we have $W_\xi\in\mathcal{T}^*(X)$ for $\xi\in A$. For $\xi\in A$ we choose $X_\xi\in\mathcal{T}^*(X)$ such that

$$x_\xi\in X_\xi\subset \text{cl}\, X_\xi\subset W_\xi,$$

and we claim, finally, that for $x\in X$ there is a neighborhood U of x such that $|\{\xi\in A: U\cap X_\xi\neq\emptyset\}|\leq 1$. Indeed if there is $\bar{\xi}\in A$ such that $x\in\text{cl}\, X_{\bar{\xi}}$, we set $U = W_{\bar{\xi}}$, and if $x\notin\cup_{\xi\in A}(\text{cl}\, X_\xi)$ we set

$$U = X\backslash\cup_{\xi\in A}(\text{cl}\, X_\xi).$$

The proof is complete.

Theorem 2.1 essentially exhausts the implications which hold among the various topological properties discussed in this chapter (but see Theorem 5.23 below, which improves Theorem 2.1(h)).

In Chapter 3 we use the Erdős–Rado theorems to investigate products of (infinitely many) spaces with the properties defined above. But the following simple result is available immediately from the definitions.

2.2 *Theorem.* Let $\alpha\geq\beta\geq\gamma\geq\delta\geq 2$ with $\beta\geq\omega$ and let X and Y be spaces.

(a) If X has calibre (α,β,δ) and Y has calibre (β,γ,δ), then $X\times Y$ has calibre (α,γ,δ).

(b) If X has compact-calibre (α,β,δ) and Y has compact-calibre (β,γ,δ), then $X\times Y$ has compact-calibre (α,γ,δ).

Proof. Let $\{U_\xi\times V_\xi:\xi<\alpha\}\subset\mathcal{T}^*(X)\times\mathcal{T}^*(Y)$. In (a) there is $B\in[\alpha]^\beta$ such that if $D\in[B]^\delta$ then $\cap_{\xi\in D}U_\xi\neq\emptyset$, and in (b) there is

$B \in [\alpha]^\beta$ such that for $D \in [B]^\delta$ there is compact $K \subset X$ such that $U_\xi \cap K \neq \emptyset$ whenever $\xi \in D$. And in (a) there is $C \in [B]^\gamma$ such that if $D \in [C]^\delta$ then $\cap_{\xi \in D} V_\xi \neq \emptyset$, and in (b) there is $C \in [B]^\gamma$ such that for $D \in [C]^\delta$ there is compact $L \subset Y$ such that $V_\xi \cap L \neq \emptyset$ whenever $\xi \in D$. It is then clear in (a) that if $D \in [C]^\delta$ then $\cap_{\xi \in D}(U_\xi \times V_\xi) \neq \emptyset$, and in (b) that for $D \in [C]^\delta$ there is compact $K \times L \subset X \times Y$ such that $(U_\xi \times V_\xi) \cap (K \times L) \neq \emptyset$ whenever $\xi \in D$.

It follows in particular (replacing α, β, γ and δ by α, α, α and β, respectively) that if $\alpha \geq \beta \geq 2$ with $\alpha \geq \omega$ then the product of a finite number of spaces with calibre (α, α, β) has calibre (α, α, β) and the product of a finite number of spaces with compact-calibre (α, α, β) has compact-calibre (α, α, β).

We note in Corollary 9.8 that for $\alpha \geq \beta \geq \omega$ the product of a finite number of pseudo-(α, β)-compact spaces, each completely regular and Hausdorff, need not be pseudo-(α, β)-compact.

The pseudoweight $\pi w(X)$ of a space X is defined in the Appendix.

2.3 *Theorem.* Let $\alpha \geq \omega$, and let X be a space.

(a) If X has calibre α then X has calibre $\mathrm{cf}(\alpha)$;

(a′) if X has calibre $\mathrm{cf}(\alpha)$ and $\pi w(X) < \alpha$, then X has calibre α;

(b) if X has compact-calibre α then X has compact-calibre $\mathrm{cf}(\alpha)$;

(b′) if X has compact-calibre $\mathrm{cf}(\alpha)$ and $\pi w(X) < \alpha$, then X has compact-calibre α;

(c) if X is pseudo-(α, α)-compact then X is pseudo-$(\mathrm{cf}(\alpha), \mathrm{cf}(\alpha))$-compact; and

(c′) if X is pseudo-$(\mathrm{cf}(\alpha), \mathrm{cf}(\alpha))$-compact and $\pi w(X) < \alpha$, then X is pseudo-(α, α)-compact.

Proof. We assume that α is singular and we choose a family $\{\alpha_\sigma : \sigma < \mathrm{cf}(\alpha)\}$ such that

$$0 = \alpha_0 < \alpha_{\sigma'} < \alpha_\sigma < \alpha \text{ for } 0 < \sigma' < \sigma < \mathrm{cf}(\alpha), \text{ and}$$
$$\alpha = \sum_{\sigma < \mathrm{cf}(\alpha)} \alpha_\sigma.$$

We first prove (a), (b) and (c). Let $\{U_\sigma : \sigma < \mathrm{cf}(\alpha)\} \subset \mathscr{T}^*(X)$, and set

$$V_\eta = U_\sigma \text{ for } \alpha_\sigma \leq \eta < \alpha_{\sigma+1}.$$

Then in (a) there is $A \in [\alpha]^\alpha$ such that $\cap_{\eta \in A} V_\eta \neq \emptyset$; in (b) there are $A \in [\alpha]^\alpha$ and compact $K \subset X$ such that $V_\eta \cap K \neq \emptyset$ whenever $\eta \in A$; and in (c) there is $x \in X$ such that $|\{\eta < \alpha : V \cap V_\eta \neq \emptyset\}| = \alpha$ for every neighborhood V of x.

Now in (a) and (b) set

$$A(\sigma) = A \cap [\alpha_\sigma, \alpha_{\sigma+1}) \text{ for } \sigma < \mathrm{cf}(\alpha),$$

and in (c) let V be a neighborhood of x and set

$$A(\sigma) = \{\eta \in [\alpha_\sigma, \alpha_{\sigma+1}) : V \cap V_\eta \neq \emptyset\} \text{ for } \sigma < \mathrm{cf}(\alpha);$$

in each case set $B = \{\sigma < \mathrm{cf}(\alpha) : A(\sigma) \neq \emptyset\}$, and for $\sigma \in B$ let $\eta(\sigma) \in A(\sigma)$. Then $B \in [\mathrm{cf}(\alpha)]^{\mathrm{cf}(\alpha)}$ (since $A = \cup_{\sigma \in B} A(\sigma)$ and $|A(\sigma)| < \alpha$ for $\sigma \in B$) and we have:

$$\bigcap_{\sigma \in B} U_\sigma = \bigcap_{\sigma \in B} V_{\eta(\sigma)} \neq \emptyset \text{ in (a)},$$

$$U_\sigma \cap K = V_{\eta(\sigma)} \cap K \neq \emptyset \quad \text{for } \sigma \in B \text{ in (b), and}$$

$$V \cap U_\sigma = V \cap N_{\eta(\sigma)} \neq \emptyset \quad \text{for } \sigma \in B \text{ in (c).}$$

as required.

We prove (a′), (b′) and (c′). Let $\{U_\xi : \xi < \alpha\} \subset \mathscr{T}^*(X)$, let $\{B_\eta : \eta < \beta\}$ be a pseudobase for X with $\beta < \alpha$, and set

$$A(\eta) = \{\xi < \alpha : B_\eta \subset U_\xi\} \text{ for } \eta < \beta.$$

Since $\alpha = \cup_{\eta < \beta} A(\eta)$ and $\beta < \alpha$, we have

$$\sup \{|A(\eta)| : \eta < \beta\} = \alpha;$$

for $\sigma < \mathrm{cf}(\alpha)$ we choose $\eta(\sigma) < \beta$ such that $|A(\eta(\sigma))| \geq \alpha_\sigma$. Then in (a′) there is $C \in [\mathrm{cf}(\alpha)]^{\mathrm{cf}(\alpha)}$ such that $\cap_{\sigma \in C} B_{\eta(\sigma)} \neq \emptyset$; in (b′) there are $C \in [\mathrm{cf}(\alpha)]^{\mathrm{cf}(\alpha)}$ and compact $K \subset X$ such that $B_{\eta(\sigma)} \cap K \neq \emptyset$ whenever $\sigma \in C$; and in (c′) there is $x \in X$ such that $|\{\sigma < \mathrm{cf}(\alpha) : V \cap B_{\eta(\sigma)} \neq \emptyset\}| = \mathrm{cf}(\alpha)$ for every neighborhood V of x. In (c′) let V be a neighborhood of x and set

$$C = \{\sigma < \mathrm{cf}(\alpha) : V \cap B_{\eta(\sigma)} \neq \emptyset\},$$

and in each case set $A = \cup_{\sigma \in C} A(\eta(\sigma))$. Then $A \in [\alpha]^\alpha$ and we have:

$$\bigcap_{\xi \in A} U_\xi \supset \bigcap_{\sigma \in C} B_{\eta(\sigma)} \neq \emptyset \text{ in (a′)},$$

$$U_\xi \cap K \supset B_{\eta(\sigma)} \cap K \neq \emptyset \quad \text{for } \xi \in A(\eta(\sigma)) \text{ in (b′), and}$$

$$V \cap V_\xi \supset V \cap B_{\eta(\sigma)} \neq \emptyset \quad \text{for } \xi \in A(\eta(\sigma)) \text{ in (c′),}$$

as required.

The proof is complete.

We show in Theorems 4.3, 6.2 and 7.1 that for α singular the assumption $\pi w(X) < \alpha$ in (a′), (b′) and (c′) cannot be omitted.

2.4 *Theorem.* Let $\alpha \geq \beta \geq \omega$ and let γ be a cardinal number.

(a) The following are equivalent: the space γ has calibre (α, β); the space γ has compact-calibre (α, β); the space γ is pseudo-(α, β)-compact.

(b) The space γ has calibre α if and only if $\gamma < \mathrm{cf}(\alpha)$.

(c) The space γ has calibre (α, β) if and only if ($\alpha = \beta$ and $\gamma < \mathrm{cf}(\alpha)$) or ($\beta < \alpha$ and $\gamma < \alpha$).

Proof. (a) For $\{U_\xi : \xi < \alpha\} \subset \mathscr{T}^*(\gamma)$ and $\eta \in \gamma$ we have ($\{\eta\}$ is a neighborhood of η and) $\eta \in U_\xi$ if and only if $\{\eta\} \cap U_\xi \neq \varnothing$. Thus the discrete space γ, if pseudo-(α, β)-compact, has calibre (α, β). The remaining required implications are given by Theorem 2.1 ((d) and (f)).

(b) ($\Rightarrow$) There is $\{A_\sigma : \sigma < \mathrm{cf}(\alpha)\}$ such that

$$\bigcup_{\sigma < \mathrm{cf}(\alpha)} A_\sigma = \alpha,$$

$$A_{\sigma'} \cap A_\sigma = \varnothing \quad \text{for } \sigma' < \sigma < \mathrm{cf}(\alpha), \text{ and}$$

$$|A_\sigma| < \alpha \quad \text{for } \sigma < \mathrm{cf}(\alpha).$$

If $\mathrm{cf}(\alpha) \leq \gamma$, then for $\xi \in A_\sigma$ we set

$$U_\xi = \{\sigma\} \subset \mathrm{cf}(\alpha) \subset \gamma.$$

Then $\{U_\xi : \xi < \alpha\} \subset \mathscr{T}^*(\gamma)$ and for $\eta \in \gamma$ the set $\{\eta\}$ is a neighborhood of η such that

$$|\{\xi < \alpha : \{\eta\} \cap U_\xi \neq \varnothing\}| < \alpha,$$

a contradiction.

($\Leftarrow$) Clearly for $\{U_\xi : \xi < \alpha\} \subset \mathscr{T}^*(\gamma)$ there is $\eta \in \gamma$ such that $|\{\xi < \alpha : \eta \in U_\xi\}| = \alpha$.

(c) ($\Rightarrow$) Suppose that γ has calibre (α, β). If $\alpha = \beta$ then $\gamma < \mathrm{cf}(\alpha)$ by (b). If $\beta < \alpha$ and $\gamma \geq \alpha$, then $\{\{\xi\} : \xi < \alpha\}$ is a family of non-empty open subsets of α (hence of γ) such that

$$\{\xi'\} \cap \{\xi\} = \varnothing \text{ for } \xi' < \xi < \alpha,$$

so that γ does not have calibre (α, β).

($\Leftarrow$) If $\beta = \alpha$ and $\gamma < \mathrm{cf}(\alpha)$, then γ has calibre (α, β) by (b). If $\beta < \alpha$

and $\gamma < \alpha$ we set $\delta = \max\{\beta, \gamma\}$; since $\gamma < \delta^+ = \mathrm{cf}(\delta^+)$ the space γ has calibre δ^+ by (b), and since $\beta < \delta^+ \le \alpha$ the space γ has calibre (α, β) by Theorem 2.1(a).

We note finally that the properties considered above are preserved under certain standard topological operations. We consider in particular the following four (possible) relationships between spaces X and Y:

(i) There is a continuous function from Y onto X; or

(ii) Y is a dense subspace of X; or

(iii) X is an open subspace of Y; or

(iv) X is the closure (in Y) of an open subspace of Y.

2.5 *Theorem.* Let $\alpha \ge \beta \ge \gamma \ge 2$ and let X and Y be spaces.

(a) If X and Y are related as in (i), (ii), (iii) or (iv) above and if Y has calibre (α, β, γ), then X has calibre (α, β, γ);

(b) if X and Y are related as in (i), (ii) or (iv) above and if Y has compact-calibre (α, β, γ) or is pseudo-(α, β)-compact, then X has the same property.

Proof. We note that statement (a)(iv) follows from (a)(ii) and (a)(iii).

We prove that if Y is pseudo-(α, β)-compact and $X = \mathrm{cl}_Y U$ with $U \in \mathscr{T}(Y)$, then X is pseudo-(α, β)-compact. If $\{U_\xi : \xi < \alpha\} \subset \mathscr{T}^*(X)$, then since $\{U \cap U_\xi : \xi < \alpha\} \subset \mathscr{T}^*(Y)$ there is $x \in Y$ such that

$$|\{\xi < \alpha : V \cap (U \cap U_\xi) \neq \emptyset\}| \ge \beta$$

for every neighborhood V (in Y) of x. If $x \in Y \backslash X$, then $Y \backslash X$ is a neighborhood in Y of x such that $(Y \backslash X) \cap (U \cap U_\xi) = \emptyset$ for all $\xi < \alpha$; hence $x \in X$, as required.

We omit the remaining routine verifications.

2.6 *Remarks.* (a) It follows from Theorem 2.5((i) and (ii)) that if $\alpha \ge \beta \ge \gamma \ge 2$ and X is a space, and if the discrete space $d(X)$ has calibre (α, β, γ) or compact-calibre (α, β, γ) or is pseudo-(α, β)-compact, then X has the same property.

(b) There is a converse to (a portion of) Theorem 2.5(a)(ii): If $\alpha \ge \beta \ge 2$ and Y is dense in X and X has calibre $(\alpha, \beta, 2)$, then Y has calibre $(\alpha, \beta, 2)$. If $\{U_\xi : \xi < \alpha\} \subset \mathscr{T}^*(Y)$, there is $\{\tilde{U}_\xi : \xi < \alpha\}$ such that $\tilde{U}_\xi \cap Y = U_\xi$ for $\xi < \alpha$, and since X has calibre $(\alpha, \beta, 2)$ there is $B \in [\alpha]^\beta$ such that $\tilde{U}_\xi \cap \tilde{U}_{\xi'} \neq \varnothing$ for $\xi, \xi' \in B$; since Y is dense in X it follows that $U_\xi \cap U_{\xi'} \neq \varnothing$ for $\xi, \xi' \in B$, as required.

(c) If $\alpha \geq \beta \geq \gamma$ and $\alpha \geq \omega$, then every compact space has compact-calibre (α, β, γ) and is pseudo-(α, β)-compact. Hence an open subspace of a space with compact-calibre (α, β, γ) need not have compact-calibre (α, β, γ), and an open subspace of a pseudo-(α, β)-compact space need not be pseudo-(α, β)-compact.

We note next that for every non-empty space X some (infinite, regular) cardinal is a calibre of X.

2.7 *Theorem.* Let α be an infinite cardinal and X a non-empty space. If $d(X) < \mathrm{cf}(\alpha)$, then X has calibre α.

Proof. Let $\{x_\eta : \eta < \beta\}$ be a dense subset of X with $\beta < \mathrm{cf}(\alpha)$, let $\{U_\xi : \xi < \alpha\} \subset \mathscr{T}^*(X)$, and set

$$A(\eta) = \{\xi < \alpha : x_\eta \in U_\xi\} \text{ for } \eta < \beta.$$

Since $\cup_{\eta < \beta} A(\eta) = \alpha$ and $\beta < \mathrm{cf}(\alpha)$, there is $\bar{\eta} < \beta$ such that $A(\bar{\eta}) \in [\alpha]^\alpha$; we have

$$x_{\bar{\eta}} \in \bigcap_{\xi \in A(\bar{\eta})} U_\xi,$$

as required.

The λ-Souslin numbers

Definition. Let X be a space. A *cellular family* in X is an indexed set $\{U_i : i \in I\}$ of non-empty open subsets of X such that

$$U_i \cap U_{i'} = \varnothing \text{ for } i, i' \in I, i \neq i'.$$

It follows from Theorem 2.7 that for X a non-empty space and λ a cardinal, $\lambda > 0$, there is a cardinal α such that X has calibre (α, λ). (One may take $\alpha = (\max\{d(X), \lambda, \omega\})^+$.) This validates the following definition.

Definition. For $\lambda > 0$ and X a non-empty space, the *λ-Souslin number* of X, denoted $S_\lambda(X)$, is the cardinal

$$S_\lambda(X) = \min\{\alpha : X \text{ has calibre } (\alpha, \lambda)\}.$$

The 2-Souslin number of X is called, simply, the *Souslin number of* X and is (usually) denoted $S(X)$.

We note some simple consequences of this definition.

2.8 *Lemma.* Let $\lambda > 0$, let α be a cardinal and let X be a non-empty space.

(a) $\lambda \leq S_\lambda(X)$;

(b) $\alpha < S_\lambda(X)$ if and only if there is a set $\{U_\xi : \xi < \alpha\}$ of non-empty, open subsets of X such that if $A \subset \alpha$ with $|A| = \lambda$ then $\cap_{\xi \in A} U_\xi = \varnothing$;

(c) if $\lambda = 1$ then $S_\lambda(X) = 1$;

(d) if $\lambda = 2$ then $S_\lambda(X)$ is the least cardinal β such that no cellular family in X is indexed by β; and,

(e) if $\lambda \geq \kappa$ then $S_\lambda(X) \geq S_\kappa(X)$.

We determine the λ-Souslin numbers of discrete spaces.

2.9 *Lemma.* Let $\lambda > 0$ and $\beta > 0$ and set

$$\begin{aligned} \alpha &= (\lambda - 1)\beta + 1 && \text{if } \lambda < \omega, \beta < \omega, \\ \alpha &= \beta^+ && \text{if } \beta \geq \omega, \beta \geq \lambda > 1, \\ \alpha &= \lambda^+ && \text{if } \lambda \geq \omega, \mathrm{cf}(\lambda) \leq \beta < \lambda, \\ \alpha &= \lambda && \text{if } \lambda \geq \omega, \beta < \mathrm{cf}(\lambda). \end{aligned}$$

Then $\alpha = S_\lambda(\beta)$.

Proof. ($\geq$) It is clear in each case that if $\{U_\xi : \xi < \alpha\} \subset \mathscr{T}^*(\beta)$ and $x(\xi) \in U_\xi$, then there is $\eta < \beta$ such that

$$|\{\xi < \alpha : x(\xi) = \eta\}| \geq \lambda.$$

($\leq$) For $\lambda \geq \omega, \beta < \mathrm{cf}(\lambda)$ we have $S_\lambda(\beta) \geq \lambda = \alpha$ from Lemma 2.8(a). To establish the required inequality in the other three cases we set

$$\begin{aligned} \gamma &= (\lambda - 1)\beta && \text{if } \lambda < \omega, \beta < \omega \\ \gamma &= \beta && \text{if } \beta \geq \omega, \beta \geq \lambda > 1 \\ \gamma &= \lambda && \text{if } \lambda \geq \omega, \mathrm{cf}(\lambda) \leq \beta < \lambda, \end{aligned}$$

and we note in each case that it is simple to find $\{x(\xi) : \xi < \gamma\} \subset \beta$ so that

$$|\{\xi < \gamma : x(\xi) = \eta\}| < \lambda \text{ for } \eta < \beta.$$

It follows that $S_\lambda(\beta) > \gamma$; hence $S_\lambda(\beta) \geq \alpha$.

The following terminology will be used (only) in the proof of Lemma 2.10 below.

A space X has *large finite cellular families* if for all $n < \omega$ there is a cellular family $\{U_i : i \in I\}$ in X such that $|I| \geq n$.

It is clear that if X has large finite cellular families and $\mathscr{U}$ is a finite family of open subsets of X such that $\cup \mathscr{U}$ is dense in X, then there is $U \in \mathscr{U}$ such that U has large finite cellular families.

2.10 *Lemma.* Let X be a space. If X has large finite cellular families, then there is a cellular family $\{U_n : n < \omega\}$ (indexed by ω) in X.

Proof. For $n < \omega$ let $\mathscr{U}_n = \{U_i(n) : i \in I(n)\}$ be a cellular family in X with $n + 2 \leq |I(n)| < \omega$. We assume without loss of generality, adjoining

$$X \backslash \mathrm{cl}(\bigcup_{i \in I(n)} U_i(n))$$

to $\mathscr{U}_n$ and reindexing, that $\cup \mathscr{U}_n$ is dense in X for $n < \omega$. We assume also, recursively replacing $\mathscr{U}_{n+1}$ by

$$\{U_i(n+1) \cap U_j(n) : i \in I(n+1), j \in I(n), U_i(n+1) \cap U_j(n) = \varnothing\}$$

and reindexing, that $\mathscr{U}_{n+1}$ is a *refinement* of $\mathscr{U}_n$ for $n < \omega$ (in the sense that for $i \in I(n+1)$ there is $j \in U(n)$ such that $U_i(n+1) \subset U_j(n)$). Finally we assume, again using finite recursion to replace $\mathscr{U}_n$ by a refinement, that if $i \in I(n)$ and $U_i(n)$ has large finite cellular families, then

$$|\{j \in I(n+1) : U_j(n+1) \subset U_i(n)\}| \geq 2.$$

It is now simple to define by finite recursion sets $\{i(n) : n < \omega\}$ and $\{j(n) : n < \omega\}$ such that

(i) $i(n), j(n) \in I(n)$ for $n < \omega$;

(ii) $U_{i(n)}(n)$ has large finite cellular families for $n < \omega$;

(iii) $i(n) \neq j(n)$ for $n < \omega$; and

(iv) $U_{i(n)}(n) \cup U_{j(n)}(n) \subset U_{i(n-1)}(n-1)$ for $0 < n < \omega$.

It is then clear that $\{U_{j(n)}(n) : n < \omega\}$ is a cellular family in X.

The proof of Lemma 2.10 is essentially the proof of the classical tree theorem of D. König (1927).

From Lemma 2.10 it follows that if X is a space then $S(X) \neq \omega$.

2.11 *Corollary.* Let X be an infinite Hausdorff space and $1 < \lambda < \omega$. Then

(a) $S_2(X) \geq \omega^+$ (i.e., X does not have calibre $(\omega, 2, 2)$);

(b) $S_2(X) = S_\lambda(X)$.

Proof. (a) It follows from Lemma 2.10 that there is a cellular family in X indexed by ω. Hence $S_2(X) \geq \omega^+$.

(b) now follows from Theorem 2.1((a) and (h)).

2.12 *Corollary.* (a) $\{0,1\}^m$ has calibre α for $0 < m < \omega, \alpha \geq \omega$; and

(b) $\{0,1\}^\omega$ does not have calibre α if $\mathrm{cf}(\alpha) = \omega$.

Proof. (a) Since $|\{0,1\}^m| < \omega$ for $0 < m < \omega$, this follows from Theorem 2.6.

(b) From Corollary 2.11(a), Theorem 2.1(a) and Theorem 2.3(a), respectively, it follows that $\{0,1\}^\omega$ does not have calibre $(\omega, 2, 2)$, hence does not have calibre ω, hence does not have calibre α if $\mathrm{cf}(\alpha) = \omega$.

2.13 *Lemma.* Let $\lambda \geq 2$ and let X be a non-empty space. If $S_\lambda(X)$ is a limit cardinal, then there is a non-empty, open subset U of X such that $S_\lambda(V) = S_\lambda(X)$ for every non-empty, open subset V of U.

Proof. From Lemma 2.8(a) we have $S_\lambda(X) \geq \lambda$. If $S_\lambda(X) = \lambda$, then for $V \in \mathscr{T}^*(X)$ we have

$$\lambda \leq S_\lambda(V) \leq S_\lambda(X) \leq \lambda,$$

and we complete the proof by setting $U = X$. In what follows we assume $S_\lambda(X) > \lambda$.

If the lemma fails, then for all $U \in \mathscr{T}^*(X)$ there is $V \in \mathscr{T}^*(U)$ such that $S_\lambda(V) < S_\lambda(X)$. We choose a maximal cellular family $\{V_i : i \in I\}$ of X such that $S_\lambda(V_i) < S_\lambda(X)$ for $i \in I$, and we set $\alpha = \sum_{i \in I} S_\lambda(V_i)$. Since $|I| < S_2(X) \leq S_\lambda(X)$, we have

$$\lambda \leq \alpha \leq |I| \cdot S_\lambda(X) = S_\lambda(X);$$

we claim $\alpha < S_\lambda(X)$.

If the claim fails, then α is the limit cardinal $S_\lambda(X)$ and there is a family $\{\alpha_\sigma : \sigma < \mathrm{cf}(\alpha)\}$ of cardinal numbers such that

$$\lambda = \alpha_0 \leq \alpha_{\sigma'} < \alpha_\sigma < \alpha \quad \text{for } \sigma' < \sigma < cf(\alpha), \text{ and}$$

$$\sum_{\sigma < \mathrm{cf}(\alpha)} \alpha_\sigma = \alpha.$$

We note that $\alpha = \sup\{S_\lambda(V_i) : i \in I\}$; indeed if there is $\beta < \alpha$ such that $S_\lambda(V_i) \leq \beta$ for all $i \in I$, then from $|I| < S_2(X) \leq S_\lambda(X) = \alpha$ we would have the contradiction

$$\alpha = \sum_{i \in I} S_\lambda(V_i) \leq |I| \cdot \beta < \alpha.$$

Thus for $\sigma < \mathrm{cf}(\alpha)$ there is $i(\sigma) \in I$ such that $S_\lambda(V_{i(\sigma)}) > \alpha_{\sigma+1} > \lambda$. We define $\{U_\xi : \alpha_\sigma \leq \xi < \alpha_{\sigma+1}\} \subset \mathscr{T}^*(V_{i(\sigma)})$ such that $\cap_{\xi \in A} U_\xi = \varnothing$ for all $A \in [\alpha_{\sigma+1} \backslash \alpha_\sigma]^\lambda$ and we note that if $A \in [\alpha \backslash \lambda]^\lambda$ then $\cap_{\xi \in A} U_\xi = \varnothing$; this is clear if there is $\sigma < \mathrm{cf}(\alpha)$ such that $A \in [\alpha_{\sigma+1} \backslash \alpha_\sigma]^\lambda$, while if there are $\zeta, \zeta' \in A$ and $\sigma, \sigma' < \mathrm{cf}(\alpha)$ such that

$$\alpha_{\sigma'} \leq \zeta' < \alpha_{\sigma'+1} \text{ and } \alpha_\sigma \leq \zeta < \alpha_{\sigma+1},$$

then $\cap_{\xi \in A} U_\xi \subset U_{\zeta'} \cap U_\zeta \subset V_{i(\sigma')} \cap V_{i(\sigma)} = \varnothing$.

This completes the proof that $S_\lambda(X) > \alpha$.

Now set $\beta = (\alpha \cdot |I|)^+$. Since $\beta < S_\lambda(X)$, there is $\{W_\xi : \xi < \beta\} \subset \mathscr{T}^*(X)$ such that $\cap_{\xi \in A} W_\xi = \varnothing$ for all $A \in [\beta]^\lambda$. For $i \in I$ we set

$$A_i = \{\xi < \beta : W_\xi \cap V_i \neq \varnothing\}.$$

From the maximality of the cellular family $\{V_i : i \in I\}$ it follows that $\cup_{i \in I} V_i$ is dense in X, so that $\cup_{i \in I} A_i = \beta$. Since $|I| < \beta$ there is $\bar{i} \in I$ such that $|A_{\bar{i}}| = \beta$, and from the relations

$$\{W_\xi \cap V_{\bar{i}} : \xi \in A_{\bar{i}}\} \subset \mathscr{T}^*(V_{\bar{i}}),$$
$$S_\lambda(V_{\bar{i}}) \leq \alpha < \beta$$

there is $A \in [A_{\bar{i}}]^\lambda$ such that $\cap_{\xi \in A}(W_\xi \cap V_{\bar{i}}) \neq \varnothing$. We have

$$\varnothing \neq \bigcap_{\xi \in A} (W_\xi \cap V_{\bar{i}}) \subset \bigcap_{\xi \in A} W_\xi = \varnothing,$$

a contradiction.

The proof is complete.

2.14 *Theorem.* Let λ be a cardinal such that either $\lambda = 2$ or λ is (infinite and) regular, and let X be an infinite Hausdorff space. Then $S_\lambda(X)$ is a regular cardinal.

Proof. We argue by contradiction. We assume the result fails, we set $\alpha = S_\lambda(X)$, we note that $\lambda < \alpha$ and we (will) define $\{V_\xi : \lambda \leq \xi < \alpha\} \subset \mathscr{T}^*(X)$ such that $\cap_{\xi \in A} V_\xi = \varnothing$ for all $A \in [\alpha \backslash \lambda]^\lambda$.

From Corollary 2.11 we have $S(X) \geq \omega^+$. Since $\alpha = S_\lambda(X) \geq S(X)$ (by Lemma 2.8(e)) the cardinal α is a singular limit cardinal and from

Lemma 2.13 there is $U \in \mathscr{T}^*(X)$ such that $S_\lambda(V) = \alpha$ for all $V \in \mathscr{T}^*(U)$. We choose a family $\{\alpha_\sigma : \sigma < \mathrm{cf}(\alpha)\}$ of cardinal numbers such that

$$\lambda = \alpha_0 \leq \alpha_{\sigma'} < \alpha_\sigma < \alpha \text{ for } \sigma' < \sigma < \mathrm{cf}(\alpha), \text{ and}$$
$$\sum_{\sigma < \mathrm{cf}(\alpha)} \alpha_\sigma = \alpha,$$

and we define $\{V_\xi : \alpha_\sigma \leq \xi < \alpha_{\sigma+1}\}$ for $\sigma < \mathrm{cf}(\alpha)$ as follows. If $\mathrm{cf}(\alpha) < \lambda$ we define $\{V_\xi : \alpha_\sigma \leq \xi < \alpha_{\sigma+1}\} \subset \mathscr{T}^*(U)$ so that $\cap_{\xi \in A} V_\xi = \varnothing$ for all $A \in [\alpha_{\sigma+1} \backslash \alpha_\sigma]^\lambda$; and if $\lambda \leq \mathrm{cf}(\alpha)$ we define $\{U_\sigma : \sigma < \mathrm{cf}(\alpha)\} \subset \mathscr{T}^*(U)$ so that $\cap_{\sigma \in A} U_\sigma = \varnothing$ for all $A \in [\mathrm{cf}(\alpha)]^\lambda$ and (using the relation $S_\lambda(U_\sigma) = \alpha$) we define $\{V_\xi : \alpha_\sigma \leq \xi < \alpha_{\sigma+1}\} \subset \mathscr{T}^*(U_\sigma)$ so that $\cap_{\xi \in A} V_\xi = \varnothing$ for all $A \in [\alpha_{\sigma+1} \backslash \alpha_\sigma]^\lambda$.

We show that the family $\{V_\xi : \lambda \leq \xi < \alpha\}$ is as required. Let $A \in [\alpha \backslash \lambda]^\lambda$, set

$$A_\sigma = A \cap [\alpha_\sigma, \alpha_{\sigma+1}) \text{ for } \sigma < \mathrm{cf}(\alpha),$$

and set $B = \{\sigma < \mathrm{cf}(\alpha) : A_\sigma \neq \varnothing\}$. If $|B| < \lambda$, then since $\lambda = 2$ or λ is regular there is $\bar{\sigma} < \mathrm{cf}(\alpha)$ such that $|A_{\bar{\sigma}}| = \lambda$ and we have

$$\bigcap_{\xi \in A} V_\xi \subset \cap \{V_\xi : \xi \in A_{\bar{\sigma}}\} = \varnothing;$$

and if $|B| = \lambda$ then ($\lambda \leq \mathrm{cf}(\alpha)$ and U_σ is defined for $\sigma < \mathrm{cf}(\alpha)$ and) we have

$$\bigcap_{\xi \in A} V_\xi \subset \bigcap_{\sigma \in B} U_\sigma = \varnothing.$$

Reduction to compact Hausdorff spaces

For use in describing the calibre properties of compact spaces we introduce the relations $\leq$ and $\equiv$ on the class of spaces and we show that every space X is $\equiv$-equivalent to its (compact) Gleason space $G(X)$.

Definition. Let X and Y be spaces. Then $X \leq Y$ if there is $\varphi : \mathscr{T}^*(X) \to \mathscr{T}^*(Y)$ such that

if $n < \omega$ and $\{U_k : k < n\} \subset \mathscr{T}^*(X)$ and $\cap_{k<n} U_k = \varnothing$, then $\cap_{k<n} \varphi(U_k) = \varnothing$.

With X, Y and φ as above we say that φ *makes* $X \leq Y$.

It is clear that if X, Y and Z are spaces, then $X \leq X$ (i.e., $\leq$ is a *reflexive* relation) and if $X \leq Y$ and $Y \leq Z$ then $X \leq Z$ (i.e., $\leq$ is a *transitive* relation).

2.15 *Lemma.* Let X and Y be spaces, $\mathscr{B}$ a pseudobase for X and φ a function from $\mathscr{B}$ into $\mathscr{T}^*(Y)$ such that

if $n < \omega$ and $\{B_k : k < n\} \subset \mathscr{B}$ and $\cap_{k<n} B_k = \emptyset$, then $\cap_{k<n} \varphi(B_k) = \emptyset$. Then $X \leq Y$.

Proof. For $U \in \mathscr{T}^*(X)$ we choose $B_U \in \mathscr{B}$ with $\emptyset \neq B_U \subset U$, and we define $\psi : \mathscr{T}^*(X) \to \mathscr{T}^*(Y)$ by the rule $\psi(U) = \varphi(B_U)$. Then ψ makes $X \leq Y$.

2.16 *Theorem.* Let X and Y be spaces.

(a) If X is an open subspace of Y, then $X \leq Y$;

(b) if X is a dense subspace of Y, then $X \leq Y$ and $Y \leq X$;

(c) if X is a continuous image of Y, then $X \leq Y$; and

(d) if there are sets $\{X_i : i \in I\}$ and $\{Y_i : i \in I\}$ of spaces with $X_i \leq Y_i$ for $i \in I$ and with $X = \prod_{i \in I} X_i$ and $Y = \prod_{i \in I} Y_i$, then $X \leq Y$.

Proof. (a) Define $\varphi : \mathscr{T}^*(X) \to \mathscr{T}^*(Y)$ by $\varphi(U) = U$. Then φ makes $X \leq Y$.

(b) For $U \in \mathscr{T}^*(X)$ let $\varphi(U)$ be an element of $\mathscr{T}^*(Y)$ such that $\varphi(U) \cap X = U$. Then φ makes $X \leq Y$.

For $V \in \mathscr{T}^*(Y)$ let $\psi(V) = V \cap X$. Then ψ makes $Y \leq X$.

(c) Let f be a continuous function from Y onto X, and define $\varphi : \mathscr{T}^*(X) \to \mathscr{T}^*(Y)$ by $\varphi(U) = f^{-1}(U)$. Then φ makes $X \leq Y$.

(d) Let φ_i make $X_i \leq Y_i$ for $i \in I$, let $\mathscr{B}$ denote the canonical base for the cartesian product topology of X, and define $\varphi : \mathscr{B} \to \mathscr{T}^*(Y)$ by $\varphi(\cap_{i \in F} \pi_i^{-1}(U_i)) = \cap_{i \in F} \pi_i^{-1}(\varphi_i(U_i))$ for $F \in \mathscr{P}^*_\omega(I)$, $U_i \in \mathscr{T}^*(X_i)$ for $i \in F$. We have $X \leq Y$ from Lemma 2.15.

It is convenient to consider several classes of spaces.

Definition. Let X be a space, α an infinite cardinal, ρ an ordinal such that $\rho \leq \alpha$, and $2 \leq n < \omega$.

(a) X has *the* c.c.c. (or, *X is a* c.c.c. *space*) if every cellular family in X is finite or countably infinite, i.e., if $S(X) \leq \omega^+$.

(b) X is *productively* c.c.c. if $X \times Y$ is a c.c.c. space for every c.c.c. space Y.

(c) X has *precalibre* (α, ρ) if for every set $\{U_\xi : \xi < \alpha\}$ of non-empty, open subsets of X there is $A \subset \alpha$ such that type $A = \rho$ and $\{U_\xi : \xi \in A\}$

has the finite intersection property; X has *precalibre* α if X has precalibre (α, α).

(d) X has *property* $\mathrm{K}_{\alpha,n}$ if X has calibre (α, α, n); X has *property* K_n if X has property $\mathrm{K}_{\omega^+,n}$; and X has *property* (K) if X has property K_2.

It follows from Theorem 2.1(a) that a space with calibre ω^+ has property (K), and a space with property (K) has the c.c.c.

The notation c.c.c. abbreviates the expression *countable chain condition*.

2.17 *Theorem.* Let X and Y be spaces with $X \leq Y$. Then

(a) $S(X) \leq S(Y)$;

(b) if Y is productively c.c.c. then X is productively c.c.c.;

(c) if α is an infinite cardinal and ρ an ordinal such that Y has precalibre (α, ρ), then X has precalibre (α, ρ);

(d) if α is an infinite cardinal, $2 \leq n < \omega$ and Y has property $\mathrm{K}_{\alpha,n}$, then X has property $\mathrm{K}_{\alpha,n}$.

Proof. Let $\varphi: \mathscr{T}^*(X) \to \mathscr{T}^*(Y)$ make $X \leq Y$.

(a) If the indexed family $\{U_i : i \in I\}$ is a cellular family in X, then $\{\varphi(U_i) : i \in I\}$ is a cellular family in Y; hence $S(X) \leq S(Y)$.

(b) Let Z be a c.c.c. space. Since $X \leq Y$ and $Z \leq Z$ we have $X \times Z \leq Y \times Z$ (from Theorem 2.16(c)) and hence

$$S(X \times Z) \leq S(Y \times Z) \leq \omega^+$$

from part (a). It follows that $X \times Z$ is a c.c.c. space, as required.

(c) For $\{U_\xi : \xi < \alpha\} \subset \mathscr{T}^*(X)$ we have $\{\varphi(U_\xi) : \xi < \alpha\} \subset \mathscr{T}^*(Y)$ and since Y has precalibre (α, ρ) there is $A \subset \alpha$ with type $A = \rho$ such that $\{\varphi(U_\xi) : \xi \in A\}$ has the finite intersection property. It follows that $\{U_\xi : \xi \in A\}$ has the finite intersection property, as required.

(d) For $\{U_\xi : \xi < \alpha\} \subset \mathscr{T}^*(X)$ we have $\{\varphi(U_\xi) : \xi < \alpha\} \subset \mathscr{T}^*(Y)$ and since Y has property $\mathrm{K}_{\alpha,n}$ there is $A \in [\alpha]^\alpha$ such that if $C \subset A$ and $|C| = n$ then $\cap_{\xi \in C} \varphi(U_\xi) \neq \emptyset$. It follows that if $C \subset A$ and $|C| = n$ then $\cap_{\xi \in C} U_\xi \neq \emptyset$, as required.

Definition. For spaces X and Y, the notation $X \equiv Y$ means that $X \leq Y$ and $Y \leq X$.

It is clear that $\equiv$ is an equivalence relation on the class of spaces.

2.18 *Corollary.* Let X and Y be spaces with $X \equiv Y$. Then

(a) $S(X) = S(Y)$;

(b) X is productively c.c.c. if and only if Y is productively c.c.c.;

(c) if α is an infinite cardinal and ρ an ordinal such that $\rho \leq \alpha$, then X has precalibre (α, ρ) if and only if Y has precalibre (α, ρ);

(d) if α is an infinite cardinal and $2 \leq n < \omega$, then X has property $K_{\alpha,n}$ if and only if Y has property $K_{\alpha,n}$; and

(e) $X \times Z \equiv Y \times Z$ for every space Z.

Proof. Parts (a), (b), (c) and (d) follow from the corresponding parts of Theorem 2.17, and (e) follows from Theorem 2.16(d).

The following simple criterion for $\equiv$-equivalence, involving pseudobases, is useful.

2.19 *Lemma.* Let X and Y be spaces with pseudobases $\mathscr{B}$ and $\mathscr{C}$ respectively, and let φ be a function from $\mathscr{B}$ onto $\mathscr{C}$ such that if $n < \omega$ and $\{B_k : k < n\} \subset \mathscr{B}$, then $\cap_{k<n} B_k = \emptyset$ if and only if $\cap_{k<n} \varphi(B_k) = \emptyset$.

Then $X \equiv Y$.

Proof. Since $\varphi[\mathscr{B} \backslash \{\emptyset\}] \subset \mathscr{T}^*(Y)$ we have $X \leq Y$ from Lemma 2.15. For $V \in \mathscr{C}$ with $V \neq \emptyset$ we choose $\psi(V) \in \mathscr{B}$ such that $\varphi(\psi(V)) = V$; then $\psi(V) \neq \emptyset$ and again from Lemma 2.15 we have $Y \leq X$.

For a space X the Boolean algebra $\mathscr{R}(X)$ of regular-open subsets of X, and the Stone space $S(\mathscr{R}(X))$ of $\mathscr{R}(X)$ (which is the Gleason space $G(X)$ of X) are defined and discussed in the Appendix.

The next result shows that for every space X there is an extremally disconnected, compact Hausdorff space (namely, the Gleason space $G(X)$ of X) which is $\equiv$-equivalent to X.

2.20 *Theorem.* If X is a space then $X \equiv G(X)$.

Proof. It follows from the Appendix B that the Boolean algebra $\mathscr{R}(X)$ is a pseudobase (indeed, a base) for $G(X)$. We define

$$\varphi : \mathscr{T}(X) \to \mathscr{R}(X) \subset \mathscr{T}(G(X))$$

by

$$\varphi(U) = \text{int}_X \, \text{cl}_X \, U \quad \text{for } U \in \mathscr{T}(X),$$

and we set $\varphi^* = \varphi|\mathscr{T}^*(X)$. From Lemma B.14 of the Appendix it follows that

φ^* is a function from $\mathscr{T}^*(X)$ onto $\mathscr{R}^*(X)$, and if $n < \omega$ and $\{U_k : k < n\} \subset \mathscr{T}^*(X)$, then $\cap_{k<n} U_k = \varnothing$ if and only if $\cap_{k<n} \varphi(U_k) = \varnothing$.

Hence from Lemma 2.19 we have $X \equiv G(X)$, as required.

We continue with two consequences of Theorem 2.20.

2.21 *Corollary*. Let $\{X_i : i \in I\}$ be a set of spaces. Then $G(\Pi_{i\in I} X_i) \equiv \Pi_{i\in I} G(X_i)$.

Proof. From Theorems 2.20 and 2.16(d) we have

$$G\left(\prod_{i\in I} X_i\right) \equiv \prod_{i\in I} X_i \equiv \prod_{i\in I} G(X_i),$$

as required.

2.22 *Corollary*. Let α be an infinite cardinal, ρ an ordinal such that $\rho \leq \alpha$, $2 \leq n < \omega$, and X a space. Then

(a) X has precalibre (α, ρ) if and only if $G(X)$ has precalibre (α, ρ); and

(b) X has property $\mathrm{K}_{\alpha,n}$ if and only if $G(X)$ has property $\mathrm{K}_{\alpha,n}$.

Proof. The statements are immediate from Corollary 2.18 and Theorem 2.20.

The following simple proof derives directly from the definitions (and does not depend on the relations $\leq$ and $\equiv$ discussed above).

For $\lambda \geq \omega$ and X a space, the space $X_{(\lambda)}$ and its elementary properties are described in Appendix B.

2.23 *Lemma*. Let $\lambda > \omega$, let X be a locally compact Hausdorff space, and U a non-empty, open subset of $X_{(\lambda)}$. Then there are a non-empty, open subset V of $X_{(\lambda)}$, a cardinal $\mu \leq \lambda$ with $\mu < \lambda$ in case λ is a regular cardinal, and a set $\{V(\zeta) : \zeta < \mu\}$ of (non-empty) open subsets of X such that

$V \subset U$,

$V = \bigcap_{\zeta<\mu} V(\zeta) = \bigcap_{\zeta<\mu} \mathrm{cl}_X V(\zeta)$,

$\mathrm{cl}_X V(\zeta)$ is compact for $\zeta < \mu$, and

$\{V(\zeta) : \zeta < \mu\}$ is closed under finite intersections.

Proof. Let $p \in U$. It follows from Appendix B that there are $\mu \leq \lambda$, with $\mu \geq \omega$ and with $\mu < \lambda$ in case λ is regular, and a set $\{U(\zeta) : \zeta < \mu\}$ of open subsets of X, such that

$$p \in \bigcap_{\zeta < \mu} U(\zeta) \subset U;$$

we assume without loss of generality that $\mathrm{cl}_X U(\zeta)$ is compact for $\zeta < \mu$.

We set $U(\zeta, 0) = U(\zeta)$, and if $n < \omega$ and $U(\zeta, n)$ has been defined we choose an open subset $U(\zeta, n+1)$ of X such that

$$p \in U(\zeta, n+1) \subset \mathrm{cl}_X U(\zeta, n+1) \subset U(\zeta, n).$$

We note that $\mathrm{cl}_X U(\zeta, n)$ is compact for all $n < \omega$. We let $\{V(\zeta) : \zeta < \mu\}$ be an indexing of the set of finite intersections of elements of the family

$$\{U(\zeta, n) : \zeta < \mu, n < \omega\},$$

and we set $V = \bigcap_{\zeta < \mu} V(\zeta)$. It is clear that V, μ and $\{V(\zeta) : \zeta < \mu\}$ are as required.

2.24 *Corollary.* Let $\alpha \geq \beta \geq \omega$ and $\lambda \geq \omega$, and let X be a locally compact Hausdorff space. If $X_{(\lambda)}$ has precalibre (α, β), then $X_{(\lambda)}$ has calibre (a, β).

Proof. Let $\{U_\xi : \xi < \alpha\}$ be a set of non-empty, open subsets of $X_{(\lambda)}$. We consider two cases.

Case 1. $\lambda = \omega$. For $\xi < \alpha$ let V_ξ be a non-empty, open subset of U_ξ such that $\mathrm{cl}_X V_\xi$ is compact. There is $B \subset \alpha$ with $|B| = \beta$ such that $\{V_\xi : \xi \in B\}$ has the finite intersection property, and we have

$$\bigcap_{\xi \in B} U_\xi \supset \bigcap_{\xi \in B} \mathrm{cl}_X V_\xi \neq \varnothing.$$

Case 2. $\lambda > \omega$. For $\xi < \alpha$ there are by Lemma 2.23 a non-empty open subset V_ξ of U_ξ, a cardinal $\mu(\xi) \leq \lambda$ and a set $\{V_\xi(\zeta) : \zeta < \mu(\xi)\}$ of open subsets of X such that

$$V_\xi = \bigcap_{\zeta < \mu(\xi)} V_\xi(\zeta) = \bigcap_{\zeta < \mu(\xi)} \mathrm{cl}_X V_\xi(\zeta).$$

There is $B \subset \alpha$ with $|B| = \beta$ such that $\{V_\xi : \xi \in B\}$ has the finite intersection property, and since $\{\mathrm{cl}_X V_\xi(\zeta) : \xi \in B, \zeta < \mu(\xi)\}$ has the finite intersection property, we have

$$\bigcap_{\xi\in B} U_\xi \supset \bigcap_{\xi\in B} V_\xi = \cap\{\mathrm{cl}_X V_\xi(\zeta) : \xi\in B, \zeta < \mu(\xi)\} \neq \varnothing,$$

as required.

2.25 *Corollary*. Let $\alpha \geq \omega$ and let X be a space. If X has precalibre α, then X has precalibre $\mathrm{cf}(\alpha)$.

Proof. From Corollary 2.18(b) and Theorem 2.20 it follows that $G(X)$ has precalibre α. Then $G(X)$ has calibre α (by Corollary 2.24), hence calibre $\mathrm{cf}(\alpha)$ (by Theorem 2.3(a)), hence precalibre $\mathrm{cf}(\alpha)$. Then from Corollary 2.18(b) and Theorem 2.20 it follows that X has precalibre $\mathrm{cf}(\alpha)$, as required.

We note that Corollary 2.25 can be proved directly, without using Lemma 2.23 or the Gleason space $G(X)$, by an argument similar to that of Theorem 2.3(a).

Notes for Chapter 2

Calibres are the creation of the Soviet mathematician Shanin (Šanin); without his highly original papers (1946a, b, c, 1948) the present monograph could not have been written or even conceived. His work is described in Chapter 3 and its Notes; for the moment we note simply that the expression calibre α is first introduced (with very nearly the definition we have adopted here) in Shanin (1946b), and that the case $\alpha = \beta = \gamma$ of Theorem 2.5(a)(i) is in Shanin (1946b, Theorem 2).

Those properties which in our terminology are denoted calibre $(\omega^+, 2, 2)$ and calibre $(\omega^+, \omega^+, 2)$ were considered by Marczewski (1941) and Knaster (1945), respectively.

In considering chain conditions satisfied by a space X, many authors (see for example Juhász (1971, 1980) and Rudin (1975) prefer to deal not with the Souslin number $S(X)$, but with the *cellular number* $c(X)$, defined by the rule:

$$c(X) = \sup\{|\mathscr{U}| : \mathscr{U} \text{ is cellular family in } X\}.$$

In our experience such discussions usually require considering separately the case in which there is, and the case in which there is not, a cellular family $\mathscr{U}$ in X such that $|\mathscr{U}| = c(X)$ ('the sup = max problem'); this explains our preference for the cardinal $S(X)$ over $c(X)$. In any event, the relation between the two is easily described. We have

$$S(X) = (c(X))^+ \text{ if there is cellular } \mathscr{U} \subset \mathscr{T}(X) \text{ with } |\mathscr{U}| = c(X)$$
$$S(X) = c(X) \text{ otherwise.}$$

Historically, the cellular number precedes what we have called the Souslin number. The cellular number was introduced and studied in the doctoral dissertation of Kurepa (1935).

So far as we are aware, the first mathematician to consider (what we have called) cellular families and a chain condition was Cantor (1882); he showed, without the concept of separability at his disposal, that the Euclidean spaces $\mathbb{R}^n$ are c.c.c. spaces. The famous problem of Souslin (1920) (discussed briefly in the Notes for Chapter 7) belongs also to the theory of chain conditions: 'Un ensemble ordonné linéairement sans sautes ni lacunes et tel que tout ensemble de ses intervalles (contenant plus qu'un élément) n'empiétant pas les uns sur les autres est au plus dénombrable, est-il nécessairement un continu linéaire (ordinaire)?'

Hewitt (1948) defined pseudocompact spaces (as those completely regular Hausdorff spaces on which every continuous real-valued function is bounded) and gave several characterizations based on the ring $C(X)$ and the inclusion $X \subset \beta X$. That Hewitt's spaces are exactly those for which every locally finite family of open subsets is finite was shown by Glicksberg (1952); see also Mardesič & Papič (1955) and Bagley, Connell & McKnight (1958). Isbell (1964) (page 135) introduced pseudo-α-compact spaces for $\alpha \geq \omega$ as those (uniform) spaces whose covering character in the fine uniformity is at most α; it is noted by Noble (1969a) (page 389) that Isbell's pseudo-α-compact spaces are exactly ours. Pseudo-(α, β)-compact spaces were introduced (for $\alpha \geq \beta \geq \omega$) by Comfort & Negrepontis (1972b).

Frolík (1960) considered those spaces which (in our terminology) have compact-calibre ω, and he showed that the product of any such space with a pseudocompact space is pseudocompact. Noble (1969a) improves some of Frolík's results and gives analogues for countably compact spaces.

The properties calibre (α, β, γ) and compact-calibre (α, β, γ) were defined in the abstract of Comfort & Negrepontis (1978a). See also Miščenko (1966a, b) for another (two-cardinal) generalization of Shanin's calibre (to a family of sets; in particular, to a topology on a set) and for applications to the question of expressing a uniformly continuous function $f: X_I \to Y$ on a product X_I in the form $f = g \circ \pi_J$ with $J \subset I, |J| \leq \omega$.

Comfort (1971, Theorem 4) and Comfort & Negrepontis (1974, Theorem 3.5) have recorded the following theorem of Haratomi (1931, Satz 21): If X is metrizable, then $S(X) = (d(X))^+$. The case $\lambda = 2$ of Theorem 2.14, a generalization of Haratomi's result due to Erdős & Tarski (1943), has been proved independently by several authors, in some cases with a view to the application of Boolean-valued models in set theory; see in this connection Vopěnka (1965), Štěpánek & Vopěnka (1967), Efimov (1968), Bukovský (1968, introduction to section 2), and, for a generalized treatment, Mostowski (1969, Chapter 13, Section 3). Our motivation for introducing the λ-Souslin numbers $S_\lambda(X)$ was the observation that the above-cited argument (the case $\lambda = 2$) sufficed with only minor modifications to prove Theorem 2.14 in its present degree of generality.

The material in the section 'Reduction to compact Hausdorff spaces', including in particular the definition of the relations $\leq$ and $\equiv$ and the useful result that $X \equiv G(X)$ (Theorem 2.20), is from Galvin (1980).

3

Chain Conditions in Products

The chain conditions introduced in Chapter 2 are studied here in cartesian products and powers, both with the usual cartesian product topology and with various strong κ-box topologies. Most of the results follow this pattern: if every 'small' subproduct (or, every 'small' power of a space) satisfies a particular chain condition, then the full 'large' product (or, every power of the space) also satisfies it. The combinatorial tool suitable for this purpose is the Erdős–Rado theorem on quasi-disjoint families and its extension to singular cardinals.

Our results are more complete and satisfactory for the usual product topology than for the κ-box topologies, and for chain conditions concerning calibre and compact-calibre than for pseudo-compactness numbers. The distinction we draw between powers and products is not artificial, since some chain conditions are preserved in the former context and not in the latter (compare, for example, Theorem 3.18(a) (of Shelah) or its extension Theorem 3.19(a) (based on an argument of Argyros) with Theorem 4.4 (of Gerlits), and compare Theorem 3.18(b) or Theorem 3.19(b) with Theorem 8.3).

The chart in section 3.24 summarizes what we know about chain conditions in products and powers, and indicates some open questions.

Notation. Let $\{X_i : i \in I\}$ be a family of sets and $X = \prod_{i \in I} X_i$. If $J \subset I$ and $J \neq \emptyset$, we set

$$X_J = \prod_{i \in J} X_i,$$

and we denote by π_J the projection function from X onto X_J given by $\pi_J(\langle p_i : i \in I \rangle) = \langle p_i : i \in J \rangle$. In particular, $X = X_I$. If $p \in X$, we sometimes write p_J for $\pi_J(p)$, though we write (as usual) π_i and p_i for $\pi_{\{i\}}$ and $p_{\{i\}}$, respectively.

A subset A of X such that

$$A = \prod_{i\in I} A_I,$$

where $A_i \subset X_i$ for $i \in I$, is called a *(generalized) rectangle* in X, and the set

$$R(A) = \{i \in I : A_i \neq X_i\}$$

is called the *restriction set* of A.

3.1 *Lemma.* Let $A = \prod_{i\in I} A_i$ and $B = \prod_{i\in I} B_i$ be non-empty generalized rectangles in the product $X = \prod_{i\in I} X_i$, and let $R(A) \cap R(B) \subset J \subset I$. Then the following are equivalent.

(a) $A \cap B = \varnothing$;

(b) there is $i \in J$ such that $A_i \cap B_i = \varnothing$;

(c) $J \neq \varnothing$ and $\pi_J[A] \cap \pi_J[B] = \varnothing$.

Proof. (a) $\Rightarrow$ (b). If (b) fails there is $x = \langle x_i : i \in I\rangle \in X$ such that

$$\begin{array}{ll} x_i \in A_i \cap B_i & \text{if } i \in J \\ x_i \in A_i & \text{if } i \in R(A)\backslash J \\ x_i \in B_i & \text{if } i \in R(B)\backslash J \\ x_i \in X_i & \text{if } i \in I\backslash(R(A) \cup R(B)). \end{array}$$

Then $x \in A \cap B$ and hence (a) fails.

(b) $\Rightarrow$ (c) is obvious.

(c) $\Rightarrow$ (a). If $x \in A \cap B$ then $x_J \in \pi_J[A] \cap \pi_J[B]$.

Definition. Let $\{X_i : i \in I\}$ be a family of spaces, let X be the set $\prod_{i\in I} X_i$ and let κ be an infinite cardinal. The *κ-box topology* on X is the topology with base (called the *canonical* base) equal to the family of all subsets U of X such that

$$U = \prod_{i\in I} U_i,$$

$$U_i \text{ is open in } X_i \text{ for } i \in I, \text{ and}$$

$$|R(U)| < \kappa.$$

It is simple to verify that the canonical base for the κ-box topology just defined is indeed a base (for a topology). The product set

$X_I = \prod_{i\in I} X_i$ with the κ-box topology is denoted by $(X_I)_\kappa$ or $(\prod_{i\in I} X_i)_\kappa$.

The ω-product topology is the usual product topology; we write simply X_I and $\prod_{i\in I} X_i$ for $(X_I)_\omega$ and $(\prod_{i\in I} X_i)_\omega$, respectively.

In discussing the κ-box topology for a product and its subsets we use, among the various bases for it that exist, only the canonical base. Thus for $\{X_i : i\in I\}$ and κ as above and $Y \subset \prod_{i\in I} X_i$ the expression 'V is basic in Y' means that there is an element U of the canonical base for $(\prod_{i\in I} X_i)_\kappa$ such that $V = U \cap Y$.

In the next lemma and elsewhere in this monograph we allow ourselves the following convenient abuse of notation: Given $\{X_i : i\in I\}$, $\varnothing \neq J \subsetneqq I$ and $\kappa \geq \omega$, we identify X_I with $X_J \times X_{I\setminus J}$ and $(X_I)_\kappa$ with $(X_J)_\kappa \times (X_{I\setminus J})_\kappa$.

3.2 *Lemma.* Let $\kappa \geq \omega$, let $\{X_i : i\in I\}$ be a set of spaces, and $J \in \mathscr{P}^*(I)$.

(a) The projection function $\pi_J : (X_I)_\kappa \to (X_J)_\kappa$ is an open, continuous function;

(b) if K is a compact subset of $(X_J)_\kappa$ and $p \in X_{I\setminus J}$, then $\{p\} \times K$ is a compact subset of $(X_I)_\kappa$.

We omit the simple proof of this lemma.

In sections 3.3–3.7 we show that for certain of the properties P we have been considering, the question whether a product $(X_I)_\kappa$ (or even certain of its dense subspaces) has P is determined by whether certain of the subproducts $(X_J)_\kappa$ have P.

In the interest of completeness we give first a particularly simple result. Its proof does not depend on the Erdős–Rado Theorem.

We note that in Theorems 3.3, 3.6, 3.9, 3.12, 3.13 and 3.25 the space $Y = (X_I)_\kappa$ satisfies the conditions imposed on Y. This is the case of principal interest.

3.3 *Theorem.* Let $\alpha \geq \beta \geq \gamma \geq 2$, $\alpha \geq \kappa \geq \omega$, let $\{X_i : i\in I\}$ be a set of spaces and Y a subspace of $(X_I)_\kappa$.

(a) If $(X_J)_\kappa$ has calibre (α, β, γ) for every non-empty $J \subset I$ with $|J| \leq \alpha$, and if $\pi_J[Y] = X_J$ for every non-empty $J \subset I$ with $|J| \leq \alpha$, then Y has calibre (α, β, γ);

(b) if $(X_J)_\kappa$ has compact-calibre (α, β, γ) for every non-empty

$J \subset I$ with $|J| \leq \alpha$, and if for every compact $K_J \subset (X_J)_\kappa$ with J non-empty, $J \subset I$ and $|J| \leq \alpha$ there is compact $K \subset Y$ such that $\pi_J[K] = K_J$, then Y has compact-calibre (α, β, γ);

(c) if $(X_J)_\kappa$ is pseudo-(α, β)-compact for every non-empty $J \subset I$ with $|J| \leq \alpha$, and if $\pi_J[Y] = X_J$ for every non-empty $J \subset I$ with $|J| \leq \alpha$, then Y is pseudo-(α, β)-compact.

Proof. We assume in what follows, the statements being obvious otherwise, that $X_i \neq \emptyset$ for $i \in I$.

Let $\{U^\xi : \xi < \alpha\}$ be a set of non-empty basic subsets of $(X_I)_\kappa$ and set $J = \cup_{\xi < \alpha} R(U^\xi)$. We note that $|J| \leq \alpha$.

(a) There is $B \in [\alpha]^\beta$ such that if $C \in [B]^\gamma$ then $\cap_{\xi \in C} \pi_J[U^\xi] \neq \emptyset$. Let $C \in [B]^\gamma$ and $p \in \cap_{\xi \in C} \pi_J[U^\xi]$ and let x be an element of Y such that $\pi_J(x) = p$. It is clear that $x \in \cap_{\xi \in C} U^\xi$.

(b) There is $B \in [\alpha]^\beta$ such that for $C \in [B]^\gamma$ there is compact $K_J \subset (X_J)_\kappa$ such that $\pi_J[U^\xi] \cap K_J \neq \emptyset$ whenever $\xi \in C$. Let $C \in [B]^\gamma$, let K_J be such a compact subspace of $(X_J)_\kappa$ and let K be a compact subspace of Y such that $\pi_J[K] = K_J$. It is clear that $U^\xi \cap K \neq \emptyset$ whenever $\xi \in C$.

(c) There is $p \in (X_J)_\kappa$ such that

$$|\{\xi < \alpha : V_J \cap \pi_J[U^\xi] \neq \emptyset\}| \geq \beta$$

for every neighborhood V_J of p. It is clear that if x is an element of Y such that $\pi_J(x) = p$, then

$$|\{\xi < \alpha : V \cap U^\xi \neq \emptyset\}| \geq \beta$$

for every neighborhood V of x.

3.4 *Corollary.* Let $\{X_i : i \in I\}$ be a set of spaces such that X_J is pseudocompact for every non-empty $J \subset I$ with $|J| \leq \omega$. Then X_I is pseudocompact.

Proof. This is the case $\alpha = \beta = \kappa = \omega$, $Y = X_I$ of Theorem 3.3(c).

We show next that for $\kappa \ll \mathrm{cf}(\alpha)$ (and without additional hypotheses) the behavior of products of fewer than $\mathrm{cf}(\alpha)$ spaces with respect to the chain conditions we are considering is determined by the behavior of subproducts of fewer than κ spaces.

3.5 *Theorem.* Let $\alpha \geq \beta \geq \gamma \geq 2$ and $\omega \leq \kappa \ll \mathrm{cf}(\alpha)$ and let $\{X_i : i \in I\}$ be a set of spaces.

(a) If $(X_J)_\kappa$ has calibre (α,β,γ) for every non-empty $J\subset I$ with $|J|<\kappa$, then $(X_J)_\kappa$ has calibre (α,β,γ) for every non-empty $J\subset I$ with $|J|<\mathrm{cf}(\alpha)$;

(b) if $(X_J)_\kappa$ has compact-calibre (α,β,γ) for every non-empty $J\subset I$ with $|J|<\kappa$, then $(X_J)_\kappa$ has compact-calibre (α,β,γ) for every non-empty $J\subset I$ with $|J|<\mathrm{cf}(\alpha)$;

(c) if $(X_J)_\kappa$ is pseudo-(α,β)-compact for every non-empty $J\subset I$ with $|J|<\kappa$, then $(X_J)_\kappa$ is pseudo-(α,β)-compact for every non-empty $J\subset I$ with $|J|<\mathrm{cf}(\alpha)$.

Proof. We assume in what follows, the statements being obvious otherwise, that $X_i\neq\emptyset$ for $i\in I$. We assume without loss of generality that $|I|<\mathrm{cf}(\alpha)$ and we show in each case that $(X_I)_\kappa$ has the property in question.

Let $\{U^\xi:\xi<\alpha\}$ be a set of non-empty, basic subsets of $(X_I)_\kappa$. Since

$$R(U^\xi)\in\mathscr{P}_\kappa(I)\text{ for }\xi<\alpha\text{ and}$$
$$|\mathscr{P}_\kappa(I)|=\sum_{\lambda<\kappa}|I|^\lambda<\mathrm{cf}(\alpha),$$

there are $A\in[\alpha]^\alpha$ and $J\subset I$ such that

$$R(U^\xi)=J\text{ for }\xi\in A.$$

If $J=\emptyset$ then $U^\xi=(X_I)_\kappa$ for all $\xi\in A$ and hence $\cap_{\xi\in A}U^\xi\neq\emptyset$, and if $J=I$ then $|I|<\kappa$; in either case the required relations follow immediately. We assume

$$\emptyset\neq J\subset I$$

and we choose $q\in X_{I\backslash J}$.

In (a) there is $B\in[A]^\beta$ such that if $C\in[B]^\gamma$ there is

$$p(C)\in\bigcap_{\xi\in S}\pi_J[U^\xi];$$

it follows with $x(C)=\langle p(C),q\rangle\in X_J$ that $x(C)\in\cap_{\xi\in C}U^\xi$.

In (b) there is $B\in[A]^\beta$ such that if $C\in[B]^\gamma$ there is compact $K(C)\subset(X_J)_\kappa$ such that $\pi_J[U^\xi]\cap K(C)\neq\emptyset$ whenever $\xi\in C$. Then $K(C)\times\{q\}$ is a compact subspace of $(X_I)_\kappa$ and

$$U^\xi\cap(K(C)\times\{q\})\neq\emptyset\text{ whenever }\xi\in C.$$

In (c) there is $p \in (X_J)_\kappa$ such that

$$|\{\xi \in A : V_J \cap \pi_J[U^\xi] \neq \varnothing\}| \geq \beta$$

for every neighborhood V_J of p. It follows with

$$x = \langle p, q \rangle \in (X_I)_\kappa$$

that

$$|\{\xi \in A : V \cap U^\xi \neq \varnothing\}| \geq \beta$$

for every neighborhood V of x.

The proof is complete.

Product theorems for regular cardinals

3.6 *Theorem.* Let $\alpha \geq \beta \geq \gamma \geq 2$, $\omega \leq \kappa \ll \alpha$ with α regular, and let $\{X_i : i \in I\}$ be a set of spaces and $Y \subset (X_I)_\kappa$.

(a) If $(X_J)_\kappa$ has calibre (α, β, γ) for every non-empty $J \subset I$ with $|J| < \kappa$, and if $\pi_J[Y] = X_J$ for every non-empty $J \subset I$ with $|J| \leq \gamma \cdot \kappa$, then Y has calibre (α, β, γ);

(b) if $(X_J)_\kappa$ has compact-calibre (α, β, γ) for every non-empty $J \subset I$ with $|J| < \kappa$, and if for every compact $K_J \subset (X_J)_\kappa$ with J non-empty, $J \subset I$ and $|J| \leq \gamma \cdot \kappa$ there is compact $K \subset Y$ such that $\pi_J[K] = K_J$, then Y has compact-calibre (α, β, γ);

(c) if $(X_J)_\kappa$ is pseudo-(α, β)-compact for every non-empty $J \subset I$ with $|J| < \kappa$, and if $\pi_J[Y] = X_J$ for every non-empty $J \subset I$ with $|J| < \kappa$, and if $\beta \geq \kappa$, then Y is pseudo-(α, β)-compact.

Proof. We assume in what follows, the statements being obvious otherwise, that $X_i \neq \varnothing$ for $i \in I$.

Let $\{U^\xi : \xi < \alpha\}$ be a set of non-empty basic subsets of $(X_I)_\kappa$. Since $|R(U^\xi)| < \kappa$ for $\xi < \alpha$, by the Erdős–Rado theorem on quasi-disjoint families (Theorem 1.4) there are $A \in [\alpha]^\alpha$ and J such that if $\xi, \xi' \in A$ and $\xi \neq \xi'$ then $R(U^\xi) \cap R(U^{\xi'}) = J$. We consider statements (a), (b) and (c) separately.

(a) We show there is $B \in [A]^\beta$ such that if $C \in [B]^\gamma$ then $(\cap_{\xi \in C} U^\xi) \cap Y \neq \varnothing$.

If $J = \varnothing$ we let B be an arbitrary element of $[A]^\beta$, and if $J \neq \varnothing$ we choose $B \in [A]^\beta$ such that if $C \in [B]^\gamma$ then $\cap_{\xi \in C} \pi_J[U^\xi] \neq \varnothing$.

Now let $C \in [B]^\gamma$, set $F = \cup_{\xi \in C} R(U^\xi)$, and if $J \neq \varnothing$ let $p \in \cap_{\xi \in C}$

$\pi_J[U^\xi]$. Since $|F\cup J|\le\gamma\cdot\kappa+\kappa=\gamma\cdot\kappa$ we have $\pi_F[Y]=X_F$; it follows that since

$$|\{\xi\in C:i\in R(U^\xi)\}|\le 1 \text{ for } i\in F\backslash J,$$

there is $x=\langle x_i:i\in I\rangle\in Y$ such that

$$\begin{aligned} x_i &= p_i \text{ if } i\in J \\ &\in U_i^\xi \text{ if } i\in F\backslash J \text{ with } i\in R(U^\xi), \xi\in C, \\ &\in X_i \text{ if } i\in I\backslash F. \end{aligned}$$

It is clear that $x\in(\bigcap_{\xi\in C}U^\xi)\cap Y$.

(b) We show there is $B\in[A]^\beta$ such that for $C\in[B]^\gamma$ there is compact $K\subset Y$ such that $U^\xi\cap K\ne\emptyset$ whenever $\xi\in C$.

If $J=\emptyset$ we let B be an arbitrary element of $[A]^\beta$, and if $J\ne\emptyset$ we choose $B\in[A]^\beta$ such that for $C\in[B]^\gamma$ there is compact $\tilde{K}\subset(X_J)_\kappa$ such that $\pi_J[U^\xi]\cap\tilde{K}\ne\emptyset$ whenever $\xi\in C$.

Now let $C\in[B]^\xi$, set $F=\bigcup_{\xi\in C}R(U^\xi)$, and if $J\ne\emptyset$ choose compact $\tilde{K}\subset(X_J)_\kappa$ such that $\pi_J[U^\xi]\cap\tilde{K}\ne\emptyset$ whenever $\xi\in C$. Since

$$|\{\xi\in C:i\in R(U^\xi)\}|\le 1 \text{ for } i\in F\backslash J,$$

there is $q\in X_{F\backslash J}$ such that

$$q_i\in U_i^\xi \text{ if } i\in F\backslash J, i\in R(U^\xi), \xi\in C.$$

We set $K_{F\cup J}=\{q\}\times\tilde{K}$. Then $K_{F\cup J}$ is by Lemma 3.2(b) a compact subspace of $(X_{F\cup J})_\kappa$, and since

$$|F\cup J|\le\gamma\cdot\kappa+\kappa=\gamma\cdot\kappa,$$

there is compact $K\subset Y$ such that $\pi_{F\cup J}[K]=K_{F\cup J}$. It is clear that $U^\xi\cap K\ne\emptyset$ whenever $\xi\in C$, as required.

(c) We show there is $x\in Y$ such that if V is a neighborhood in $(X_I)_\kappa$ of x then

$$|\{\xi\in A:V\cap U^\xi\ne\emptyset\}|\ge\beta;$$

this is sufficient, since from $V\cap U^\xi\ne\emptyset$ we have $V\cap U^\xi\cap Y\ne\emptyset$ (because Y is dense in $(X_I)_\kappa$).

If $J=\emptyset$ we let x be an arbitrary element of Y, and if $J\ne\emptyset$ we use the fact that $(X_J)_\kappa$ is pseudo-(α,β)-compact and $\pi_J[Y]=X_J$ to find $x\in Y$ such that

$$|\{\xi\in A:V_J\cap\pi_J[U^\xi]\ne\emptyset\}|\ge\beta$$

for every neighborhood V_J of x_J in $(X_J)_\kappa$.

Now let V be a basic neighborhood in $(X_I)_\kappa$ of x, and set $V_J = \pi_J[V]$ and

$$B = \{\xi \in A : V_J \cap \pi_J[U^\xi] \neq \varnothing\}.$$

If $\xi \in B$ and $V \cap U^\xi = \varnothing$, then from Lemma 3.1 we have $(R(V) \cap R(U^\xi)) \backslash J \neq \varnothing$. Since

$$|\{\xi \in B : i \in R(U^\xi) \backslash J\}| \leq 1 \text{ for } i \in I,$$

we have

$$|\{\xi \in B : V \cap U^\xi = \varnothing\}| \leq |R(V)| < \kappa \leq \beta,$$

and from $|B| \geq \beta$ we have

$$|\{\xi \in B : V \cap U^\xi \neq \varnothing\}| \geq \beta,$$

as required.

The proof is complete.

In Theorem 3.6(c) (and Theorem 3.12(c) below) the condition $\beta \geq \kappa$ is superfluous in case $\kappa = \omega$ and the spaces X_i are regular, Hausdorff spaces (cf. in this connection Theorem 2.1(i)).

We recall that a space X is said to have precalibre (α, β) if for every set $\{U_\xi : \xi < \alpha\}$ of non-empty, open subsets of X there is $B \subset \alpha$ with $|B| = \beta$ such that $\{U_\xi : \xi \in B\}$ has the finite intersection property. We note now that the argument in the proof of Theorem 3.6(a) shows the following statement.

Theorem. Let $\alpha \geq \beta \geq \omega, \omega \leq \kappa \ll \alpha$ with α regular and let $\{X_i : i \in I\}$ be a set of spaces such that $(X_J)_\kappa$ has precalibre (α, β) for every non-empty $J \subset I$ with $|J| < \kappa$. Then $(X_I)_\kappa$ has precalibre (α, β).

3.7 *Corollary.* Let α be an uncountable, regular cardinal.

(a) The product of every set of spaces with calibre α has calibre α;

(b) the product of every set of spaces with compact-calibre α has compact-calibre α.

Proof. This follows from Theorem 2.2 and (the case $\alpha = \beta = \gamma$, $\kappa = \omega$, $Y = (X_I)_\kappa$ of) Theorem 3.6.

3.8 *Remark.* Let α, β and κ be cardinals with $\alpha \geq \beta$. The notation

$$\alpha \rightarrow \Delta(\kappa, \beta)$$

means that if $\{S_\xi : \xi < \alpha\}$ is a family of sets such that $|S_\xi| < \kappa$ for all $\xi < \alpha$, then there are $A \in [\alpha]^\beta$ and J such that

$$S_\xi \cap S_{\xi'} = J \text{ for } \xi, \xi' \in A, \xi \neq \xi'.$$

In this notation the Erdős–Rado theorem on quasi-disjoint sets (Theorem 1.4) is the statement that if $\omega \leq \kappa \ll \alpha$ with α regular then $\alpha \to \Delta(\kappa, \alpha)$. Since our concern in this work (as to cardinals) is chiefly with pairs $\langle \alpha, \kappa \rangle$ such that $\omega \leq \kappa \ll \alpha$ and with the extension to singular cardinals of the Erdős–Rado theorem, we do not consider the general relation $\alpha \to \Delta(\kappa, \beta)$. Nevertheless we note here in passing that the proof of Theorem 3.6 becomes, with only minor modifications (not given in detail here), a proof of the following statement.

Theorem. Let $\alpha \geq \beta \geq \gamma \geq \delta \geq 2, \kappa \geq \omega, \alpha \to \Delta(\kappa, \beta)$, let $\{X_i : i \in I\}$ be a set of spaces and $Y \subset (X_I)_\kappa$.

(a) If $(X_J)_\kappa$ has calibre (β, γ, δ) for every non-empty $J \subset I$ with $|J| < \kappa$, and if $\pi_J[Y] = X_J$ for every non-empty $J \subset I$ with $|J| \leq \delta \cdot \kappa$, then Y has calibre (α, γ, δ);

(b) if $(X_J)_\kappa$ has compact-calibre (β, γ, δ) for every non-empty $J \subset I$ with $|J| < \kappa$, and if for every compact $K_J \subset (X_J)_\kappa$ with J non-empty, $J \subset I$ and $|J| \leq \delta \cdot \kappa$ there is compact $K \subset Y$ such that $\pi_J[K] = K_J$, then Y has compact-calibre (α, γ, δ);

(c) if $(X_J)_\kappa$ is pseudo-(β, γ)-compact for every non-empty $J \subset I$ with $|J| < \kappa$ and if $\pi_J[Y] = X_J$ for every non-empty $J \subset I$ with $|J| < \kappa$, and if $\gamma \geq \kappa$, then Y is pseudo-(α, γ)-compact.

The methods of 3.6–3.8 are not applicable to the cardinal number $\alpha = \omega$. Indeed we have noted already in Theorem 2.10 that no infinite Hausdorff space has calibre ω, and we show below (the case $\alpha = \beta = \omega$ of Theorem 9.7(a)) that the product of a finite number of completely regular, Hausdorff, pseudocompact spaces need not be pseudocompact. We show now, in contrast, by a special argument not dependent on the Erdős–Rado theorem on quasi-disjoint sets, that the property compact-calibre ω is productive.

3.9 *Theorem.* Let $\{X_i : i \in I\}$ be a set of non-empty spaces, each with compact-calibre ω, and let $Y \subset X_I$. If for every compact $K_J \subset$

X_J with J non-empty, $J \subset I$ and $|J| \le \omega$ there is compact $K \subset Y$ such that $\pi_J[K] = K_J$, then Y has compact-calibre ω.

Proof. Let $\{U^n : n < \omega\}$ be a set of non-empty basic subsets of X_I. It is enough to show there are $A \in [\omega]^\omega$ and compact $K \subset Y$ such that $U^n \cap K \neq \emptyset$ whenever $n \in A$. We set $J = \cup_{n<\omega} R(U^n)$, we assume without loss of generality that $J \neq \emptyset$, we note that $|J| \le \omega$ since $|R(U^n)| < \omega$ for $n < \omega$, and we consider two cases.

Case 1. $|J| < \omega$. From Theorem 2.2(b) there are $A \in [\omega]^\omega$ and compact $K_J \subset X_J$ such that $\pi_J[U^n] \cap K_J \neq \emptyset$ whenever $n \in A$. We choose compact $K \subset Y$ such that $\pi_J[K] = K_J$; it is clear that A and K are as required.

Case 2. $|J| = \omega$. We write $J = \{i(k) : k < \omega\}$ with $i(k) \neq i(k')$ when $k \neq k'$, and we define

$$\{A_k : k < \omega\} \subset \mathscr{P}(\omega) \text{ and } \{K_k : k < \omega\} \text{ so that}$$

(i) $A_0 = \omega$;
(ii) $A_{k+1} \in [A_k]^\omega$ for $k < \omega$;
(iii) K_k is a compact subspace of $X_{i(k)}$ for $k < \omega$; and
(iv) $U^n_{i(k)} \cap K_k \neq \emptyset$ whenever $n \in A_{k+1}$.

We proceed by recursion. We define A_0 by (i), and for $k < \omega$ we note that since $\{U^n_{i(k)} : n \in A_k\} \subset \mathscr{T}^*(X_{i(k)})$ and $X_{i(k)}$ has compact-calibre ω, there are A_{k+1} and K_k satisfying (ii), (iii) and (iv).

Since $A_{k+1} \in [A_k]^\omega$ for $k < \omega$, there is $\{n(m) : m < \omega\}$ such that

$$n(m) \in A_m \text{ for } m < \omega, \text{ and}$$
$$n(m') < n(m) \text{ for } m' < m < \omega.$$

We set $A = \{n(m) : m < \omega\}$, we choose

$$x_{k,m} \in U^{n(m)}_{i(k)} \text{ for } k, m < \omega,$$

we set

$$\tilde{K}_k = K_k \cup \{x_{k,m} : m \le k\} \text{ for } k < \omega \text{ and}$$
$$K_J = \prod_{k<\omega} \tilde{K}_k,$$

we note that K_J is a compact subset of X_J and we choose compact $K \subset Y$ such that $\pi_J[K] = K_J$.

To show that $U^{n(m)} \cap K \neq \emptyset$ for $m < \omega$ it is sufficient, since $R(U^{n(m)}) \subset J$, to show that

$$\pi_J[U^{n(m)}] \cap K_J \neq \emptyset,$$

i.e.,

$$U_i^{n(m)} \cap \pi_i[K] \neq \varnothing \text{ for } i \in J.$$

If $i = i(k) \in J$ with $k < m$ then from $n(m) \in A_m \subset A_{k+1}$ we have

$$U_i^{n(m)} \cap \pi_i[K] \supset U_i^{n(m)} \cap K_k \neq \varnothing;$$

and if $i = i(k) \in J$ with $k \geq m$ then

$$x_{k,m} \in U_{i(k)}^{n(m)} \cap \tilde{K}_k = U_i^{n(m)} \cap \pi_i[K].$$

The proof is complete.

Product theorems for singular cardinals

We show first that the statements of the various parts of Theorem 3.6 are all false for singular cardinals $\alpha > \beta$; the correct analogue of Theorem 3.6 (requiring additional hypotheses) is given in Theorem 3.13 and its corollaries.

3.10 *Theorem.* Let α be an (infinite) singular cardinal. There is a set $\{X_i : i \in I\}$ of completely regular, Hausdorff spaces such that

(i) if $\alpha > \beta \geq 2, \omega \leq \kappa \ll \alpha$ and $\kappa \leq \mathrm{cf}(\alpha)$, then $(X_J)_\kappa$ has calibre (α, β) for all $J \subset I$ with $|J| < \kappa$; and

(ii) if $\omega \leq \kappa$ then $(X_I)_\kappa$ is not pseudo-$(\alpha, 2)$-compact.

Proof. There is a set $\{\alpha_\sigma : \sigma < \mathrm{cf}(\alpha)\}$ of cardinal numbers such that

$$\alpha_0 = \mathrm{cf}(\alpha),$$
$$\alpha_{\sigma'} < \alpha_\sigma < \alpha \text{ for } \sigma' < \sigma < \mathrm{cf}(\alpha), \text{ and}$$
$$\sum_{\sigma < \mathrm{cf}(\alpha)} \alpha_\sigma = \alpha.$$

We set $I = \mathrm{cf}(\alpha)$ and $X_\sigma = \alpha_\sigma$ for $\sigma \in I$. We verify (i) and (ii).

(i) Since $|J| < \kappa$, the space $(X_J)_\kappa$ is discrete. We set $\gamma = |(X_J)_\kappa|$ and we note from Theorem A.5(a) of the Appendix that $\gamma < \alpha$. It follows from Theorem 2.4(c) that the space $(X_J)_\kappa$, which is the discrete space γ, has calibre (α, β) for all β such that $\alpha > \beta \geq \omega$. In particular $(X_J)_\kappa$ has calibre (α, ω) and hence calibre (α, β) for $\beta < \omega$.

(ii) It is enough to show that there is a (faithfully indexed) cellular family $\{U(s) : s \in S\}$ in X_I with $|S| = \alpha$ such that

$$X_I = (X_I)_\omega = \bigcup_{s \in S} U(s).$$

We set

$$U(0) = \{x \in X_I : x_0 = 0\}, \text{ and}$$
$$U(\sigma, \eta) = \{x \in X_I : x_0 = \sigma, x_\sigma = \eta\} \text{ for } 0 < \sigma < \mathrm{cf}(\alpha), \eta < \alpha_\sigma.$$

It is clear that the family

$$\{U(\sigma, \eta) : 0 < \sigma < \mathrm{cf}(\alpha), \eta < \alpha_\sigma\} \cup \{U(0)\},$$

indexed by the set $\cup\{\{\sigma\} \times \alpha_\sigma : 0 < \sigma < \mathrm{cf}(\alpha)\} \cup \{0\}$, is as required.

3.11 *Corollary*. Let α be an (infinite) singular cardinal, $\alpha > \beta \geq 2$, $\omega \leq \kappa \ll \alpha$ and $\kappa \leq \mathrm{cf}(\alpha)$. There is a set $\{X_i : i \in I\}$ of completely regular, Hausdorff spaces such that

(i) $(X_J)_\kappa$ has calibre (α, β) for every non-empty $J \subset I$ with $|J| < \kappa$, and $(X_I)_\kappa$ does not have calibre (α, β);

(ii) $(X_J)_\kappa$ has compact-calibre (α, β) for every non-empty $J \subset I$ with $|J| < \kappa$, and $(X_I)_\kappa$ does not have compact-calibre (α, β) if $\beta \geq \omega$; and

(iii) $(X_J)_\kappa$ is pseudo-(α, β)-compact for every non-empty $J \subset I$ with $|J| < \kappa$, and $(X_I)_\kappa$ is not pseudo-(α, β)-compact.

Proof. It follows from Theorem 2.1 that the set $\{X_i : i \in I\}$ of Theorem 3.10 has the required properties.

We remark that it follows from Theorem 3.5 that among all sets I which may index a set of spaces with the properties of those in 3.10 and 3.11, the present set I (satisfying $|I| = \mathrm{cf}(\alpha)$) is of minimal cardinality.

We note that for $J \subset I$ with $0 < |J| < \kappa$ the (discrete) space $(X_J)_\kappa$ of Theorem 3.10 satisfies $|(X_J)_\kappa| \geq \mathrm{cf}(\alpha)$; it follows from Theorem 2.4 that this space does not have calibre α, does not have compact-calibre α, and is not pseudo-(α, α)-compact.

The questions whether, with α a singular cardinal and with $\alpha = \beta$, calibre (α, β) and compact-calibre (α, β) are determined in product spaces by 'small' subproducts, are settled (in the negative, by counterexamples) in Theorems 4.4 and 8.3, respectively. Unexpectedly, the analogous question for pseudo-(α, β)-compactness has a positive answer (cf. Theorem 3.16 below).

In the interest of completeness we give first a particularly simple positive result for singular cardinals, not needed later, whose proof parallels that of Theorem 3.3.

3.12 *Theorem.* Let α be a singular cardinal with $\mathrm{cf}(\alpha) > \omega$, let $\omega \leq \kappa \ll \alpha, \kappa \ll \mathrm{cf}(\alpha)$ and $\alpha \geq \beta \geq \gamma \geq 2$, let $\{X_i : i \in I\}$ be a set of spaces and $Y \subset (X_I)_\kappa$.

(a) If $(X_J)_\kappa$ has calibre (α, β, γ) for all non-empty $J \subset I$ with $|J| \leq \mathrm{cf}(\alpha)$, and if $\pi_J[Y] = X_J$ for all non-empty $J \subset I$ with $|J| \leq \gamma \cdot \mathrm{cf}(\alpha)$, then Y has calibre (α, β, γ);

(b) if $(X_J)_\kappa$ has compact-calibre (α, β, γ) for all non-empty $J \subset I$ with $|J| \leq \mathrm{cf}(\alpha)$, and if for every compact $K_J \subset (X_J)_\kappa$ with J non-empty, $J \subset I$ and $|J| \leq \gamma \cdot \mathrm{cf}(\alpha)$ there is compact $K \subset Y$ such that $\pi_J[K] = K_J$, then Y has compact-calibre (α, β, γ);

(c) if $(X_J)_\kappa$ is pseudo-(α, β)-compact for every non-empty $J \subset I$ with $|J| \leq \mathrm{cf}(\alpha)$, and if $\pi_J[Y] = X_J$ for every non-empty $J \subset I$ with $|J| \leq \mathrm{cf}(\alpha)$, and if $\beta \geq \kappa$, then Y is pseudo-(α, β)-compact.

Proof. We assume in what follows, the statements being otherwise obvious, that $X_i \neq \emptyset$ for $i \in I$.

It follows from Theorem A.7 that there is a set $\{\alpha_\sigma : \sigma < \mathrm{cf}(\alpha)\}$ of cardinal numbers such that

$$\begin{aligned}
&\alpha_\sigma \text{ is regular for } \sigma < \mathrm{cf}(\alpha),\\
&\kappa \ll \alpha_\sigma \text{ for } \sigma < \mathrm{cf}(\alpha),\\
&\mathrm{cf}(\alpha) < \alpha_{\sigma'} < \alpha_\sigma < \alpha \text{ for } \sigma' < \sigma < \mathrm{cf}(\alpha), \text{ and}\\
&\sum_{\sigma < \mathrm{cf}(\alpha)} \alpha_\sigma = \alpha.
\end{aligned}$$

Let $\{U^\xi : \xi < \alpha\}$ be a set of non-empty basic subsets of $(X_I)_\kappa$. It follows from the extension of the Erdős–Rado theorem to singular cardinals (Theorem 1.10) that there are $\Gamma, J, \{A_\sigma : \sigma \in \Gamma\}$ and $\{J_\sigma : \sigma \in \Gamma\}$ such that

$$\begin{aligned}
&\Gamma \in [\mathrm{cf}(\alpha)]^{\mathrm{cf}(\alpha)},\\
&A_\sigma \in [\alpha]^{\alpha_\sigma} \text{ for } \sigma \in \Gamma,\\
&A_\sigma \cap A_{\sigma'} = \emptyset \text{ for } \sigma, \sigma' \in \Gamma, \sigma \neq \sigma', \text{ and}\\
&R(U^\xi) \cap R(U^{\xi'}) = J_\sigma \text{ for } \xi, \xi' \in A_\sigma, \xi \neq \xi', \sigma \in \Gamma\\
&\qquad\qquad\qquad\quad\;\; = J \text{ for } \xi \in A_\sigma, \xi' \in A_{\sigma'}, \sigma, \sigma' \in \Gamma, \sigma \neq \sigma'.
\end{aligned}$$

We set

$$A = \bigcup_{\sigma \in \Gamma} A_\sigma \text{ and } \tilde{J} = \bigcup_{\sigma \in \Gamma} J_\sigma,$$

and we note that $|A| = \alpha$ and $|\tilde{J}| \leq \mathrm{cf}(\alpha)$.

(a) There is $B \in [A]^\beta$ such that if $C \in [B]^\gamma$ then $\cap_{\xi \in C} \pi_{\tilde{J}}[U^\xi] \neq \emptyset$. Let $C \in [B]^\gamma$ and $p \in \cap_{\xi \in C} \pi_{\tilde{J}}[U^\xi]$, and set $F = \cup_{\xi \in C} R(U^\xi)$. Since

$$|\{\xi \in C : i \in R(U^\xi)\}| \leq 1 \text{ for } i \in F \backslash \tilde{J} \text{ and}$$
$$|F \cup \tilde{J}| \leq \gamma \cdot \kappa + \mathrm{cf}(\alpha) = \gamma \cdot \mathrm{cf}(\alpha),$$

there is $x = \langle x_i : i \in I \rangle \in Y$ such that

$$x_i = p_i \text{ if } i \in \tilde{J}$$
$$x_i \in U_i^\xi \text{ if } i \in F \backslash \tilde{J}, i \in R(U^\xi)$$
$$x_i \in X_i \text{ if } i \in I \backslash (F \cup \tilde{J}).$$

It is clear that $x \in \cap_{\xi \in C} U^\xi$.

(b) There is $B \in [A]^\beta$ such that for $C \in [B]^\gamma$ there is compact $K_{\tilde{J}} \subset (X_{\tilde{J}})_\kappa$ such that $\pi_{\tilde{J}}[U^\xi] \cap K_{\tilde{J}} \neq \emptyset$ whenever $\xi \in C$. Let $C \in [B]^\gamma$, let $K_{\tilde{J}}$ be such a compact subspace of $(X_{\tilde{J}})_\kappa$, and set $F = \cup_{\xi \in C} R(U^\xi)$. Since

$$|\{\xi \in C : i \in R(U^\xi)\} \leq 1 \text{ for } i \in F \backslash \tilde{J},$$

there is $q \in X_{F \backslash \tilde{J}}$ such that

$$q_i \in U_i^\xi \text{ if } i \in F \backslash \tilde{J}, i \in R(U^\xi).$$

We set $K_{F \cup \tilde{J}} = \{q\} \times K_{\tilde{J}}$. Then $K_{F \cup \tilde{J}}$ is by Lemma 3.2(b) a compact subspace of $(X_{F \cup \tilde{J}})_\kappa$, and since

$$|F \cup \tilde{J}| \leq \gamma \cdot \kappa + \mathrm{cf}(\alpha) = \gamma \cdot \mathrm{cf}(\alpha),$$

there is compact $K \subset Y$ such that $\pi_{F \cup \tilde{J}}[K] = K_{F \cup \tilde{J}}$. It is clear that $U^\xi \cap K \neq \emptyset$ whenever $\xi \in C$, as required.

(c) There is $p \in X_{\tilde{J}}$ such that

$$|\{\xi < \alpha : V_{\tilde{J}} \cap \pi_{\tilde{J}}[U^\xi] \neq \emptyset\}| \geq \beta$$

for every neighborhood $V_{\tilde{J}}$ of p, and since $|\tilde{J}| \leq \mathrm{cf}(\alpha)$ there is $x \in Y$ such that $x_J = p$. Let V be a basic neighborhood in $(X_I)_\kappa$ of x, and set $V_{\tilde{J}} = \pi_{\tilde{J}}[V]$,

$$C = \{\xi < \alpha : V_{\tilde{J}} \cap \pi_{\tilde{J}}[U^\xi] \neq \emptyset\} \text{ and } F = \bigcup_{\xi \in C} R(U^\xi).$$

If $\xi \in C$ and $V \cap U^\xi = \emptyset$, then from Lemma 3.1 we have $(R(V) \cap R(U^\xi)) \backslash \tilde{J} \neq \emptyset$. Since

$$|\{\xi \in C : i \in R(U^\xi) \backslash \tilde{J}\}| \leq 1 \text{ for } i \in F,$$

we have

$$|\{\xi \in C : V \cap U^\xi = \emptyset\}| \leq |R(V)| < \kappa \leq \beta,$$

and from $|C| \geq \beta$ we have

$$|\{\xi \in C : V \cap U^\xi \neq \emptyset\}| \geq \beta,$$

as required.

In the proof of part (b) of Theorem 3.13 we consider (implicitly) a set $\{X_i : i \in I\}$ of spaces, a set $\{\tilde{J}_\sigma : \sigma \in \Delta\}$ of pairwise disjoint, non-empty subsets of I, and a subspace $(\prod_{\sigma \in \Delta} K_\sigma)_\kappa$ of $(\prod_{i \in \tilde{I}} X_i)_\kappa$ (with $\tilde{I} = \cup_{\sigma \in \Delta} \tilde{J}_\sigma$) with K_σ a compact subspace of $(X_{\tilde{J}_\sigma})_\kappa$. Except in trivial cases, the product $(\prod_{\sigma \in \Delta} K_\sigma)_\kappa$ is not compact when $\kappa > \omega$. This explains the condition $\kappa = \omega$ in the statement of Theorem 3.13(b).

3.13 *Theorem*. Let α be singular with $\mathrm{cf}(\alpha) > \omega$, let $\omega \leq \kappa \ll \alpha$, $\kappa \ll \mathrm{cf}(\alpha)$, $\alpha \geq \beta \geq \omega$ and $\alpha \geq \beta \geq \gamma \geq 2$, and let $\{\alpha_\sigma : \sigma < \mathrm{cf}(\alpha)\}$ be a set of cardinal numbers such that

$$\begin{aligned} &\alpha_\sigma \text{ is regular for } \sigma < \mathrm{cf}(\alpha), \\ &\kappa \ll \alpha_\sigma \text{ for } \sigma < \mathrm{cf}(\alpha), \\ &\mathrm{cf}(\alpha) < \alpha_{\sigma'} < \alpha_\sigma < \alpha \text{ for } \sigma' < \sigma < \mathrm{cf}(\alpha), \text{ and} \\ &\sum_{\sigma < \mathrm{cf}(\alpha)} \alpha_\sigma = \alpha. \end{aligned}$$

Let $\{\beta_\sigma : \sigma < \mathrm{cf}(\alpha)\}$ be a set of cardinal numbers such that

$$\begin{aligned} &\alpha_\sigma \geq \beta_\sigma \geq 2 \text{ for } \sigma < \mathrm{cf}(\alpha), \\ &\beta_{\sigma'} \leq \beta_\sigma \leq \beta \text{ for } \sigma' < \sigma < \mathrm{cf}(\alpha), \text{ and} \\ &\beta = \sum_{\sigma < \mathrm{cf}(\alpha)} \beta_\sigma, \end{aligned}$$

and let $\gamma_\sigma = \min\{\gamma, \beta_\sigma\}$ for $\sigma < \mathrm{cf}(\alpha)$.

Let $\{X_i : i \in I\}$ be a set of non-empty spaces and $Y \subset (X_I)_\kappa$.

(a) If $(X_J)_\kappa$ has calibre (β, β, γ) and calibre $(\alpha_\sigma, \beta_\sigma, \gamma_\sigma)$ for every non-empty $J \subset I$ with $|J| < \kappa$, and if $\pi_J[Y] = X_J$ for every non-empty $J \subset I$ with $|J| \leq \gamma \cdot \kappa$, then Y has calibre (α, β, γ);

(b) if $\kappa = \omega$ and X_J has compact-calibre (β, β, γ) and compact-calibre $(\alpha_\sigma, \beta_\sigma, \gamma_\sigma)$ for every non-empty $J \subset I$ with $|J| < \omega$, and if for every compact $K_J \subset X_J$ with J non-empty, $J \subset I$ and $|J| \leq \gamma \cdot \omega$ there is compact $K \subset Y$ such that $\pi_J[K] = K_J$, then Y has compact calibre (α, β, γ).

Proof. We assume in what follows, the statements being obvious otherwise, that $X_i \neq \emptyset$ for $i \in I$.

Let $\{U^\xi : \xi < \alpha\}$ be a set of non-empty basic subsets of $(X_I)_\kappa$. We show in (a) there is $B \in [\alpha]^\beta$ such that if $C \in [B]^\gamma$ then $(\cap_{\xi \in C} U^\xi) \cap Y \neq \varnothing$, and in (b) there is $B \in [\alpha]^\beta$ such that if $C \in [B]^\gamma$ there is compact $K \subset Y$ such that $U^\xi \cap K \neq \varnothing$ whenever $\xi \in C$.

From the extension to singular cardinals of the Erdős–Rado theorem on quasi-disjoint families (Theorem 1.10), there are Γ, J, $\{A_\sigma : \sigma \in \Gamma\}$ and $\{J_\sigma : \sigma \in \Gamma\}$ such that

$$\begin{aligned}
&\Gamma \in [\mathrm{cf}(\alpha)]^{\mathrm{cf}(\alpha)},\\
&A_\sigma \in [\alpha]^{\alpha_\sigma} \text{ for } \sigma \in \Gamma,\\
&A_\sigma \cap A_{\sigma'} = \varnothing \text{ for } \sigma, \sigma' \in \Gamma, \sigma \neq \sigma' \text{ and}\\
&R(U^\xi) \cap R(U^{\xi'}) = J_\sigma \text{ for } \xi, \xi' \in A_\sigma, \xi \neq \xi'\\
&\qquad\qquad\qquad = J \text{ for } \xi \in A_\sigma, \xi' \in A_{\sigma'}, \sigma \neq \sigma'.
\end{aligned}$$

We set $\tilde{J}_\sigma = J_\sigma \backslash J$ for $\sigma \in \Gamma$. If $\tilde{J}_\sigma = \varnothing$ we let B_σ be an arbitrary element of $[A_\sigma]^{\beta_\sigma}$, and if $\tilde{J}_\sigma \neq \varnothing$ then: in (a), using the fact that $(X_{\tilde{J}_\sigma})\kappa$ has calibre $(\alpha_\sigma, \beta_\sigma, \gamma_\sigma)$, we choose $[B_\sigma] \in [A_\sigma]^{\beta_\sigma}$ such that if $C_\sigma \in [B_\sigma]^{\gamma_\sigma}$ then $\cap_{\xi \in C} \pi_{\tilde{J}_n}[U^\xi] \neq \varnothing$; and in (b), using the fact that $X_{\tilde{J}_\sigma}$ has compact-calibre $(\alpha_\sigma, \beta_\sigma, \gamma_\sigma)$, we choose $B_\sigma \in [A_\sigma]^{\beta_\sigma}$ such that if $C_\sigma \in [B_\sigma]^{\gamma_\sigma}$ then there is compact $K_\sigma \subset X_{\tilde{J}_\sigma}$ such that $\pi_{\tilde{J}_\sigma}[U^\xi] \cap K_\sigma \neq \varnothing$ whenever $\xi \in C_\sigma$.

We set $\tilde{B} = \cup_{\sigma \in \Gamma} B_\sigma$. Since $\beta_{\sigma'} \leq \beta_\sigma$ for $\sigma' < \sigma < \mathrm{cf}(\alpha)$ and $\beta = \sum_{\sigma < \mathrm{cf}(\alpha)} \beta_\sigma$, we have $|\tilde{B}| = \beta$. If $J = \varnothing$ we let B be an arbitrary element of $|\tilde{B}|^\beta$, and if $J \neq \varnothing$ then: in (a), using the fact that $(X_J)_\kappa$ has calibre (β, β, γ), we choose $B \in |\tilde{B}|^\beta$ such that if $C \in [B]^\gamma$ then $\cap_{\xi \in C} \pi_J[U^\xi] \neq \varnothing$; and in (b), using the fact that X_J has compact-calibre (β, β, γ), we choose $B \in [\tilde{B}]^\beta$ such that if $C \in [B]^\gamma$ then there is compact $K \subset X_J$ such that $\pi_J[U^\xi] \cap K \neq \varnothing$ whenever $\xi \in C$.

To show that B is as required we let $C \in [B]^\gamma$ and we conclude separately the arguments for (a) and (b).

(a) If $\tilde{J}_\sigma \neq \varnothing$, then since

$$|C \cap B_\sigma| \leq \min\{\gamma, \beta_\sigma\} = \gamma_\sigma,$$

there is $p(\sigma) \in \cap \{\pi_{\tilde{J}_\sigma}[U^\xi] : \xi \in C \cap B_\sigma\}$; and if $J \neq \varnothing$ there is $p \in \cap_{\xi \in C} \pi_J[U^\xi]$. We set $F = \cup_{\xi \in C} R(U^\xi)$; since $|F| \leq \gamma \cdot \kappa$ we have $\pi_F[Y] = X_F$, and since

$$|\{\xi \in C : i \in R(U^\xi)\}| \leq 1 \text{ for } i \in F \backslash \bigcup_{\sigma \in \Gamma} J_\sigma,$$

there is $x = \langle x_i : i \in I \rangle \in Y$ such that

$$\begin{aligned} x_i &= p_i \quad \text{if } i \in J \\ x_i &= p(\sigma)_i \text{ if } i \in \tilde{J}_\sigma, \sigma \in \Gamma \\ x_i &\in U_i^\xi \quad \text{if } i \in F \setminus \bigcup_{\sigma \in \Gamma} J_\sigma, i \in R(U^\xi), \xi \in C \\ x_i &\in X_i \quad \text{if } i \in I \setminus F. \end{aligned}$$

It is clear that $x \in (\cap_{\xi \in C} U^\xi) \cap Y$.

(b) If $\tilde{J}_\sigma \neq \varnothing$, then since

$$|C \cap B_\sigma| \leq \min\{\gamma, \beta_\sigma\} = \gamma_\sigma,$$

there is compact $K_\sigma \subset X_{J_\sigma}$ such that $\pi_{J_\sigma}[U^\xi] \cap K_\sigma \neq \varnothing$ whenever $\xi \in C \cap B_\sigma$; and if $J \neq \varnothing$ there is compact $K_J \subset X_J$ such that $\pi_J[U^\xi] \cap K_J \neq \varnothing$ whenever $\xi \in C$. We set $F = \cup_{\xi \in C} R(U^\xi)$, we note that

$$|\{\xi \in C : i \in R(U^\xi)\}| \leq 1 \text{ for } i \in F \setminus \bigcup_{\sigma \in \Gamma} J_\sigma,$$

and we choose

$$x_i \in U_i^\xi \text{ for } i \in F \setminus \bigcup_{\sigma \in \Gamma} J_\sigma, i \in R(U^\xi), \xi \in C.$$

Since $|F| \leq \gamma \cdot \omega$, there is compact $K \subset Y$ such that

$$\begin{aligned} \pi_J[K] &= K_J \text{ if } J \neq \varnothing \\ \pi_{\tilde{J}_\sigma}[K] &= K_\sigma \text{ if } \tilde{J}_\sigma \neq \varnothing \\ \pi_i[K] &= \{x_i\} \text{ if } i \in F \setminus \bigcup_{\sigma \in \Gamma} J_\sigma, i \in R(U^\xi). \end{aligned}$$

It is clear that $U^\xi \cap K \neq \varnothing$ whenever $\xi \in C$.

The proof is complete.

3.14 *Corollary*. Let α be singular with $\mathrm{cf}(\alpha) > \omega$, let $\{\alpha_\sigma : \sigma < \mathrm{cf}(\alpha)\}$ be a set of cardinal numbers such that

α_σ is regular for $\sigma < \mathrm{cf}(\alpha)$,

$\mathrm{cf}(\alpha) < \alpha_{\sigma'} < \alpha_\sigma < \alpha$ for $\sigma' < \sigma < \mathrm{cf}(\alpha)$, and

$$\sum_{\sigma < \mathrm{cf}(\alpha)} \alpha_\sigma = \alpha,$$

and let $\{X_i : i \in I\}$ be a set of spaces.

(a) If X_J has calibre α and calibre $(\alpha_{\sigma+1}, \alpha_\sigma)$ for all non-empty $J \subset I$ with $|J| < \omega$, then X_I has calibre α;

(b) if X_J has compact-calibre α and compact-calibre $(\alpha_{\sigma+1}, \alpha_\sigma)$ for all non-empty $J \subset I$ with $|J| < \omega$, then X_I has compact-calibre α.

Proof. Theorem 3.13 reduces to this when $\kappa, \alpha, \beta, \gamma, \langle \alpha_\sigma, \beta_\sigma, \gamma_\sigma \rangle$ and Y are replaced by $\omega, \alpha, \alpha, \alpha, \langle \alpha_{\sigma+1}, \alpha_\sigma, \alpha_\sigma \rangle$ and X_I, respectively.

3.15 *Corollary.* Let α be singular with $\mathrm{cf}(\alpha) > \omega$ and let $\{X_i : i \in I\}$ be a set of spaces.

(a) If each of the spaces X_i has calibre α and calibre β for all (infinite) regular $\beta < \alpha$, then X_I has calibre α;

(b) if each of the spaces X_i has compact-calibre α and compact-calibre β for all (infinite) regular $\beta < \alpha$, then X_I has compact-calibre α.

Proof. (a) It follows from Theorem 2.2(a) that for all non-empty $J \subset I$ with $|J| < \omega$ the space X_J has calibre α and calibre β for all (infinite) regular $\beta < \alpha$. Part (a) then follows from Corollary 3.14(a).

The proof of (b) is similar.

3.16 *Theorem.* Let $\omega \leq \kappa \ll \alpha, \kappa \ll \mathrm{cf}(\alpha)$, let $\{X_i : i \in I\}$ be a set of spaces and $Y \subset (X_I)_\kappa$. If $(X_J)_\kappa$ is pseudo-(α, α)-compact for every non-empty $J \subset I$ with $|J| < \kappa$, and if $\pi_J[Y] = X_J$ for every non-empty $J \subset I$ with $|J| < \kappa$, then Y is pseudo-(α, α)-compact.

Proof. We assume in what follows, the statement being obvious otherwise, that $X_i \neq \varnothing$ for $i \in I$.

If α is regular the theorem is the case $\beta = \alpha$ of Theorem 3.6. We assume that α is singular, using Theorem A.7 we choose a set $\{\alpha_\sigma : \sigma < \mathrm{cf}(\alpha)\}$ of regular cardinal numbers such that

$$\kappa \ll \alpha_\sigma < \alpha \text{ for } \sigma < \mathrm{cf}(\alpha) \text{ and}$$
$$\sum_{\sigma < \mathrm{cf}(\alpha)} \alpha_\sigma = \alpha,$$

we let $\{U^\xi : \xi < \alpha\}$ be a set of non-empty basic subsets of $(X_I)_\kappa$, and from the extension to singular cardinals of the Erdős–Rado theorem we choose $\Gamma, J, \{A_\sigma : \sigma \in \Gamma\}, A$ and $\{J_\sigma : \sigma \in \Gamma\}$ such that

$$\Gamma \in [\mathrm{cf}(\alpha)]^{\mathrm{cf}(\alpha)},$$
$$A_\sigma \in [\alpha]^{\alpha_\sigma} \text{ for } \sigma < \mathrm{cf}(\alpha),$$
$$A_\sigma \cap A_{\sigma'} = \varnothing \text{ for } \sigma, \sigma' \in \Gamma, \sigma \neq \sigma',$$
$$A = \bigcup_{\sigma < \mathrm{cf}(\alpha)} A_\sigma, \text{ and}$$
$$R(U^\xi) \cap R(U^{\xi'}) = J_\sigma \text{ for } \xi, \xi' \in A_\sigma, \xi \neq \xi'$$
$$= J \text{ for } \xi \in A_\sigma, \xi' \in A_{\sigma'}, \sigma, \sigma' \in \Gamma, \sigma \neq '.$$

If $J = \emptyset$ we let x be an arbitrary element of Y, and if $J \neq \emptyset$ we use the facts that $(X_J)_\kappa$ is pseudo-(α, α)-compact and $\pi_J[Y] = X_J$ to find $x \in Y$ such that

$$|\{\xi \in A : V_J \cap \pi_J[U^\xi] \neq \emptyset\}| = \alpha$$

for every neighborhood V_J in $(X_J)_\kappa$ of x_J.

Let V be a basic neighborhood in $(X_I)_\kappa$ of x and set $V_J = \pi_J[V]$ and $B = \{\xi \in A : V_J \cap \pi_J[U^\xi] \neq \emptyset\}$.

For $i \in I$ we set $C(i) = \{\xi \in B : i \in R(U^\xi)\}$. We note that if $i \in I$ and $\sigma \in \Gamma$ and $|C(i) \cap A_\sigma| \geq 2$ then $i \in J_\sigma$. Since $J_\sigma \cap J_{\sigma'} = J$ for $\sigma, \sigma' \in \Gamma$ with $\sigma \neq \sigma'$ it follows that

$$|\{\sigma \in \Gamma : |C(i) \cap A_\sigma| \geq 2\}| \leq 1 \text{ for } i \in I \backslash J.$$

Since $C(i) = \cup_{\sigma \in \Gamma}(C(i) \cap A_\sigma)$ and $|A_\sigma| = \alpha_\sigma < \alpha$ and $|\Gamma| = \text{cf}(\alpha) < \alpha$, we have $|C(i)| < \alpha$ for $i \in I \backslash J$.

We set $C = \cup\{C(i) : i \in R(V) \backslash J\}$. Since $|R(V)| < \kappa < \text{cf}(\alpha)$ we have $|C| < \alpha$, and from $|B| = \alpha$ it follows that $|B \backslash C| = \alpha$.

If $\xi \in B$ and $V \cap U^\xi = \emptyset$, then since $V_J \cap \pi_J[U^\xi] \neq \emptyset$ there is $i \in I$ such that

$$i \in (R(V) \cap R(U^\xi)) \backslash J,$$

and hence $\xi \in C(i) \subset C$. We have

$$\{\xi \in B : V \cap U^\xi \neq \emptyset\} \supset B \backslash C$$

and $|B \backslash C| = \alpha$, as required.

Chain conditions in powers

3.17 *Lemma*. Let $\omega \leq \text{cf}(\alpha) < \text{cf}(\beta) \leq \beta < \alpha$ and let X be a space.

(a) If X has calibre α then there is γ such that $\beta < \gamma < \alpha$ and X has calibre (γ, β); and

(b) if X has compact-calibre α then there is γ such that $\beta < \gamma < \alpha$ and X has compact-calibre (γ, β).

Proof. We prove (b). If the result fails there is a set $\{\alpha_\sigma : \sigma < \text{cf}(\alpha)\}$ of cardinal numbers such that

$$\beta = \alpha_0 < \alpha_{\sigma'} < \alpha_\sigma < \alpha \quad \text{for } 0 < \sigma' < \sigma < \text{cf}(\alpha),$$

$$\alpha = \sum_{\sigma < \text{cf}(\alpha)} \alpha_\sigma, \text{ and}$$

X does not have compact-calibre (α_σ, β) for $\sigma < \text{cf}(\alpha)$.

There is $\{U_\xi : \alpha_\sigma \leq \xi < \alpha_{\sigma+1}\} \subset \mathscr{T}^*(X)$ such that $|\{\xi : \alpha_\sigma \leq \xi < \alpha_{\sigma+1}, U^\xi \cap K \neq \varnothing\}| < \beta$ for all compact $K \subset X$; since $\{U_\xi : \beta \leq \xi < \alpha\} \subset \mathscr{T}^*(X)$ and X has compact-calibre (α, β), there are $B \in [\alpha \backslash \beta]^\beta$ and compact $K \subset X$ such that $U_\xi \cap K \neq \varnothing$ whenever $\xi \in B$. We set

$$B(\sigma) = B \cap [\alpha_\sigma, \alpha_{\sigma+1}) \text{ for } \alpha < \text{cf}(\alpha).$$

Since $B = \cup_{\sigma < \text{cf}(\alpha)} B(\sigma)$ and $\text{cf}(\alpha) < \text{cf}(\beta)$, there is $\bar{\sigma} < \text{cf}(\alpha)$ such that $|B(\bar{\sigma})| = \beta$. This contradiction completes the proof.

We omit the (similar, simpler) proof of (a).

A non-empty finite space has calibre α for all $\alpha \geq \omega$; but from Corollary 2.10 and Theorem 2.3(a) it follows that if $\text{cf}(\alpha) = \omega$ then for $\gamma \geq \omega$ the (infinite) space $\{0, 1\}^\gamma$ does not have calibre α. We show next, in contrast, that for $\text{cf}(\alpha) > \omega$ every power of a space with calibre α has calibre α.

3.18 *Theorem.* Let $\text{cf}(\alpha) > \omega$ and let X be a space.

(a) If X has calibre α then X^I has calibre α for every non-empty set I; and

(b) if X has compact-calibre α then X^I has compact-calibre α for every non-empty set I.

Proof. For α regular this is a special case of Corollary 3.7. We assume in what follows that α is singular.

There is a set $\{\beta_\sigma : \sigma < \text{cf}(\alpha)\}$ of cardinal numbers such that

$$\beta_{\sigma'} < \beta_\sigma < \alpha \text{ for } \sigma' < \sigma < \text{cf}(\alpha) \text{ and}$$
$$\sum_{\sigma < \text{cf}(\alpha)} \beta_\sigma = \alpha.$$

We define sets $\{\alpha_\sigma(n) : \sigma < \text{cf}(\alpha), n < \omega\}$ and $\{\alpha_\sigma : \sigma < \text{cf}(\alpha)\}$ of cardinal numbers such that

(i) $\alpha_0 = \alpha_0(n) = (\text{cf}(\alpha))^+$ for $n < \omega$;

(ii) $\alpha_\sigma = \alpha_\sigma(n) = (\max\{\beta_\sigma, \sup_{\sigma' < \sigma} \alpha_{\sigma'}\})^+$ for non-zero limit ordinals $\sigma < \text{cf}(\alpha)$;

(iii) $\alpha_{\sigma+1}(n) > \alpha_\sigma$ for $\sigma < \text{cf}(\alpha), n < \omega$;

(iv)(a) in (a): X^n has calibre $(\alpha_{\sigma+1}(n), \alpha_\sigma)$ for $\sigma < \text{cf}(\alpha), n < \omega$;

(iv)(b) in (b): X^n has compact-calibre $(\alpha_{\sigma+1}(n), \alpha_\sigma)$ for $\sigma < \text{cf}(\alpha)$, $n < \omega$; and

(v) $\alpha_{\sigma+1} = (\max\{\beta_{\sigma+1}, \sup_{n<\omega} \alpha_{\sigma+1}(n)\})^+$ for $\sigma < \text{cf}(\alpha)$.

We proceed by transfinite recursion. We define $\alpha_0, \alpha_0(n)$ for $n < \omega$

by (i), and for non-zero limit ordinals $\sigma < \mathrm{cf}(\alpha)$ we define α_σ, $\alpha_\sigma(n)$ for $n < \omega$ by (ii). For $\sigma < \mathrm{cf}(\alpha)$ we note that α_σ is a regular cardinal, we note for $n < \omega$ from Theorem 2.2 that X^n has calibre α in (a) and X^n has compact-calibre α in (b), and we use Lemma 3.17 to define $\alpha_{\sigma+1}(n)$ for $n < \omega$ so that (iii), and (iv)(a) or (iv)(b), are satisfied; and we define $\alpha_{\sigma+1}$ by (v).

The definition of the families is complete. We note that:

α_σ is regular for $\sigma < \mathrm{cf}(\alpha)$;

$\mathrm{cf}(\alpha) < \alpha_{\sigma'} < \alpha_\sigma < \alpha$ for $\sigma' < \sigma < \mathrm{cf}(\alpha)$;

$\sum_{\sigma < \mathrm{cf}(\alpha)} \alpha_\sigma = \alpha$; and

X_J has calibre α and calibre $(\alpha_{\sigma+1}, \alpha_\sigma)$ for all $J \in \mathscr{P}^*_\omega(I)$ in (a), and

X_J has compact-calibre α and compact-calibre $(\alpha_{\sigma+1}, \alpha_\sigma)$ for all $J \in \mathscr{P}^*_\omega(I)$ in (b).

The required conclusions now follow from Corollary 3.14.

Remark. We have seen in Corollary 3.7(a) that for $\alpha > \omega$ with α regular the product of a set of spaces with calibre α has calibre α. For singular α the analogous statement is false (cf. Theorem 4.4); thus in Theorem 3.18(a) the restriction to powers of a fixed space cannot be waived, and Theorem 3.18(a) is, in this sense, the best possible. For compact spaces, however, a strong positive result is available (Corollary 5.8): Assuming the generalized continuum hypothesis, the product of a set of compact spaces with calibre α (with $\mathrm{cf}(\alpha) > \omega$) has calibre α.

In Theorem 3.19 we generalize the results of Theorem 3.18.

3.19 *Theorem.* Let $\omega \leq \kappa \ll \alpha, \kappa \ll \mathrm{cf}(\alpha)$, $\alpha \geq \beta \geq \gamma \geq 2$, and let X be a space.

(a) If $(X^J)_\kappa$ has calibre (α, β, γ) for all J with $|J| < \kappa$, then $(X^I)_\kappa$ has calibre (α, β, γ) for all I;

(b) if ($\kappa = \omega$ and) X^n has compact-calibre (α, β, γ) for all $n < \omega$, then X^I has compact-calibre (α, β, γ) for all I;

(c) if ($\kappa = \omega$ and) X^n is pseudo-(α, β)-compact for all $n < \omega$, then X^I is pseudo-(α, β)-compact for all I.

Proof. The statements are clear if $X = \varnothing$; we assume in what follows that $X \neq \varnothing$.

Let $\{U^\xi : \xi < \alpha\}$ be a set of non-empty, basic open subsets of $(X^I)_\kappa$.

For notational convenience we let $R(\xi)$ denote $R(U^\xi)$ for $\xi < \alpha$. Since $|R(\xi)| < \kappa$ for $\xi < \alpha$ and $\kappa \ll \mathrm{cf}(\alpha)$ we may suppose without loss of generality that there is $\lambda < \kappa$ such that $|R(\xi)| = \lambda$ for all $\xi < \alpha$; for $\xi < \alpha$ we let φ_ξ be a one-to-one function from $R(\xi)$ onto λ and, using Corollary 1.11, we choose $A \in [\alpha]^\alpha$ such that if $\xi, \xi' \in A$ and $i \in R(\xi) \cap R(\xi')$ then $\varphi_\xi(i) = \varphi_{\xi'}(i)$.

For $\xi \in A$ we define a function $h_\xi : X^{R(\xi)} \to X^\lambda$ by the rule

$$(h_\xi(x))_\zeta = x_i \quad \text{for } \zeta = \varphi_\xi(i) < \lambda.$$

It is clear that h_ξ is a homeomorphism of $(X^{R(\xi)})_\kappa$ onto $(X^\lambda)_\kappa$.

For $\xi \in A$ we set $V^\xi = \pi_{R(\xi)}[U^\xi]$. We complete the proofs of (a) and (b) separately.

(a) Since $h_\xi[V^\xi]$ is open in $(X^\lambda)_\kappa$ and $(X^\lambda)_\kappa$ has calibre (α, β, γ), there is $B \in [A]^B$ such that if $C \in [B]^\gamma$ then $\cap_{\xi \in C} h_\xi[V^\xi] \neq \emptyset$. We claim that if $C \in [B]^\gamma$ then $\cap_{\xi \in C} U^\xi \neq \emptyset$. Let $C \in [B]^\gamma$ and $p \in \cap_{\xi \in C} h_\xi[V^\xi]$ and define $x \in X^I$ by the rule

$$x_{R(\xi)} = h_\xi^{-1}(\{p\}) \quad \text{for } \xi \in C,$$
$$x_i \in X_I \quad \text{if } i \in I \backslash \bigcup_{\xi \in C} R(\xi).$$

(If $i \in R(\xi) \cap R(\xi')$ with $\xi, \xi' \in C$, there is $\zeta < \lambda$ such that

$$\varphi_\xi(i) = \varphi_{\xi'}(i) = \zeta;$$

hence $x_i = p_\zeta$. It follows that x is a well-defined element of X^I.)

For $\xi \in C$ we have $h_\xi(x_{R(\xi)}) = p \in h_\xi[V^\xi]$ and hence $x_{R(\xi)} \in V^\xi$. Since $U_i^\xi = X$ for $i \in I \backslash R(\xi)$ we have $x \in U^\xi$ for $\xi \in C$, as required.

(b) Since $h_\xi[V^\xi]$ is open in X^λ and X^λ has compact-calibre (α, β, γ), there is $B \in [A]^\beta$ such that if $C \in [B]^\gamma$ there is compact $\tilde{K} \subset X^\lambda$ such that $h_\xi[V^\xi] \cap \tilde{K} \neq \emptyset$ whenever $\xi \in C$. We claim that if $C \in [B]^\gamma$ there is compact $K \subset X^I$ such that $U^\xi \cap K \neq \emptyset$ whenever $\xi \in C$. Let $C \in [B]^\gamma$ and let $\tilde{K}$ be a compact subspace of X^λ such that $h_\xi[V^\xi] \cap \tilde{K} \neq \emptyset$ whenever $\xi \in C$; we suppose without loss of generality that $\tilde{K} = \prod_{k < \lambda} \pi_k[\tilde{K}]$.

We choose $x \in X$ and we define $K = \prod_{i \in I} K_i \subset X^I$ by the rule

$$\prod_{i \in R(U^\xi)} K_i = h_\xi^{-1}(\tilde{K}) \quad \text{for } \xi \in C,$$

$$K_i = \{x\} \quad \text{if } i \in I \backslash \bigcup_{\xi \in C} R(\xi).$$

(If $i \in R(\xi) \cap R(\xi')$ with $\xi, \xi' \in C$, there is $k < \lambda$ such that

$$\varphi_\xi(i) = \varphi_{\xi'}(i) = k;$$

hence $K_i = \tilde{K}_k$. It follows that K is a well-defined (compact) subspace of X^I.)

For $\xi \in C$ we have

$$h_\xi[V^\xi] \cap h_\xi[\pi_{R(\xi)}[K]] = h_\xi[V_\xi] \cap \tilde{K} \neq \varnothing$$

and hence $V^\xi \cap \pi_{R(\xi)}[K] \neq \varnothing$. Since $U_i^\xi = X$ for $i \in I \backslash R(\xi)$, we have $U^\xi \cap K \neq \varnothing$ for $\xi \in C$, as required.

(c) Since $h_\xi[V^\xi]$ is open in X^λ and X^λ is pseudo-(α, β)-compact, there is $p \in X^\lambda$ such that

$$|\{\xi \in A : V \cap h_\xi[V^\xi] \neq \varnothing\}| \geq \beta$$

for every neighborhood V of p. We define $x \in X^I$ by the rule

$$X_{R(\xi)} = h_\xi^{-1}(\{p\}) \quad \text{for } \xi \in A,$$
$$x_i \in X_i \quad \text{if } i \in I \backslash \bigcup_{\xi \in A} R(\xi)$$

and we show

$$|\{\xi \in A : W \cap U^\xi \neq \varnothing\}| \geq \beta$$

for every neighborhood W of x. (That x is a well-defined element of X^I follows as in the proof above of part (a).)

Let W be a basic neighborhood of x and let F be a finite subset of I such that

$$R(W) \cap (\bigcup_{\xi \in A} R(\xi)) \subset \bigcup_{\eta \in F} R(\eta).$$

Since $\pi_{R(\eta)}[W]$ is a neighborhood of $x_{R(\eta)}$ and h_η is a homeomorphism of $X^{R(\eta)}$ onto X^λ, the set $h_\eta[\pi_{R(\eta)}[W]]$ is a neighborhood of $h_\eta(x_{R(\eta)}) = p$. We set

$$V = \bigcap_{\eta \in F} h_\eta[\pi_{R(\eta)}[W]].$$

Since V is a neighborhood of p we have

$$|\{\xi \in A : V \cap h_\xi[V^\xi] \neq \varnothing\}| \geq \beta.$$

To complete the proof it is enough to show that if $V \cap h_\xi[V^\xi] \neq \varnothing$ with $\xi \in A$ then $W \cap U^\xi \neq \varnothing$, and for this (since $W_i = X$ for $i \in I \backslash \bigcup_{\eta \in F} R(\eta)$) it is enough to show that

$$\pi_{R(\eta)}[W] \cap \pi_{R(\eta)}[U^\xi] \neq \varnothing \quad \text{for } n \in F.$$

We have

$$\begin{aligned}
h_\xi \circ \pi_{R(\xi)}[U^\xi] &= \prod_{\zeta<\lambda} U^\xi_{\varphi_\xi^{-1}(\{\zeta\})} \\
&= \prod_{\zeta\in\varphi_\xi[R(\xi)\cap R(\eta)]} U^\xi_{\varphi_\xi^{-1}(\{\zeta\})} \times \prod_{\xi\in\varphi_\xi[R(\xi)\setminus R(\eta)]} U^\xi_{\varphi_\xi^{-1}(\{\zeta\})} \\
&\subset \prod_{\zeta\in\varphi_\xi[R(\xi)\cap R(\eta)]} U^\xi_{\varphi_\xi^{-1}(\{\zeta\})} \times \prod_{\zeta\in\varphi_\xi[R(\xi)\setminus R(\eta)]} X \\
&= \prod_{\zeta\in\varphi_\eta[R(\xi)\cap R(\eta)]} U^\xi_{\varphi_\eta^{-1}(\{\zeta\})} \times \prod_{\zeta\in\varphi_\eta[R(\eta)\setminus R(\xi)]} X \\
&= \prod_{\zeta<\lambda} U^\xi_{\varphi_\eta^{-1}(\{\zeta\})} = h_\eta \circ \pi_{R(\eta)}[U^\xi],
\end{aligned}$$

and from the relations

$$\begin{aligned}
\varnothing = V\cap h_\xi[V^\xi] &= V\cap h_\xi[\pi_{R(\xi)}[U^\xi]] \\
&\subset V\cap h_\eta[\pi_{R(\eta)}[U^\xi]] \subset h_\eta[\pi_{R(\eta)}[W]]\cap h_\eta[\pi_{R(\eta)}[U^\xi]] \\
&= h_\eta[\pi_{R(\eta)}[W]\cap \pi_{R(\eta)}[U^\xi]]
\end{aligned}$$

we have $\pi_{R(\eta)}[W]\cap\pi_{R(\eta)}[U^\xi]\neq\phi$, as required.

3.20 *Remark.* We note that although Theorems 3.18 and 3.19 concern powers (of the form X^I or $(X^I)_\kappa$), similar results are available for products X_I or $(X_I)_\kappa$ which are sufficiently closely related to powers. We illustrate with a simple modification of Theorem 3.19(a).

Theorem. Let $\omega\leq\kappa\ll\alpha$, $\kappa\ll\mathrm{cf}(\alpha)$, $\alpha\geq\beta\geq\gamma\geq 2$ and let $\{X_i : i\in I\}$ be a set of spaces. If there are a space Y and for $i\in I$ a continuous function from Y onto a dense subspace of X_i, and if $(Y^J)_\kappa$ has calibre (α,β,γ) for all J with $|J|<\kappa$, then $(X_I)_\kappa$ has calibre (α,β,γ).

Proof. The space $(Y^I)_\kappa$ has calibre (α,β,γ) by Theorem 3.19(a), and there is a continuous function from $(Y^I)_\kappa$ onto a dense subspace of $(X_I)_\kappa$. That $(X_I)_\kappa$ has calibre (α,β,γ) follows from Theorem 2.5(a)(i) and (ii).

3.21 *Remark.* The methods of Theorem 3.19 do not apply when $\mathrm{cf}(\alpha)=\omega$, and it is natural to ask the following three questions.

Let α be singular with $\mathrm{cf}(\alpha)=\omega$ and $\alpha>\beta$ and let P be one of the properties: calibre (α,β); compact-calibre (α,β); pseudo-(α,β)-compact. Is there a completely regular, Hausdorff space X such that

X^n has P for all $n < \omega$, and
X^ω does not have P?
(The corresponding question is also unanswered for compact-calibre (α, α) with α singular, $\mathrm{cf}(\alpha) = \omega$.)

3.22 *Theorem.* Let κ and λ be cardinal numbers with $\kappa > \lambda \geq \omega$ and let $\{X_i : i \in I\}$ be a set of regular, Hausdorff spaces with $|I| \geq \lambda$ and with $|X_i| > 1$ for $i \in I$. Then $(X_I)_\kappa$ is not pseudo-$(2^\lambda, 2)$-compact.

Proof. Let $\{i(\eta) : \eta < \lambda\} \subset I$ with $i(\eta) \neq i(\eta')$ for $\eta' < \eta < \lambda$, for $\eta < \lambda$ let $U_\eta, V_\eta, \tilde{U}_\eta, \tilde{V}_\eta$ be non-empty open subsets of $X_{i(\eta)}$ such that

$$\mathrm{cl}\, U_\eta \subset \tilde{U}_\eta, \mathrm{cl}\, V_\eta \subset \tilde{V}_\eta, \tilde{U}_\eta \cap \tilde{V}_\eta = \varnothing,$$

and set

$$W(A) = (\bigcap_{\eta \in A} \pi_{i(\eta)}^{-1}(U_\eta)) \cap (\bigcap_{\eta \in \lambda \backslash A} \pi_{i(\eta)}^{-1}(V_\eta)) \quad \text{for } A \subset \lambda.$$

Since $\kappa > \lambda$ the sets $W(A)$ are open in $(X_I)_\kappa$.

We show that for $x \in (X_I)_\kappa$ there is a neighborhood W of x such that

$$|\{A \subset \lambda : W \cap W(A) \neq \varnothing\}| \leq 1.$$

Case 1. $x_{i(\eta)} \in \mathrm{cl}\, U_\eta \cup \mathrm{cl}\, V_\eta$ for $\eta < \lambda$. We set

$$B = \{\eta \in \lambda : x_{i(\eta)} \in \mathrm{cl}\, U_\eta\}, \text{ and}$$

$$W = (\bigcap_{\eta \in B} \pi_{i(\eta)}^{-1}(\tilde{U}_\eta)) \cap (\bigcap_{\eta \in \lambda \backslash B} \pi_{i(\eta)}^{-1}(\tilde{V}_\eta)).$$

Then for $A \subset \lambda$ we have $W \cap W(A) \neq \varnothing$ if and only if $A = B$.

Case 2. There is $\bar{\eta} < \lambda$ such that $x_{i(\bar{\eta})} \notin \mathrm{cl}\, U_\eta \cup \mathrm{cl}\, V_\eta$. We set

$$W = \pi_{i(\bar{\eta})}^{-1}(X_{i(\bar{\eta})} \backslash (\mathrm{cl}\, U_{\bar{\eta}} \cup \mathrm{cl} V_{\bar{\eta}})).$$

Then $W \cap W(A) = \varnothing$ for all $A \subset \lambda$.

The proof is complete.

It follows from Theorem 2.1((d) and (e)) that a space $(X_I)_\kappa$ as in Theorem 3.22 does not have calibre $(2^\lambda, 2)$. For this latter conclusion it is not necessary to assume that the Hausdorff spaces X_i are regular spaces, since the procedure of the proof of Theorem 3.22 furnishes a cellular family in $(X_I)_\kappa$ faithfully indexed by $\{A : A \subset \lambda\}$.

3.23 *Corollary.* Let α, κ and λ be infinite cardinals with $\kappa > \lambda$

Chain Conditions in Products

		calibre (α,β,γ)				compact-calibre (α,β,γ)				pseudo-(α,β)-compact			
		α regular		α singular		α regular		α singular		α regular		α singular	
		$\alpha=\omega$	$\alpha>\omega$	$\mathrm{cf}(\alpha)=\omega$	$\mathrm{cf}(\alpha)>\omega$	$\alpha=\omega$	$\alpha>\omega$	$\mathrm{cf}(\alpha)=\omega$	$\mathrm{cf}(\alpha)>\omega$	$\alpha=\omega$	$\alpha>\omega$	$\mathrm{cf}(\alpha)=\omega$	$\mathrm{cf}(\alpha)>\omega$
Products	$\alpha=\beta$	N 2.10	Y 3.6	N 4.4	N 4.4	Y 3.9	Y 3.6	N 8.3	N 8.3	N 9.9	Y 3.6, 3.16	N 9.9	Y 3.16
	$\alpha>\beta$	N 2.10	Y 3.6	N 3.11	N 3.11	Y 2.1(*g*)	Y 3.6	$\beta\geqq\omega$ N 3.11 $\beta<\omega$ Y 2.1 (g)	$\beta\geqq\omega$ N 3.11 $\beta<\omega$ Y 2.1(g)	N 9.9 + 2.1(i)	Y 3.6	N 3.11	N 3.11
Powers	$\alpha=\beta$	N 2.10	Y 3.6	N 2.12	Y 3.18, 3.19	Y 3.9	Y 3.6	?	Y 3.18, 3.19	N 9.10	Y 3.6, 3.16	N 9.10	Y 3.16, 3.19
	$\alpha>\beta$	N 2.10	Y 3.6	?	Y 3.19	Y 2.1(g)	Y 3.6	$\beta\geqq\omega$? $\beta<\omega$ Y 2.1 (g)	Y 3.19	N 9.10 + 2.1 (i)	Y 3.6	?	Y 3.19

and let $\{X_i : i \in I\}$ be a set of regular, Hausdorff spaces with $|I| \geq \lambda$ and with $|X_i| > 1$ for $i \in I$.

(a) If $2 \leq \beta \leq \alpha \leq 2^\lambda$ then $(X_I)_\kappa$ is not pseudo-(α, β)-compact.

(b) If $\mathrm{cf}(\alpha) \leq 2^\lambda$ then $(X_I)_\kappa$ is not pseudo-(α, α)-compact.

Proof. (a) follows from Theorems 3.22 and 2.1(c).

(b) follows from (a) (with α and β replaced by $\mathrm{cf}(\alpha)$) and Theorem 2.3(c).

We note that from Theorem 2.1((d) and (e)) it follows that a space $(X_I)_\kappa$ as in Corollary 3.23(b) does not have compact-calibre α or calibre α.

We note further that the assumptions $\alpha \leq 2^\lambda$ and $\mathrm{cf}(\alpha) \leq 2^\lambda$ of Corollary 3.23 are incompatible with the conditions $\kappa \ll \alpha$ and $\kappa \ll \mathrm{cf}(\alpha)$, respectively, which have occurred frequently as hypotheses in this chapter.

3.24 *Summary.* Some of the results of this chapter and of later chapters are summarized in the chart opposite. For each of the indicated chain conditions P (with $\alpha \geq \beta \geq \gamma \geq 2$) the question asked is this: If $\{X_i : i \in I\}$ is a set of spaces such that $(X_J)_\kappa$ has P for all non-empty $J \subset I$ with $|J| < \kappa$, must $(X_I)_\kappa$ have P?

Some of the results cited and some of the examples given relate only to the case $\kappa = \omega$. It is understood in each case that $\alpha \geq \kappa$, and frequently for $\kappa > \omega$ the theorems cited require in addition one or both of the hypotheses $\kappa \ll \alpha, \kappa \ll \mathrm{cf}(\alpha)$. Numbers refer to the relevant passage in this text. Y indicates that the answer to the question is Yes; N indicates No.

The numbers $S_\lambda((X_I)_\kappa)$

3.25 *Theorem.* Let $\lambda \geq 1$, $\kappa \geq \omega$, let $\{X_i : i \in I\}$ be a set of non-empty spaces and

$$\alpha = \sup\{S_\lambda((X_J)_\kappa) : \varnothing \neq J \subset I, |J| < \kappa\},$$

and let β be a regular cardinal such that $\beta \geq \alpha$ and $\kappa \ll \beta$. Then

(a) $\alpha \leq S_\lambda((X_I)_\kappa) \leq \beta$; and

(b) if the spaces X_i are Hausdorff spaces and $\alpha < \omega$, then $S_\lambda((X_I)_\kappa) = \alpha$.

Proof. (a) For $J \in \mathscr{P}^*_\kappa(I)$ the space $(X_J)_\kappa$ is the continuous image of $(X_I)_\kappa$, and from Theorem 2.5(a)(i) we have $S_\lambda((X_J)_\kappa) \le S_\lambda((X_I)_\kappa)$. It follows that $\alpha \le S_\lambda((X_I)_\kappa)$.

That $S_\lambda((X_I)_\kappa) \le \beta$ follows from Theorem 3.6(a) (with α, β, λ and Y replaced by β, λ, λ and $(X_I)_\kappa$, respectively).

(b) If $\lambda = 1$ then from Lemma 2.8(c) we have $S_\lambda((X_J)_\kappa) = 1$ for all $J \in \mathscr{P}^*_\kappa(I)$ and hence

$$\alpha = 1 = S_\lambda((X_I)_\kappa).$$

If $\lambda > 1$ then from Corollary 2.10 we have $|(X_J)_\kappa| < \omega$ for all $J \in \mathscr{P}^*_\kappa(I)$. If for all $n < \omega$ there is $J(n) \in \mathscr{P}^*_\kappa(I)$ such that $|(X_{J(n)})_\kappa| \ge n$, then $\alpha \ge S_\lambda((X_{J(n)})_\kappa) \ge S_2((X_{J(n)})_\kappa) \ge S_2(n) = n + 1$ and hence $\alpha \ge \omega$, a contradiction. It follows that there are $n < \omega$ and $\bar{J} \in P^*_\kappa(I)$ such that $|\bar{J}| = n$ and $|X_i| = 1$ for $i \in I \backslash \bar{J}$. Then $(X_I)_\kappa$ is homeomorphic to $(X_{\bar{J}})_\kappa$ and

$$S_\lambda((X_I)_\kappa) = S_\lambda((X_{\bar{J}})_\kappa) = \alpha,$$

as required.

3.26 *Remarks.* We continue the notation of Theorem 3.25.

(a) If α is regular and $\kappa \ll \alpha$ then $S_\lambda((X_I)_\kappa) = \alpha$.

(b) If $\alpha \ge \omega$ and $\gamma = ((\alpha^\kappa)^\kappa)^+$ then γ is a regular cardinal such that $\gamma > \alpha$ and $\kappa \ll \gamma$; hence $S_\lambda((X_I)_\kappa) \le \gamma$. (That γ is regular is clear. For $\delta < \gamma, \mu < \kappa$ we have

$$\delta^\mu \le ((\alpha^\kappa)^\kappa)^\mu \le ((\alpha^\kappa)^\kappa)^\kappa = (\alpha^\kappa)^\kappa < \gamma$$

from Theorem A.3(c) of the Appendix; hence $\kappa \ll \omega$.)

(c) It follows from Theorem 3.25 that if $\lambda = 2$ or λ is (infinite and) regular, and if X_I is infinite and the spaces X_i are Hausdorff spaces, then $S_\lambda((X_I)_\kappa)$ is regular. But even when the spaces X_i are completely regular, Hausdorff spaces the relation $\kappa \ll S_\lambda((X_I)_\kappa)$ can fail. For a trivial instance of this phenomenon let γ be an uncountable regular cardinal such that $\gamma \ge \lambda$ and such that $\kappa \ll \gamma$ fails, using Theorem 3.30 below let X_0 be a space such that $S_\lambda(X_0) = \lambda$, let I be a set such that $0 \in I$ and let $|X_i| = 1$ for $i \in I, i \ne 0$; then

$$S_\lambda((X_I)_\kappa = S_\lambda(X_0) = \lambda.$$

(d) We give a natural condition sufficient to ensure that $\kappa \ll \alpha$. We begin with a simple lemma.

Lemma. Let $\kappa \geq \omega$, let I be a set, $\{J(\eta): \eta < \kappa\} \subset \mathscr{P}_\kappa(I)$ and $\mu < \kappa$. Then there is $A \in [\kappa]^\mu$ such that $|\cup_{\eta \in A} J(\eta)| < \kappa$.

Proof. If κ is regular then $|\cup_{\eta \in A} J(\eta)| < \kappa$ for all $A \in [\kappa]^\mu$. We assume that κ is singular. There is a set $\{\kappa_\sigma : \sigma < \mathrm{cf}(\kappa)\}$ of cardinal numbers such that

$$\kappa_{\sigma'} < \kappa_\sigma < \kappa \quad \text{for } \sigma' < \sigma < \mathrm{cf}(\kappa), \text{ and}$$
$$\sum_{\sigma < \mathrm{cf}(\kappa)} \kappa_\sigma = \kappa.$$

We set $A(\sigma) = \{\eta < \kappa : |J(\eta)| \leq \kappa_\sigma\}$ for $\sigma < \mathrm{cf}(\kappa)$ and we consider two cases.

Case 1. There is $\sigma < \mathrm{cf}(\kappa)$ such that $|A(\sigma)| \geq \mu$. We choose $A \in [A(\sigma)]^\mu$ and we note that $|\cup_{\eta \in A} J(\eta)| \leq \mu \cdot \kappa_\sigma < \kappa$, as required.

Case 2. $|A(\sigma)| < \mu$ for all $\sigma < \mathrm{cf}(\kappa)$. Then since $\kappa = \cup_{\sigma < \mathrm{cf}(\kappa)} A(\sigma)$ we have

$$\kappa \leq \sum_{\sigma < \mathrm{cf}(\kappa)} |A(\sigma)| \leq (\mathrm{cf}(\kappa)) \cdot \mu < \kappa,$$

a contradiction.

Theorem. Let $\kappa \geq \omega$ and let λ be a cardinal such that either

(i) $\lambda = 2$, or

(ii) $\kappa \leq \mathrm{cf}(\lambda)$, and if $\bar{\lambda} < \lambda$ and $\mu < \kappa$ then $\bar{\lambda}^\mu < \lambda$.

Let $\{X_i : i \in I\}$ be a set of non-empty spaces and

$$\alpha = \sup\{S_\lambda((X_J)_\kappa) : J \in \mathscr{P}_\kappa^*(I)\},$$

and suppose for all $\gamma < \alpha$ there is $\{J(\eta): \eta < \kappa\} \subset \mathscr{P}_\kappa^*(I)$ such that

$$J(\eta') \cap J(\eta) = \varnothing \quad \text{for } \eta' < \eta < \kappa \text{ and}$$
$$S_\lambda((X_{J(\eta)})_\kappa) > \gamma.$$

If $\gamma < \alpha$ and $\mu < \kappa$ then $\gamma^\mu < \alpha$.

Proof. Let $\gamma < \alpha$ and $\mu < \kappa$ and let $\{J(\eta): \eta < \kappa\} \subset \mathscr{P}_\kappa^*(I)$ satisfy

$$J(\eta') \cap J(\eta) = \varnothing \quad \text{for } \eta' < \eta < \kappa \text{ and}$$
$$S_\lambda((X_{J(\eta)})_\kappa) > \gamma \quad \text{for } \eta < \kappa.$$

For $\eta < \kappa$ there is $\{U^\xi(\eta): \xi < \gamma\} \subset \mathscr{T}^*((X_{J(\eta)})_\kappa)$ such that $\cap_{\xi \in C} U^\xi(\eta) = \varnothing$ for all $C \in [\gamma]^\lambda$; from the lemma there is $A \in [\kappa]^\mu$ such that $|\cup_{\eta \in A} J(\eta)| < \kappa$.

We set $\bar{J} = \cup_{\eta \in A} J(\eta)$. For $f \in \gamma^A$ we set $U(f) = \prod_{\eta \in A} U^{f(\eta)}(\eta)$ and we note that $U(f) \in \mathscr{T}^*((X_{\bar{J}})_\kappa)$.

For $C \in [\gamma^A]^\lambda$ and $\eta \in A$ we set

$$C(\eta) = \{f(\eta) : f \in C\}.$$

We claim that if $C \in [\gamma^A]^\lambda$ there is $\bar{\eta} \in A$ such that $|C(\bar{\eta})| = \lambda$. We consider two cases.

Case 1. $\lambda = 2$. Let $C = \{f_0, f_1\}$. Since $f_0 \neq f_1$ there is $\bar{\eta} \in A$ such that $f_0(\bar{\eta}) \neq f_1(\bar{\eta})$.

Case 2. $\kappa \leq \mathrm{cf}(\lambda)$, and if $\bar{\lambda} < \lambda$ and $\mu < \kappa$ then $\bar{\lambda}^\mu < \lambda$. If $|C(\eta)| < \lambda$ for all $\eta \in A$ we set $\bar{\lambda} = \sum_{\eta \in A} |C(\eta)|$ and we note that $\bar{\lambda} < \lambda$ (since $|A| = \mu < \kappa \leq \mathrm{cf}(\lambda)$); hence we have

$$\lambda = |C| \leq \prod_{\eta \in A} |C(\eta)| \leq \bar{\lambda}^\mu < \lambda,$$

a contradiction.

The claim is proved.

We show next that $S_\lambda((X_J)_\kappa) > \gamma^\mu$. Since $|A| = \gamma$ and $\{U(f) : f \in \gamma^A\} \subset \mathscr{T}^*((X_J)_\kappa)$, it is enough to show that if $C \in [\gamma^A]^\lambda$ then $\cap_{f \in C} U(f) = \emptyset$. Let $C \in [\lambda^A]^\lambda$ and choose $\bar{\eta} \in A$ such that $|C(\bar{\eta})| = \lambda$. Since $\pi_{J(\bar{\eta})}[U(f)] = U^{f(\bar{\eta})}(\bar{\eta})$ for $f \in C$ we have

$$\bigcap_{f \in C} \pi_{J(\bar{\eta})}[U(f)] = \bigcap_{f \in C} U^{f(\bar{\eta})}(\bar{\eta}) = \bigcap_{\xi \in C(\bar{\eta})} U^\xi(\bar{\eta}) = \emptyset$$

and hence $\cap_{f \in C} U(f) = \emptyset$.

We note finally that

$$\alpha \geq S_\lambda((X_J)_\kappa) > \gamma^\mu.$$

as required.

The proof is complete.

We sharpen the estimates of Theorem 3.25 in the case $\kappa = \omega$.

3.27 *Theorem.* Let λ be a cardinal such that either $\lambda = 2$ or λ is (infinite and) regular, let $\{X_i : i \in I\}$ be a set of non-empty, Hausdorff spaces, and

$$\alpha = \sup\{S_\lambda(X_J) : \emptyset \neq J \subset I, |J| < \omega\}.$$

Then

$$\begin{aligned} S_\lambda(X_I) = \alpha \text{ if } & \alpha < \omega, \text{ or} \\ & \alpha = \omega \text{ and } |X_I| < \omega, \text{ or} \\ & \alpha > \omega \text{ and } \alpha \text{ is regular} \\ = \alpha^+ & \text{ in all other cases.} \end{aligned}$$

Proof. If $\alpha < \omega$ we have $S_\lambda(X_I) = \alpha$ from Theorem 3.25(b), and if $\alpha > \omega$ and α is regular we have $S_\lambda(X_I) = \alpha$ from Theorem 3.25(a). If $\alpha = \omega$ and $|X_I| < \omega$ then from Lemma 2.8 the relations $\lambda = 2$ and $\lambda > \omega$ both fail; hence $\lambda = \omega$ and from Lemma 2.8 we have $S_\lambda(X_I) = \omega = \alpha$. From Theorem 3.25(a) we have

$$\alpha \leq S_\lambda(X_I) \leq \alpha^+.$$

If $\alpha = \omega$ and $|X_I| \geq \omega$ then from Corollary 2.10 we have $S_\lambda(X_I) > \alpha$ and hence $S_\lambda(X_I) = \alpha^+$; and if $\alpha > \omega$ and α is singular, then $|X_I| \geq \omega$ and from Theorem 2.12 we have $S_\lambda(X_I) \neq \alpha$ and hence $S_\lambda(X_I) = \alpha^+$.

The proof is complete.

3.28 *Theorem.* If $\omega \leq \kappa < \alpha$ with α regular, then the following are equivalent.

(a) $\kappa \ll \alpha$;

(b) if $\{X_i : i \in I\}$ is a set of spaces with $d(X_i) < \alpha$ for $i \in I$, then $S((X_I)_\kappa) \leq \alpha$.

Proof. (a) $\Rightarrow$ (b). It follows from Theorem 3.25(a) (with β and λ replaced by α and 2, respectively) that it is sufficient to show that if $J \in \mathscr{P}^*_\kappa(I)$ then $S((X_J)_\kappa) \leq \alpha$. For $i \in J$ let D_i be dense in X_i with $|D_i| < \alpha$ and set $D = \prod_{i \in J} D_i$. It is clear that D is dense in $(X_I)_\kappa$. From Theorem A.5(c) we have

$$|D| = \prod_{i \in J} |D_i| < \alpha;$$

hence $S((X_J)_\kappa) \leq \alpha$.

(b) $\Rightarrow$ (a). If $\beta < \alpha$ and $\lambda < \kappa$ then the space $(\beta^\lambda)_\kappa$ is discrete, and from Lemma 2.8 we have $S((\beta^\lambda)_\kappa) = (\beta^\lambda)^+$. It follows from (b) that $(\beta^\lambda)^+ \leq \alpha$, i.e., $\beta^\lambda < \alpha$.

The P_κ-space $X_{(\kappa)}$ determined by a space X is defined in the Appendix.

3.29 *Corollary.* Let $\omega \leq \kappa \ll \alpha$ with κ and α regular and let $\{X_i : i \in I\}$ be a set of spaces such that $w(X_i) < \alpha$ for $i \in I$. Then $S((X_I)_{(\kappa)}) \leq \alpha$.

Proof. From the fact that κ is regular, it follows that if $\mathscr{B}$ is a base for X_i then $\{\cap \mathscr{A} : \mathscr{A} \in \mathscr{P}_\kappa(\mathscr{B})\}$ is a base for $(X_i)_{(\kappa)}$. Thus for

$i \in I$ we have

$$d((X_i)_{(\kappa)}) \leq w((X_i)_{(\kappa)}) \leq (w(X_i))^\kappa < \alpha,$$

and from Theorem 3.28 we have

$$S((\prod_{i \in I} (X_i)_{(\kappa)})_\kappa) \leq \alpha.$$

Since κ is regular the topology of $(\prod_{i \in I}(X_i)_{(\kappa)})_\kappa$ is κ-complete; it is clearly the smallest κ-complete topology on the set X_I containing the product topology. Thus the spaces $(\prod_{i \in I}(X_i)_{(\kappa)})_\kappa$ and $(X_I)_{(\kappa)}$ are equal and we have $S((X_I)_{(\kappa)}) \leq \alpha$, as required.

For every infinite cardinal λ there is by Lemma 2.8 a discrete space X such that $S_\lambda(X) = \lambda$. (For $\mathrm{cf}(\lambda) = \omega$ and X a Hausdorff space, the relation $S_\lambda(X) = \lambda$ holds if and only if $0 < |X| < \omega$.) We note now that for λ (infinite and) regular every cardinal number not proscribed by the result of Chapters 2 and 3 does arise in practice as $S_\lambda(X)$ for an appropriate completely regular, Hausdorff space X.

3.30 *Theorem*. Let λ and α be cardinals such that $\lambda \leq \alpha$, α is uncountable and regular, and either $\lambda = 2$ or λ is (infinite and) regular. There is a completely regular, Hausdorff space X such that if I is a non-empty set then

(i) $S_\lambda(X^I) = \alpha$, and

(ii) every regular cardinal γ with $\gamma > \alpha$ is a calibre for X^I.

Proof. We show first there is X such that $S_\lambda(X) = \alpha$ and every regular cardinal γ with $\gamma \geq \alpha$ is a calibre for X. We consider two cases.

Case 1. α is a successor cardinal. Let $\alpha = \beta^+$ and let X be the (discrete) space β. Then $S_\lambda(X) = \alpha$ by Lemma 2.8, and X has calibre γ by Theorem 2.4(b) (with γ and α replaced by β and γ, respectively).

Case 2. α is a limit cardinal. We choose a set $\{\alpha_\sigma : \sigma < \alpha\}$ of cardinal numbers such that

$$\omega \leq \alpha_{\sigma'} < \alpha_\sigma < \alpha \text{ for } \sigma' < \sigma < \alpha,$$

$$\sum_{\sigma < \alpha} \alpha_\sigma = \alpha, \text{ and}$$

$$\lambda < \alpha_\sigma \text{ for } \sigma < \alpha \text{ if } \lambda < \alpha,$$

and we set $X = \prod_{\sigma < \alpha} \alpha_\sigma$. For $J \in \mathscr{P}^*_\omega(\alpha)$ we set $X_J = \prod_{\sigma \in J} \alpha_\sigma$ and we note that with $\sigma = \max J$ the space X_J is (homeomorphic to) the discrete space α_σ. It follows from Theorem 3.27 and Lemma 2.8 that

$$S_\lambda(X) = \sup\{S_\lambda(X_J): J \in \mathscr{P}^*_\omega(I)\} = \sup\lambda = \lambda = \alpha \qquad \text{if } \lambda = \alpha$$
$$= \sup\{\alpha_\sigma^+ : \sigma < \alpha\} = \alpha \text{ if } \lambda < \alpha,$$

as required. Further by Theorem 2.5(b) the (discrete) spaces α_σ have calibre γ; hence X has calibre γ by Theorem 3.18(a).

The definition of X, and the proofs that $S_\lambda(X) = \alpha$ and X has calibre γ for every regular $\gamma \geq \alpha$, are complete in each case.

For $0 < n < \omega$ the space X^n is homeomorphic to X and hence $S_\lambda(X^n) = \alpha$. Thus

$$\sup\{S_\lambda(X_J): J \in \mathscr{P}^*_\omega(I)\} = \alpha$$

and statement (i) follows from Theorem 3.27. Statement (ii) again follows from Theorem 3.18(a).

Notes for Chapter 3

The first mathematician to recognize that chain conditions in spaces X_i are reflected in (possibly weaker) properties of X_I was E. Marczewski ($\equiv$ E. Szpilrajn), who showed (Marczewski, 1941, Theorem 4(i)) that the product of second-countable spaces has calibre $(\omega^+, 2, 2)$ and (Marczewski, 1947, Theorem 3.3) that the class of spaces with calibre $(\omega^+, \omega^+, 2)$ is closed under the formation of Cartesian products. The proof of the former is self-contained and uses no principle of infinite combinatorics more complicated than the pigeon-hole principle of Dedekind and Dirichlet, but in the proof of the latter one can discern (p. 140) all ingredients necessary for (the appropriate case of) the Erdős–Rado theorem.

Marczewski (1941, 1(i)) noted Lemma 3.1 explicitly and he was apparently the first to raise the question (Marczewski, 1945, p. 304) whether the class of spaces with calibre $(\omega^+, 2, 2)$ is closed under the formation of Cartesian products.

Shanin proved (1946a) a version of the Erdős–Rado theorem for $\kappa = \omega$, he defined (1946b) the expression calibre α, and he showed (1946b, 1948) that for regular $\alpha > \omega$ the class of spaces with calibre α is closed under formation of Cartesian products.

We discussed Theorem 3.6(a, b) in connection with the abstract of

Comfort & Negrepontis (1978a); Theorem 3.6(c) is from Comfort & Negrepontis (1972b). Corollary 3.4 is a remark of Glicksberg (1959, p. 370), and (the case $Y = X_I$ of) Theorem 3.9 is due to Noble (1969b, Theorem 3.4).

Quasi-disjoint families were introduced (and named Δ-systems) by Erdős & Rado (1960); the notation $\alpha \to \Delta(\kappa, \beta)$ of Remark 3.8 is from Erdős & Rado (1969).

The device used in the proof of Theorem 3.10 to define a cover by pairwise disjoint open subsets of a product of discrete spaces has served related purposes in the past (see for example Mycielski, 1964; Engelking 1968, Exercise 3.2F(b); and Kemperman & Maharam, 1970). The κ-box product space $(X_I)_\kappa$ of Theorem 3.10, closely related to the example of Gerlits in Theorem 4.4 below, was suggested in correspondence by Eric K. van Douwen (in connection with Corollary 3.11 (iii)); see also Comfort (1979, Theorem 3.5).

In formulating the results of 3.13–3.15 we were motivated by Shelah (1977), who proved Lemma 3.17(a) and Theorem 3.18(a). Theorem 3.16 was communicated to us by Argyros (May, 1978), who found in addition a new proof (based on Corollary 1.11) of Shelah's Theorem 3.18(a); in Theorem 3.19 we have applied Argyros' basic argument to formulate Shelah-type theorems of considerable generality.

The case $\lambda = 2$ of Theorem 3.26(a) has been noted by Comfort & Negrepontis (1972a, Corollary 2.2) and (for $\kappa = \omega$) by Juhász (1971, Remark, p. 53). The results of 3.25 through 3.30 for $\lambda = 2$ are taken from Comfort & Negrepontis (1972a, 1974, 3.9–3.12); the case $\alpha = \mathfrak{m}^+, \kappa = \mathfrak{n}^+, \mathfrak{m}^\mathfrak{n} = \mathfrak{m}$ of Corollary 3.29 is due to Engelking & Karłowicz (1965, Theorem 6) (see also Engelking, 1965, Theorem 7), and the case $\lambda = 2$ of Theorem 3.30(i) was noted by Efimov (1968).

It is proved by Shanin (1946c, Theorem 5) that if α is a regular, uncountable cardinal and $d(X_i) < \alpha$ for $i \in I$, then $S(X_I) \leq \alpha$; this is the case $\kappa = \omega$ of Theorem 3.28. In Comfort & Negrepontis (1974, Remark, p. 78) we note that Theorem 3.28 can be proved, without any appeal to the Erdős–Rado theorem, by invoking the following (generalization of the) so-called Hewitt–Marczewski–Pondiczery theorem: if $\omega \leq \kappa \leq \alpha = \alpha^\kappa$ and $d(X_i) \leq \alpha$ for $i \in I$ with $|I| \leq 2^\alpha$, then

$d((X_I)_\kappa) \leq \alpha$. For a measure-theoretic proof (noted by Oxtoby, 1961) that every product X of separable spaces satisfies $S(X) \leq \omega^+$, see Theorem 6.7 and Corollary 6.8 below.

For an extensive treatment of cardinal numbers associated with topological spaces, including a number of results concerning calibres in product spaces and elsewhere, we refer the reader to Juhász (1971, 1980).

We note that from Corollary 3.14(a) it follows that if $\mathrm{cf}(\alpha) > \omega$ and $\{X_i : i \in I\}$ is a set of spaces such that there is a set B of regular cardinal numbers satisfying

$$\beta < \alpha \quad \text{for } \beta \in B,$$
$$\sup B = \alpha, \text{ and}$$
$$X_i \text{ has calibre } \beta \text{ and calibre } \alpha \text{ for } \beta \in B, i \in I,$$

then X_I has calibre α. The same conclusion is given by Juhász (1980, Theorem 5.11) under the (stronger) hypotheses that $\mathrm{cf}(\alpha) > \omega$ and $d(X_i) < \mathrm{cf}(\alpha)$ for $i \in I$.

4

Classes of Calibres, Using Σ-products

We investigate the extent of classes of calibres of spaces; i.e., we characterize those classes of infinite cardinals which arise, for some space X, as the class of calibres of X. (For technical reasons it turns out to be more convenient to determine classes of non-calibres than classes of calibres. The resulting description (Corollary 4.7) is satisfactory in that those restrictions which are clearly necessary on the basis of the preceding chapters turn out to be sufficient.) The spaces we define with calibre properties prescribed in advance are products of Σ-products of the generalized Cantor powers 2^I. These are dense subspaces of the powers 2^I, and accordingly their Souslin number is equal to ω^+ and they are not compact; a consequence of (the proof of) Corollary 4.8 is that if for a class $\mathbf{T}$ of infinite cardinals there is a Hausdorff space X such that X does not have calibre α if and only if $\alpha \in \mathbf{T}$, then X may be chosen to be a completely regular, Hausdorff space such that $S(X) = \omega^+$.

The question of the extent of calibres of compact spaces is more delicate and is treated in detail in Chapter 5.

We have seen in Chapter 3 that for regular cardinals the calibre properties of a product space X_I are determined by the products X_J with $J \in \mathscr{P}^*_\omega(I)$. An argument given by Gerlits shows (Theorem 4.4) that the analogous statement fails for every (infinite) singular cardinal.

Definition. Let κ be an infinite cardinal, $\{X_i : i \in I\}$ a set of sets and

$$p = \langle p_i : i \in I \rangle \in X_I = \prod_{i \in I} X_i.$$

For $x \in X_I$ the *exceptional set of* x (with respect to p), denoted $E_p(x)$, is the set

$$E_p(x) = \{i \in I : x_i \neq p_i\};$$

and the *κ-Σ-product of* X_I *based at* p, denoted $\Sigma_\kappa(p, X_I)$, is the set

$$\{x \in X_I : |E_p(x)| < \kappa\}.$$

The set $\Sigma_\kappa(p, X_I)$ is a *generalized Σ-product*. When ambiguity is unlikely we write $E(x)$ in place of $E_p(x)$ and $\Sigma_\kappa(p)$ in place of $\Sigma_\kappa(p, X_I)$.

If $\lambda \geq \omega$ and p is the point of $\{0, 1\}^\lambda$ defined by

$$p_\xi = 0 \quad \text{for } \xi < \lambda,$$

we write $\Sigma_\kappa\{0, 1\}^\lambda$ in place of $\Sigma_\kappa(p, \{0, 1\}^\lambda)$.

We note that if $\kappa > |I|$ then $\Sigma_\kappa(p) = X_I$ for all $p \in X_I$.

The basic projection properties of generalized Σ-products are given in this simple result.

4.1 *Theorem.* Let $\kappa \geq \omega$, let $\{X_i : i \in I\}$ be a set of sets, $p \in X_I$ and $Y = \Sigma_\kappa(p)$. Then

(a) $\pi_J[Y] = X_J$ for every non-empty $J \subset I$ with $|J| < \kappa$; and

(b) if the sets X_i are spaces then for every compact $K_J \subset X_J$ with J non-empty, $J \subset I$ and $|J| < \kappa$ there is compact $K \subset Y$ such that $\pi_J[K] = K_J$.

Proof. (a) For $x \in X_J$ we set

$$y = \langle x, p_{I \setminus J} \rangle \in X_J \times X_{I \setminus J} = X_I;$$

it is clear that $y \in Y$ and $\pi(y) = x$.

(b) We set $K = K_J \times \{p_{I \setminus J}\}$. Then K is a compact subspace of Y such that $\pi_J[K] = K_J$.

It follows from Theorem 4.1 that certain generalized Σ-products satisfy the hypotheses, and hence the conclusions, of Theorems 3.3, 3.5 and 3.8.

We have noted already that an infinite Hausdorff space does not have calibre α when $\text{cf}(\alpha) = \omega$. Since there are arbitrarily large cardinals α such that $\text{cf}(\alpha) = \omega$, for every infinite space X the class of cardinals α such that X does not have calibre α is a proper class. On the other hand it follows from Theorem 2.6 that every regular cardinal α satisfying $\alpha > d(X)$ is a calibre for X; thus for every space X the class

$$\{\alpha : \alpha \text{ is a regular cardinal and } X \text{ does not have calibre } \alpha\}$$

is a set. Thus it is more natural, given a space X, to specify those infinite cardinals α which are not calibres for X than to identify those that are.

We do this now for certain generalized Σ-products.

4.2 *Theorem.* Let α, κ and λ be infinite cardinals and $X = \Sigma_\kappa \{0, 1\}^\lambda$. The following statements are equivalent.

(a) $\mathrm{cf}(\alpha) = \omega$ or $\kappa \leq \alpha \leq \lambda$ or $\kappa \leq \mathrm{cf}(\alpha) \leq \lambda$;

(b) X does not have calibre α.

Proof. (a)$\Rightarrow$(b). We have $|X| \geq \omega$; from Theorems 2.10 and 2.3(a) it follows that X does not have calibre α if $\mathrm{cf}(\alpha) = \omega$.

We note next that X does not have calibre (λ, κ). Indeed let

$$U^\eta = \{x \in X : x_\eta = 1\} \quad \text{for } \eta < \lambda;$$

then $\{U^\eta : \eta < \lambda\} \subset \mathscr{T}^*(X)$, and $\cap_{\eta \in A} U^\eta = \varnothing$ for $A \in [\lambda]^\kappa$.

Now if either $\kappa \leq \alpha \leq \lambda$ or $\kappa \leq \mathrm{cf}(\alpha) \leq \lambda$, then (from Theorem 2.1(a)) either X does not have calibre α or X does not have calibre $\mathrm{cf}(\alpha)$. It follows from Theorem 2.3(a) that X does not have calibre α.

(b)$\Rightarrow$(a). We assume $\mathrm{cf}(\alpha) > \omega$ and we consider four cases.

Case 1. $\kappa > \lambda$. Then $X = \{0, 1\}^\lambda$ and X has calibre α by Theorem 3.18.

Case 2. $\kappa \leq \lambda$ and $\alpha < \kappa$. Let $\{U^\xi : \xi < \alpha\}$ be a set of basic subsets of X and set $J = \cup_{\xi < \alpha} R(U^\xi)$. Since $\{0, 1\}^J$ has calibre α (by Theorem 3.18) there are $A \in [\alpha]^\alpha$ and $p \in \{0, 1\}^J$ such that

$$p \in \bigcap_{\xi \in A} \pi_J[U^\xi].$$

We define $x \in \{0, 1\}^\lambda$ by

$$\begin{aligned} x_\eta &= p_\eta \quad \text{if } \eta \in J \\ &= 0 \quad \text{if } \eta \in \lambda \backslash J; \end{aligned}$$

then $x \in \cap_{\xi \in A} U^\xi$, and from $|J| \leq \alpha < \kappa$ we have $x \in X$.

Case 3. $\kappa \leq \lambda$ and $\lambda < \mathrm{cf}(\alpha)$. There is a pseudobase $\mathscr{B}$ for X such that $|\mathscr{B}| = \lambda$. Let $\{U^\xi : \xi < \alpha\} \subset \mathscr{T}^*(X)$ and set

$$A(B) = \{\xi < \alpha : B \subset U^\xi\} \quad \text{for } B \in \mathscr{B}.$$

Then $\cap_{\xi \in A(B)} U^\xi \supset B \neq \varnothing$ for $B \in \mathscr{B}$, and since $\lambda < \mathrm{cf}(\alpha)$ there is $\bar{B} \in \mathscr{B}$ such that $|A(\bar{B})| = \alpha$.

Case 4. $\kappa \leq \lambda$ and $\mathrm{cf}(\alpha) < \kappa$ and $\lambda < \alpha$. It follows from Case 2 (applied to $\mathrm{cf}(\alpha)$ in place of α) and X has calibre $\mathrm{cf}(\alpha)$.

Since $w(\{0,1\}^{\lambda}) = \lambda$ we have $\pi w(X) < \alpha$ and hence from Theorem 2.3(a′) the space X has calibre α.

In each of the four cases the space X has calibre α.

The proof is complete.

In Theorem 8.11 we determine those (infinite) cardinals α for which the spaces $\Sigma_{\kappa}\{0,1\}^{\lambda}$ have compact-calibre α and are pseudo-(α, α)-compact.

We see next that in Theorem 2.3(a′) the condition $\pi w(X) < \alpha$ cannot be omitted.

4.3 *Theorem.* Let α be a singular cardinal such that $\mathrm{cf}(\alpha) > \omega$. There is a completely regular, Hausdorff space X such that X has calibre $\mathrm{cf}(\alpha)$, X does not have calibre α, and $\pi w(X) = \alpha$.

Proof. We set $X = \Sigma_{(\mathrm{cf}(\alpha))^+}\{0,1\}^{\alpha}$. It is immediate from Theorem 4.2 that X has the required properties.

We saw in Corollary 3.7(a) that if α is an uncountable regular cardinal and $\{X_i : i \in I\}$ a set of spaces such that X_i has calibre α for all $i \in I$, then X_I has calibre α. We see next that the analogous statement fails for singular cardinals α. Theorem 4.4 supplements and complements Corollary 3.11(i), which treats products of spaces with calibre (α, β) with α singular and $\alpha > \beta$.

4.4 *Theorem.* Let α be an infinite singular cardinal. There is a set $\{X_i : i \in I\}$ of completely regular, Hausdorff spaces such that

X_i has calibre α for all $i \in I$, and

X_I does not have calibre α.

Proof. If $\mathrm{cf}(\alpha) = \omega$ we establish the statement by setting $I = \omega$ and $X_n = \{0,1\}$ for $n < \omega$. We assume in what follows that $\mathrm{cf}(\alpha) > \omega$.

There is a set $\{A_\sigma : \sigma < \mathrm{cf}(\alpha)\}$ of sets such that

$$\mathrm{cf}(\alpha) < |A_\sigma| < \alpha \text{ for } \sigma < \mathrm{cf}(\alpha),$$

$$A_{\sigma'} \cap A_\sigma = \varnothing \text{ for } \sigma' < \sigma < \mathrm{cf}(\alpha), \text{ and}$$

$$\bigcup_{\sigma < \mathrm{cf}(\alpha)} A_\sigma = \alpha.$$

We set

$$X_\sigma = \Sigma_{(\mathrm{cf}(\alpha))^+}\{0,1\}^{A_\sigma} \text{ for } \sigma < \mathrm{cf}(\alpha), \text{ and}$$

$$X = \prod_{\sigma < \mathrm{cf}(\alpha)} X_\sigma.$$

From Theorem 4.2 (with κ and λ replaced by $(\text{cf}(\alpha))^+$ and $|A_\sigma|$, respectively) it follows that X_σ has calibre α. Then from Theorem 2.2(a) it follows that X_J has calibre α for all $J \in \mathscr{P}^*_\omega(I)$.

We are to show that X does not have calibre α. We show that in fact X does not have calibre $(\alpha, (\text{cf}(\alpha))^+)$.

We have

$$X \subset \prod_{\sigma < \text{cf}(\alpha))^+} \{0,1\}^{A_\sigma} = \{0,1\}^\alpha$$

and in fact

$$X = \Sigma_{(\text{cf}(\alpha))^+}\{0,1\}^\alpha.$$

We set

$$U_\xi = \{x \in X : x_\xi = 1\} \text{ for } \xi < \alpha;$$

then $\{U_\xi : \xi < \alpha\} \subset \mathscr{T}^*(X)$, and

$$\bigcap_{\xi \in A} U_\xi = \varnothing \quad \text{for } A \in [\alpha]^{(\text{cf}(\alpha))^+},$$

as required.

We note that for $\text{cf}(\alpha) > \omega$ the space X_I of Theorem 4.1 does have calibre $\text{cf}(\alpha)$.

We give now a consequence of Theorem 4.2, and a theorem which allows us to determine those classes of infinite cardinal numbers which can arise as the class of non-calibres of a space.

4.5 *Corollary*. Let λ and α be infinite cardinals and $X = \Sigma_\lambda\{0,1\}^\lambda$. Then X has calibre α if and only if $\alpha \neq \lambda$, $\text{cf}(\alpha) \neq \lambda$, and $\text{cf}(\alpha) \neq \omega$.

4.6 *Theorem*. Let $\mathbf{S}$ be a non-empty set of infinite cardinal numbers, $X(\lambda) = \Sigma_\lambda\{0,1\}^\lambda$ for $\lambda \in \mathbf{S}$ and $X(\mathbf{S}) = \prod_{\lambda \in \mathbf{S}} X(\lambda)$.

For $\alpha \geq \omega$, the following statements are equivalent.

(a) α is not a calibre for $X(\mathbf{S})$;

(b) $\alpha \in \mathbf{S}$ or $\text{cf}(\alpha) \in \mathbf{S}$ or $\text{cf}(\alpha) = \omega$.

Proof. (a) $\Rightarrow$ (b). We suppose that $\alpha \notin \mathbf{S}$, $\text{cf}(\alpha) \notin \mathbf{S}$ and $\text{cf}(\alpha) > \omega$. It follows from Theorem 4.2 that $X(\lambda)$ has calibre α for all $\lambda \in \mathbf{S}$.

We show that $X(\mathbf{S})$ has calibre α. If α is regular this follows from Corollary 3.7(a). We assume in what follows that α is singular.

Let $\{\alpha_\sigma : \sigma < \mathrm{cf}(\alpha)\}$ be a set of cardinal numbers such that

α_σ is regular for $\sigma < \mathrm{cf}(\alpha)$,

$\mathrm{cf}(\alpha) < \alpha_{\sigma'} < \alpha_\sigma < \alpha$ for $\sigma' < \sigma < \mathrm{cf}(\alpha)$ and

$$\sum_{\sigma < \mathrm{cf}(\alpha)} \alpha_\sigma = \alpha.$$

We claim that if $J \in \mathscr{P}^*_\omega(\mathbf{S})$ then X_J has calibre α and calibre $(\alpha_{\sigma+1}, \alpha_\sigma)$. That X_J has calibre α follows from Theorem 2.2(a); to prove that X_J has calibre $(\alpha_{\sigma+1}, \alpha_\sigma)$ we consider two cases.

Case 1. $\alpha_\sigma \notin J$. Then $X(\lambda)$ has calibre α_σ for all $\lambda \in J$. It follows that X_J has calibre $(\alpha_{\sigma+1}, \alpha_\sigma)$.

Case 2. $\alpha_\sigma \in J$. If $J \backslash \{\alpha_\sigma\} = \emptyset$ then X_J, which is $X(\alpha_\sigma)$, has calibre $\alpha_{\sigma+1}$ and hence calibre $(\alpha_{\sigma+1}, \alpha_\sigma)$. If $J \backslash \{\alpha_\sigma\} \neq \emptyset$ then, since $X_{J \backslash \{\alpha_\sigma\}}$ has calibre $(\alpha_{\sigma+1}, \alpha_\sigma)$ (by Case 1) and $X(\alpha_\sigma)$ has calibre $\alpha_{\sigma+1}$, the space X_J, which is $X_{J \backslash \{\alpha_\sigma\}} \times X(\alpha_\sigma)$, has calibre $(\alpha_{\sigma+1}, \alpha_\sigma)$.

The proof of the claim is complete. It follows from Corollary 3.11(a) that $X(\mathbf{S})$ has calibre α, as required.

(b) $\Rightarrow$ (a). From Corollary 4.5 it follows that there is $\lambda \in \mathbf{S}$ such that $X(\lambda)$ does not have calibre α. It is then clear that $X(\mathbf{S})$ does not have calibre α.

4.7 *Corollary*. Let $\mathbf{T}$ be a class of infinite cardinal numbers. Then statements (a) and (b) are equivalent.

(a) There is an infinite Hausdorff space X such that X does not have calibre α if and only if $\alpha \in \mathbf{T}$.

(b) (i) $\omega \in \mathbf{T}$,

(ii) if $\mathrm{cf}(\alpha) \in \mathbf{T}$ then $\alpha \in \mathbf{T}$,

(iii) $\{\mathrm{cf}(\alpha) : \alpha \in \mathbf{T}\}$ is a set, and

(iv) if κ is a regular cardinal and $\kappa \notin \mathbf{T}$, then $\{\alpha \in \mathbf{T} : \mathrm{cf}(\alpha) = \kappa\}$ is a set.

Proof. (a) $\Rightarrow$ (b). (i) follows from Corollary 2.10.

(ii) follows from Theorem 2.3(a).

(iii) If α is an infinite cardinal and $\mathrm{cf}(\alpha) > d(X)$, then it follows from Theorem 2.6 that X has calibre α. Hence

$$\{\mathrm{cf}(\alpha) : \alpha \in \mathbf{T}\} \subset (d(X))^+.$$

(iv) From Theorem 2.3(a′) we have

$$\{\alpha \in \mathbf{T} : \mathrm{cf}(\alpha) = \kappa\} \subset (\pi w(X))^+ \quad \text{for } \kappa \notin \mathbf{T}.$$

(b) ⇒ (a). For $\kappa \in \{\mathrm{cf}(\alpha) : \alpha \in \mathbf{T}\}$ we set

$$\begin{aligned} \mathbf{S}(\kappa) &= \{\kappa\} && \text{if } \kappa \in \mathbf{T} \\ &= \{\alpha \in \mathbf{T} : \mathrm{cf}(\alpha) = \kappa\} && \text{if } \kappa \notin \mathbf{T}, \end{aligned}$$

we set

$$\mathbf{S} = \cup \{\mathbf{S}(\kappa) : \kappa \in \{\mathrm{cf}(\alpha) : \alpha \in \mathbf{T}\}\},$$

we note that $\mathbf{S}$ is a set and $\mathbf{S} \subset \mathbf{T}$, and we set

$$X(\lambda) = \Sigma_\lambda \{0, 1\}^\lambda \text{ for } \lambda \in \mathbf{S}, \text{ and}$$

$$X = \prod_{\lambda \in \mathbf{S}} X(\lambda).$$

We show first that if α is an infinite cardinal and X does not have calibre α then $\alpha \in \mathbf{T}$. From Theorem 4.6 we have $\alpha \in \mathbf{S}$ or $\mathrm{cf}(\alpha) \in \mathbf{S}$ or $\mathrm{cf}(\alpha) = \omega$. Since $\mathbf{S} \subset \mathbf{T}$, from (ii) we have $\alpha \in \mathbf{T}$ if $\alpha \in \mathbf{S}$ or $\mathrm{cf}(\alpha) \in \mathbf{S}$; and from (i) and (ii) we have $\alpha \in \mathbf{T}$ if $\mathrm{cf}(\alpha) = \omega$.

We show next that if $\alpha \in \mathbf{T}$ then X does not have calibre α. If $\mathrm{cf}(\alpha) \in \mathbf{T}$ then

$$\mathrm{cf}(\alpha) \in \mathbf{S}(\mathrm{cf}(\alpha)) \subset \mathbf{S},$$

and if $\mathrm{cf}(\alpha) \notin \mathbf{T}$ then $\alpha \in \mathbf{S}(\mathrm{cf}(\alpha)) \subset \mathbf{S}$; it follows from Theorem 4.6 that X does not have calibre α.

4.8 *Remark.* We note that the space X defined in the proof of the implication (b) ⇒ (a) in Corollary 4.7 is a dense subspace of a generalized Cantor power 2^I. The space X is then a completely regular, Hausdorff space, and from Theorem 3.25 and Remark 2.6(b) it follows that $S(X) = \omega^+$. This shows that if for a class $\mathbf{T}$ of infinite cardinal numbers there is a Hausdorff space X such that X does not have calibre α if and only if $\alpha \in \mathbf{T}$, then X may be chosen to be a completely regular, Hausdorff space such that $S(X) = \omega^+$.

Notes for Chapter 4

The term Σ-product was introduced by Corson (1959) and applied to spaces denoted (in our notation) $\Sigma_{\omega^+}(p, X_I)$.

Theorem 4.2 is noted by Comfort (1980, Theorem 4.1(a)).

Comfort (1971, p. 175) raised the question of 'characterizing those collections of cardinal numbers which can arise as the totality of calibres possessed by some space'. Theorem 4.6, responding to this question, is from Shelah (1977, Theorem 3.3), though the spaces defined by Shelah are not regular spaces. Our proof of Theorem 4.6, systematically exploiting the calibre properties of the generalized Σ-products $\Sigma_\kappa\{0,1\}^\lambda$ (Theorem 4.2), closely parallels the treatment (obtained independently) of Broverman, Ginsburg, Kunen & Tall (1978); these authors consider in addition a number of properties (e.g., normality and 'Baireness'), and their behavior under special set-theoretic assumptions, of spaces of the form

$$\Sigma(\mathscr{I}) = \{x \in \{0,1\}^{(\omega^+)} : \{\xi < \omega^+ : x_\xi = 1\} \in \mathscr{I}\}$$

with $\mathscr{I}$ a σ-ideal in ω^+.

The example of Theorem 4.4 was communicated in conversation by J. Gerlits (February, 1978).

5

Calibres of Compact Spaces

In this chapter we study the calibres of compact spaces (and the precalibres of all spaces); the surprising fact is that an extensive class of cardinals, depending only on the Souslin number of the compact space in question, is a class of calibres for the space. This is the content of Theorems 5.6 and 5.25, the two main results of this section. (When the generalized continuum hypothesis is assumed, these results assume a particularly simple form, given in Corollary 5.26.) The techniques used for the proof of these results are combinatorial, and are contained in Lemmas 5.1 (a ramification method for regular cardinals), 5.4 (a method that deals with singular cardinals), and 5.22 (dealing with the 'exceptional' cardinals). It is noted that the combinatorial techniques used here involve ramification methods more general and more powerful than those given by the Erdős–Rado theorem on quasi-disjoint families of sets and its extension to singular cardinals; as to the determination of cardinal invariants, these latter results are useful essentially only for product spaces or for spaces with a product-like structure, while we in the present section are able to examine spaces quite general in nature. On the basis of examples given in Theorems 5.28 and 6.18, we can claim that the results are essentially all that can be proved for general spaces (see the diagrams in 7.19).

For purposes of motivation we note that the results of this section are closely related to a number of results, also due to Argyros and Tsarpalias and stated in some detail in the Notes, deriving from the following question concerning Banach spaces: When can the Banach space l_α^1 of absolutely summable real functions defined on (a set of cardinality) α be embedded isomorphically into every closed, linear α-dimensional subspace of the Banach space $C(X)$ of real-valued, continuous functions on the compact space X? The appropriate condition, roughly speaking, is that α and the Souslin number κ of X

are in the cardinal relation given in 5.6 (and analogously for 5.25). While these functional analytic proofs are more intricate, they also rely on the same combinatorial methods, together with the reduction (given by Rosenthal) of the embeddability of l_α^1 to a combinatorial question on the existence of certain (Boolean) independent families of pairs of open subsets of X; this question in turn is related to (but more delicate than) the question of the existence of a family of open subsets of X with non-empty intersection appearing in the concept of calibre for X.

Byproducts of Theorem 5.6 are: (a) the impossibility of finding, for a given class **T** of cardinals satisfying the natural conditions given in the statement of Corollary 4.7, a *compact* space whose class of non-calibres is exactly **T**; (b) assuming the generalized continuum hypothesis, the reduction of the question of determining all calibres of a compact space to the consideration of regular cardinals only (Corollary 5.9); and (c) again assuming the generalized continuum hypothesis, a satisfactory product theorem for singular (as well as regular) cardinals (Corollary 5.10).

These results have analogues concerning calibres of compact spaces X with λ^+-complete topologies for $\lambda < S(X)$ (Theorem 5.16) and concerning calibres of λ^+-box topologies on product spaces $\prod_{i\in I} X_i$ with λ a cardinal less than a regular upper bound for the numbers $S(X_i)$, $i \in I$ (Theorem 5.21); the regular cases of these theorems strengthen results given earlier by Juhász (Corollary 5.13) and by Kurepa and Hedrlín (Corollary 5.18), respectively.

Strongly $S(X)$-inaccessible cardinals as calibres

Definition. Let X be a space and $\{U_\xi : \xi < \alpha\}$ a set of non-empty, open subsets of X. An index $\bar{\xi} < \alpha$ is a *representative* of $\{U_\xi : \xi < \alpha\}$ if the following condition is satisfied:

If $\beta < \alpha$ and $\{W_\zeta : \zeta < \beta\}$ is a set of non-empty, open subsets of $U_{\bar{\xi}}$, then there is $A \subset \alpha$ with $|A| = \alpha$ such that $\{U_\xi \cap W_\zeta : \xi \in A\}$ has the finite intersection property for all $\zeta < \beta$.

5.1 *Lemma.* Let $\omega \leq \kappa \ll \alpha$ with α regular and let X be a space such that $S(X) \leq \kappa$. Then every set $\{U_\xi : \xi < \alpha\}$ of non-empty, open subsets of X has a representative.

Proof. Suppose that $\{U_\xi : \xi < \alpha\}$ is a set of non-empty, open subsets of X with no representative. For $A \subset \alpha$ with $|A| = \alpha$ we choose $\xi_A \in A$. There are

a cardinal $\beta_A < \alpha$, and

a set $\{W^A_\zeta : \zeta < \beta_A\}$ of non-empty, open subsets of U_{ξ_A}, such that

if $B \subset A$ and $\{U_\xi \cap W^A_\zeta : \xi \in B\}$ has the finite intersection property for all $\zeta < \beta_A$, then $|B| < \alpha$.

For $A \subset \alpha$ with $|A| = \alpha$ we set

$$\mathscr{B}_A = \{B \subset A : \{U_\xi \cap W^A : \xi \in B\} \text{ has the finite intersection property for all } \zeta < \beta_A\}.$$

Since the set $\mathscr{B}_A$ partially ordered by inclusion is inductive, there is a maximal element B_A of $\mathscr{B}_A$. We have $|B_A| < \alpha$, and since $\{\xi_A\} \in \mathscr{B}_A$ we have $B_A \neq \varnothing$.

For $\xi \in A \backslash B_A$ it follows from the maximality of B_A that there are a finite, non-empty subset $F(\xi)$ of B_A and $\zeta(\xi) < \beta_A$ such that

$$\Big(\bigcap_{i \in F(\xi)} U_i\Big) \cap W^A_{\zeta(\xi)} \cap U_\xi = \varnothing.$$

We define

$$\varphi_A : A \backslash B_A \to \mathscr{P}^*_\omega(B_A) \times \beta_A$$

by the rule

$$\varphi_A(\xi) = (F(\xi), \zeta(\xi)),$$

and we set

$$\mathscr{P}_A = \{B_A\} \cup \{\varphi_A^{-1}(\{(F, \zeta)\}) : F \in \mathscr{P}^*_\omega(B_A), \zeta < \beta_A\}.$$

Then $\mathscr{P}_A$ is a partition of A, and since

$$|\mathscr{P}^*_\omega(B_A)| \leq \sup\{\omega, |B_A|\} < \alpha \text{ and } \beta_A < \alpha,$$

we have $|\mathscr{P}_A| < \alpha$. It follows from (the case $\beta = \alpha$ of) Lemma 1.1 that there is a family $\{A_\eta : \eta < \kappa\}$ of subsets of α such that

$$|A_\eta| = \alpha \quad \text{for } \eta < \kappa,$$
$$A_{\eta+1} \in \mathscr{P}_{A_\eta} \quad \text{for } \eta < \kappa, \text{ and}$$
$$A_{\eta'} \subset A_\eta \quad \text{for } \eta < \eta' < \kappa.$$

For $\eta < \kappa$ there are $\bar{F}(\eta) \in \mathscr{P}^*_\omega(B_{A_\eta}), \bar{\zeta}(\eta) < \beta_{A_\eta}$ such that

$$A_{\eta+1} = \varphi^{-1}_{A_\eta}(\{(\bar{F}(\eta), \bar{\zeta}(\eta))\}).$$

For $\eta < \kappa$ we set

$$V_\eta = (\bigcap_{i \in \bar{F}(\eta)} U_i) \cap W^{A_\eta}_{\bar{\zeta}(\eta)},$$

and we note that $V_\eta \neq \varnothing$ (since $\bar{F}(\eta)$ is a finite subset of B_{A_η} and $\{U_i \cap W^{A_\eta}_{\bar{\zeta}(\eta)} : i \in B_{A_\eta}\}$ has the finite intersection property).

We claim further that

$$V_\eta \cap V_{\eta'} = \varnothing \quad \text{for } \eta < \eta' < \kappa.$$

Indeed since $\bar{F}(\eta') \neq \varnothing$, there is $\xi \in \bar{F}(\eta')$, and from

$$\xi \in \bar{F}(\eta') \subset A_{\eta+1} = \varphi^{-1}_{A_\eta}(\{(\bar{F}(\eta), \bar{\zeta}(\eta))\})$$

we have

$$\varphi_{A_\eta}(\xi) = (\bar{F}(\eta), \bar{\zeta}(\eta))$$

and hence

$$\varnothing = (\bigcap_{i \in \bar{F}(\eta)} U_i) \cap W^{A_\eta}_{\bar{\zeta}(\eta)} \cap U_\xi = V_\eta \cap U_\xi;$$

then from $V_{\eta'} \subset U_\xi$ we have $V_\eta \cap V_{\eta'} = \varnothing$, as claimed.

It follows that $S(X) > \kappa$. This contradiction completes the proof.

5.2 *Corollary.* Let $\omega \leq \kappa \ll \alpha$ with α regular and let X be a space such that $S(X) \leq \kappa$. Then α is a precalibre for X.

Proof. Let $\{U_\xi : \xi < \alpha\}$ be a family of non-empty, open subsets of X, and let $\bar{\xi}$ be the ordinal number given by Lemma 5.1. With $\beta = 1$ and with

$$\{W_\zeta : \zeta < \beta\} = \{W_0\} = \{U_{\bar{\xi}}\},$$

it follows from Lemma 5.1 that there is $A \subset \alpha$ with $|A| = \alpha$ such that the family $\{U_\xi \cap U_{\bar{\xi}} : \xi \in A\}$ has the finite intersection property.

The proof is complete.

5.3 *Remarks.* (a) The power and usefulness of Lemma 5.1 will become apparent in 5.5 below, but for the moment we note that a simple (and similar) argument is sufficient to prove Corollary 5.2.

Suppose that α is not a precalibre for X, and let $\{U_\xi : \xi < \alpha\}$ be a set of non-empty, open subsets of X such that if $B \subset \alpha$ and $\{U_\xi : \xi \in B\}$ has the finite intersection property, then $|B| < \alpha$.

For $A \subset \alpha$ with $|A| = \alpha$ we set

$$\mathscr{B}_A = \{B \subset A : \{U_\xi : \xi \in B\} \text{ has the finite intersection property}\}.$$

Since the set $\mathscr{B}_A$ partially ordered by inclusion is inductive, there is a maximal element B_A of $\mathscr{B}_A$. We have $|B_A| < \alpha$, and since $\{\xi\} \in \mathscr{B}_A$ for all $\xi \in A$ we have $B_A \neq \varnothing$.

For $\xi \in A \backslash B_A$ it follows from the maximality of B_A that there is a finite, non-empty subset $F(\xi)$ of B_A such that

$$\left(\bigcap_{i \in F(\xi)} U_i\right) \cap U_\xi = \varnothing.$$

We define

$$\varphi_A : A \backslash B_A \to \mathscr{P}^*_\omega(B_A)$$

by the rule

$$\varphi_A(\xi) = F(\xi),$$

and we set

$$\mathscr{P}_A = \{B_A\} \cup \{\varphi_A^{-1}(\{F\}) : F \in \mathscr{P}^*_\omega(B_A)\}.$$

Then $\mathscr{P}_A$ is a partition of A, with $|\mathscr{P}_A| < \alpha$. It follows from (the case $\beta = \alpha$ of) Lemma 1.1 that there is a family $\{A_\eta : \eta < \kappa\}$ of subsets of α such that

$$|A_\eta| = \alpha \quad \text{for } \eta < \kappa,$$
$$A_{\eta+1} \in \mathscr{P}_{A_\eta} \quad \text{for } \eta < \kappa, \text{ and}$$
$$A_{\eta'} \subset A_\eta \quad \text{for } \eta < \eta' < \kappa.$$

For $\eta < \kappa$ there is $\bar{F}(\eta) \in \mathscr{P}^*_\omega(B_{A_\eta})$ such that

$$A_{\eta+1} = \varphi_{A_\eta}^{-1}(\{\bar{F}(\eta)\}).$$

We set $V_\eta = \bigcap_{i \in \bar{F}(\eta)} U_i$ for $\eta < \kappa$ and we prove, just as in the proof of Lemma 5.1, that the family $\{V_\eta : \eta < \kappa\}$ is a faithfully indexed family of non-empty, pairwise disjoint open subsets of X. (Indeed $V_\eta \neq \varnothing$ since $\bar{F}(\eta)$ is a finite subset of B_{A_η} and $\{U_i : i \in B_{A_\eta}\}$ has the finite intersection property; and for $\eta < \eta' < \kappa$ there is $\xi \in \bar{F}(\eta')$, and from

$$\xi \in \bar{F}(\eta') \subset A_{\eta+1} = \varphi_{A_\eta}^{-1}(\{\bar{F}(\eta)\})$$

we have $\varphi_{A_\eta}(\xi) = \bar{F}(\eta)$ and hence

$$V_\eta \cap V_{\eta'} \subset V_\eta \cap U_\xi = \emptyset.)$$

This is a contradiction to the fact that $S(X) \leq \kappa$, and the proof is complete.

(b) The argument just given applies also to prove the following result. Here we use Lemma 1.2 instead of Lemma 1.1 with $\beta = \alpha$; and we replace κ throughout with α.

Theorem. Let α be a weakly compact cardinal and let X be a space such that $S(X) \leq \alpha$. Then α is a precalibre for X.

We use the following lemma in the proofs of Theorems 5.5 and 5.25.

5.4 *Lemma.* Let α be a singular cardinal, $\{\alpha_\sigma : \sigma < \mathrm{cf}(\alpha)\}$ a set of regular cardinals such that

$$\mathrm{cf}(\alpha) < \alpha_{\sigma'} < \alpha_\sigma < \alpha \quad \text{for } \sigma' < \sigma < \mathrm{cf}(\alpha), \text{ and}$$

$$\alpha = \sum_{\sigma < \mathrm{cf}(\alpha)} \alpha_\sigma,$$

and let X be a space and $\{U_\xi : \xi < \alpha\}$ a set of non-empty open subsets of X. If there is a set $\{\bar{\xi}(\sigma) : \sigma < \mathrm{cf}(\alpha)\}$ such that

$$\bar{\xi}(\sigma) \text{ is a representative of } \{U_\xi : \xi < \alpha_\sigma\} \text{ and}$$

$$\{U_{\bar{\xi}(\sigma)} : \sigma < \mathrm{cf}(\alpha)\} \text{ has the finite intersection property,}$$

then there is $A \subset \alpha$ with $|A| = \alpha$ such that

$$\{U_\xi : \xi \in A\} \text{ has the finite intersection property.}$$

Proof. We define a family $\{A_\sigma : \sigma < \mathrm{cf}(\alpha)\}$ such that

$$A_\sigma \subset \alpha_\sigma \quad \text{for } \sigma < \mathrm{cf}(\alpha),$$

$$|A_\sigma| = \alpha_\sigma \quad \text{for } \sigma < \mathrm{cf}(\alpha), \text{ and}$$

$$\{U_\xi : \xi \in \bigcup_{\sigma < \mathrm{cf}(\alpha)} A_\sigma\} \cup \{U_{\bar{\xi}(\sigma)} : \sigma < \mathrm{cf}(\alpha)\}$$

has the finite intersection property.

We proceed by recursion. We note that

$$|\mathscr{P}^*_\omega(\mathrm{cf}(\alpha))| = \mathrm{cf}(\alpha) < \alpha_0,$$

we set

$$\beta_0 = |\mathscr{P}^*_\omega(\mathrm{cf}(\alpha))|,$$

and we let $\{W_\zeta : \zeta < \beta_0\}$ be an enumeration of

$$\{(\bigcap_{\tau \in F} U_{\bar{\xi}(\tau)}) \cap U_{\bar{\xi}(0)} : F \in \mathscr{P}^*_\omega(\mathrm{cf}(\alpha))\}.$$

Then $\{W_\zeta : \zeta < \beta_0\}$ is a set of non-empty, open subsets of $U_{\bar{\xi}(0)}$, and since $\bar{\xi}(0)$ is a representative of $\{U_\xi : \xi < \alpha_0\}$ there is $A_0 \subset \alpha_0$ with $|A_0| = \alpha_0$ such that $\{U_\xi \cap W_\zeta : \xi \in A_0\}$ has the finite intersection property for all $\zeta < \beta_0$.

Now let $\sigma < \mathrm{cf}(\alpha)$, and suppose that $A_{\sigma'}$ has been defined for $\sigma' < \sigma$ so that

$$A_{\sigma'} \subset \alpha_{\sigma'},$$
$$|A_{\sigma'}| = \alpha_{\sigma'}, \text{ and}$$
$$\{U_\xi : \xi \in \bigcup_{\sigma' < \sigma} A_{\sigma'}\} \cup \{U_{\bar{\xi}(\sigma)} : \sigma < \mathrm{cf}(\alpha)\}$$

has the finite intersection property.

Again we have $|\mathscr{P}^*_\omega(\mathrm{cf}(\alpha))| = |\mathrm{cf}(\alpha)| = \mathrm{cf}(\alpha) < \alpha_\sigma$. Further, since

$$|\sigma| < \mathrm{cf}(\alpha) < \alpha_\sigma \text{ and } \alpha_\sigma \text{ is regular}$$

we have

$$|\mathscr{P}^*_\omega(\bigcup_{\sigma' < \sigma} A_{\sigma'})| = |\bigcup_{\sigma' < \sigma} A_{\sigma'}| < \alpha_\sigma.$$

We set

$$\beta_\sigma = |\mathscr{P}^*_\omega(\mathrm{cf}(\alpha)) \cup \mathscr{P}^*_\omega(\bigcup_{\sigma' < \sigma} A_{\sigma'})|,$$

we note that $\beta_\sigma < \alpha_\sigma$, and we let $\{W_\xi : \zeta < \beta_\sigma\}$ be an enumeration of

$$\{(\bigcap_{\tau \in F} U_{\xi(\tau)}) \cap (\bigcap_{\xi \in G} U_\xi) \cap U_{\bar{\xi}(\sigma)} : F \in \mathscr{P}^*_\omega(\mathrm{cf}(\alpha)), G \in \mathscr{P}^*_\omega(\bigcup_{\sigma' < \sigma} A^{\sigma'})\}.$$

Then $\{W_\zeta : \zeta < \beta_\sigma\}$ is a set of non-empty, open subsets of $U_{\bar{\xi}(\sigma)}$, and since $\bar{\xi}(\sigma)$ is a representative of $\{U_\xi : \xi < \alpha_\sigma\}$, there is $A_\sigma \subset \alpha_\sigma$ with $|A_\sigma| = \alpha_\sigma$ such that $\{U_\xi \cap W_\zeta : \xi \in A_\sigma\}$ has the finite intersection property for all $\zeta < \beta_\sigma$.

The inductive definition of the family $\{A_\sigma : \sigma < \mathrm{cf}(\alpha)\}$ is complete. We set

$$A = \bigcup_{\sigma < \mathrm{cf}(\alpha)} A_\sigma.$$

Then $|A| = \alpha$, and

$\{U_\xi : \xi \in A\} \cup \{U_{\xi(\sigma)} : \sigma < \text{cf}(\alpha)\}$ has the finite intersection property.

In particular

$$\{U_\xi : \xi \in A\} \text{ has the finite intersection property,}$$

as required.

The proof is complete.

5.5 *Theorem.* Let α be an (infinite) singular cardinal and let X be a space with precalibre cf(α). Let $S(X) \leq \kappa$, and suppose there is a set $\{\alpha_\sigma : \sigma < \text{cf}(\alpha)\}$ such that

$$\begin{aligned} &\alpha_\sigma \text{ is a regular cardinal for } \sigma < \text{cf}(\alpha), \\ &\kappa \ll \alpha_\sigma \quad \text{for } \sigma < \text{cf}(\alpha), \\ &\text{cf}(\alpha) < \alpha_{\sigma'} < \alpha_\sigma < \alpha \quad \text{for } \sigma' < \sigma < \text{cf}(\alpha), \text{ and} \\ &\alpha = \sum_{\sigma < \text{cf}(\alpha)} \alpha_\sigma. \end{aligned}$$

Then X has precalibre α.

Proof. Let $\{U_\xi : \xi < \alpha\}$ be a set of non-empty, open subsets of X.

For $\sigma < \text{cf}(\alpha)$ the set $\{U_\xi : \xi < \alpha_\sigma\}$ is a set of non-empty, open subsets of X. Since $\omega \leq \kappa \ll \alpha_\sigma$ and α_σ is regular, it follows from Lemma 5.1 (with α_σ in place of α) that there is $\bar{\xi}(\sigma) < \alpha_\sigma$ such that $\bar{\xi}(\sigma)$ is a representative of $\{U_\xi : \xi < \alpha_\sigma\}$. Without loss of generality we assume, using the fact that X has precalibre cf(α), that $\{U_{\bar{\xi}(\sigma)} : \sigma < \text{cf}(\alpha)\}$ has the finite intersection property. It then follows from Lemma 5.4 that there is $A \subset \alpha$ with $|A| = \alpha$ such that

$$\{U_\xi : \xi \in A\} \text{ has the finite intersection property.}$$

The proof is complete.

5.6 *Theorem.* Let $\omega \leq \kappa \ll \alpha$ and $\kappa \ll \text{cf}(\alpha)$ and let X be a space such that $S(X) \leq \kappa$. Then X has precalibre cf(α) and X has precalibre α.

Proof. It follows from Corollary 5.2 (with α replaced by cf(α)) that X has precalibre cf(α). If α is singular, then from

$$\text{cf}(\kappa) \leq \kappa < \text{cf}(\alpha)$$

we have cf$(\kappa) \neq$ cf(α) and hence from Theorem A.7 of the Appendix there is a set $\{\alpha_\sigma : \sigma < \mathrm{cf}(\alpha)\}$ as in the statement of Theorem 5.5. The required conclusion then follows from Theorem 5.5.

5.7 *Theorem.* Let $\omega \le \kappa \ll \alpha$ and let X be a space such that $S(X) \le \kappa$ and X has precalibre $\mathrm{cf}(\alpha)$. If either

κ is a successor cardinal, or
$\mathrm{cf}(\kappa) \ne \mathrm{cf}(\alpha)$, or
α is a strong limit cardinal, or
$S(X) < \kappa$,

then X has precalibre α.

Proof. It is enough to show that there is a set $\{\alpha_\sigma : \sigma < \mathrm{cf}(\alpha)\}$ as in the statement of Theorem 5.5. In the first three cases this follows from Theorem A.7 of the Appendix; in case $S(X) < \kappa$ we set $\kappa' = (S(X))^+$ and we note that $\omega \le \kappa' \ll \alpha$ and $S(X) \le \kappa'$ and κ' is a successor cardinal.

We note some consequences of Theorem 5.7.

5.8. *Corollary.* Let α be a strong limit cardinal and X a space. Then X has precalibre α if and only if X has precalibre $\mathrm{cf}(\alpha)$.

In particular, if X is a compact Hausdorff space then X has calibre α if and only if X has calibre $\mathrm{cf}(\alpha)$.

Proof. The first equivalence follows from Corollary 2.25 and Theorem 5.7, the second from Corollary 2.24.

5.9 *Corollary.* Assume the generalized continuum hypothesis. Let α be an infinite cardinal and X a space. Then X has precalibre α if and only if X has precalibre $\mathrm{cf}(\alpha)$.

In particular, if X is a compact Hausdorff space then X has calibre α if and only if X has calibre $\mathrm{cf}(\alpha)$.

Proof. If α is regular there is nothing to prove. If α is singular then α is a limit cardinal, hence (from the generalized continuum hypothesis) a strong limit cardinal, and the required conclusion follows from Corollary 5.8.

5.10 *Corollary.* Assume the generalized continuum hypothesis. Let α be an (infinite) cardinal with $\mathrm{cf}(\alpha) > \omega$, let $\{X_i : i \in I\}$ be a set of non-empty spaces, and $X = \prod_{i\in I} X_i$. Then α is a precalibre for X if and only if α is a precalibre for X_i for all $i\in I$.

In particular, if X_i is a compact Hausdorff space for all $i\in I$ then α is a calibre for X if and only if α is a calibre for X_i for all $i\in I$.

Proof. Of the following seven statements, each of the first six is equivalent to its successor: α is a precalibre for X; $\text{cf}(\alpha)$ is a precalibre for X; $\text{cf}(\alpha)$ is a precalibre for $G(X)$; $\text{cf}(\alpha)$ is a precalibre for $\prod_{i\in I} G(X_i)$; $\text{cf}(\alpha)$ is a precalibre for $G(X_i)$ for all $i\in I$; $\text{cf}(\alpha)$ is a precalibre for X_i for all $i\in I$; α is a precalibre for X_i for all $i\in I$. (The indicated equivalences are given respectively by 5.9; 2.22; 2.18(b) and 2.21; 3.7(a); 2.18(b); and 5.9.)

The second statement of the corollary follows from Corollary 2.24.

It follows from Corollary 3.6(a) that for α a singular cardinal with $\text{cf}(\alpha) > \omega$ the space $X = X_I$ of Theorem 4.4 is a product of (non-compact) Hausdorff spaces each with calibre α (hence with calibre $\text{cf}(\alpha)$) such that X has calibre $\text{cf}(\alpha)$ and X does not have calibre α. Thus the analogues of the second paragraphs of Corollaries 5.9 and 5.10 for non-compact spaces both fail (cf. Theorems 4.3 and 4.4).

The results above suggest some natural questions.

5.11 *Problems.* (a) Assume the generalized continuum hypothesis and let κ, β and γ be infinite cardinals with κ regular, $\text{cf}(\beta) < \kappa$, $\text{cf}(\gamma) < \kappa$, and $\gamma > \beta$. Is there a compact space X such that $S(X) = \kappa$ and X has calibre β^+ and X does not have calibre γ^+?

Even the following (simplest) case of this question is unsolved: Assuming the generalized continuum hypothesis, is there a compact c.c.c. space for which ω^+ is, and $(\aleph_\omega)^+$ is not, a calibre?

We note that from an answer (whether positive or negative) to (a) there will follow, assuming the generalized continuum hypothesis, a characterization (analogous to Corollary 4.7) of those classes of infinite cardinal numbers which can arise as the class of calibres of a compact Hausdorff space.

(b) Let $\omega \leq \kappa \ll \alpha$ with $\kappa = \text{cf}(\alpha)$ and let X be a space such that $S(X) = \kappa$ and X has precalibre $\text{cf}(\alpha)$. Does it follow (without additional set-theoretic assumptions, or hypotheses on α or κ) that X must have precalibre α?

In connection with problem (b) we mention the following purely combinatorial question.

(c) Let $\omega \leq \kappa \ll \alpha$ and $\kappa \ll \mathrm{cf}(\alpha)$ with κ regular. Does it follow that $\alpha \to (\alpha, \kappa)^2$?

(Pre-)calibres for P_{λ^+}-spaces and λ^+-box products

In this section we indicate how several of the results above generalize (without additional assumptions) from spaces X to spaces of the form $X_{(\lambda^+)}$ and to λ^+-box products (of the form $(X_I)_{\lambda^+}$). The arguments required in the two cases have many similarities (and are, in addition, similar to much of what has appeared in the proofs of 5.1–5.10). To avoid unnecessary repetition we give complete proofs below relating to the spaces $X_{(\lambda^+)}$ in question (cf. 5.12–5.16) and we give fewer details concerning the products $(X_I)_{\lambda^+}$ (cf. 5.17–5.21).

We begin with a generalization of Corollary 5.2.

5.12 *Theorem.* Let $\omega \leq \lambda < \kappa \ll \alpha$ with α and κ regular and let X be a compact Hausdorff space such that $S(X) \leq \kappa$. Then $X_{(\lambda^+)}$ has calibre α.

Proof. It is, by Corollary 2.24, sufficient to show that $X_{(\lambda^+)}$ has precalibre α. Let $\{U_\xi : \xi < \alpha\}$ be a set of non-empty, open subsets of $X_{(\lambda^+)}$. We assume without loss of generality, using Lemma 2.23, that for $\xi < \alpha$ there is a set $\{U_\xi(\zeta) : \zeta < \lambda\}$ of (non-empty) open subsets of X such that

$$U_\xi = \bigcap_{\zeta < \lambda} U_\xi(\zeta) = \bigcap_{\zeta < \lambda} \mathrm{cl}_X U_\xi(\zeta),$$

$\mathrm{cl}_X U_\xi(\zeta)$ is compact for $\zeta < \lambda$, and

$\{U_\xi(\zeta) : \zeta < \lambda\}$ is closed under finite intersections.

Suppose that if $B \subset \alpha$ and $\{U_\xi : \xi \in B\}$ has the finite intersection property then $|B| < \alpha$, and for $A \subset \alpha$ with $|A| = \alpha$ set

$$\mathscr{B}_A = \{B \subset A : \{U_\xi : \xi \in B\} \text{ has the finite intersection property}\}.$$

Since the set $\mathscr{B}_A$ partially ordered by inclusion is inductive, there is a maximal element B_A of $\mathscr{B}_A$. We have $|B_A| < \alpha$, and since $\{\xi\} \in \mathscr{B}_A$ for all $\xi \in A$ we have $B_A \neq \varnothing$.

For $\xi \in A \backslash B_A$ it follows from the maximality of B_A that there is a finite, non-empty subset $F(\xi)$ of B_A such that

$$(\bigcap_{i \in F(\xi)} U_i) \cap U_\xi = \varnothing.$$

From the conditions on the families $\{U_\xi(\zeta) : \zeta < \lambda\}$ it follows that there are $f_\xi \in \lambda^{F(\xi)}$ and $\zeta(\xi) < \lambda$ such that

$$\bigcap_{i \in F(\xi)} U_i(f_\xi(i)) \cap U_\xi(\zeta(\xi)) = \varnothing.$$

We define

$$\varphi_A : A \backslash B_A \to \mathscr{P}^*_\omega(B_A) \times \cup \{\lambda^F : F \in \mathscr{P}^*_\omega(B_A)\} \times \lambda$$

by the rule

$$\varphi_A(\xi) = (F(\xi), f_\xi, \zeta(\xi)),$$

and we set

$$\mathscr{P}_A = \{B_A\} \cup \{\varphi_A^{-1}(\{(F, f, \zeta)\}) : F \in \mathscr{P}^*_\omega(B_A), f \in \lambda^F, \zeta < \lambda\}.$$

Then $\mathscr{P}_A$ is a partition of A, and since $|B_A| < \alpha$ and $\lambda < \alpha$ we have $|\mathscr{P}_A| < \alpha$. It follows from (the case $\beta = \alpha$ of) Lemma 1.1 that there is a family $\{A_\eta : \eta < \kappa\}$ of subsets of α such that

$$|A_\eta| = \alpha \quad \text{for } \eta < \kappa,$$
$$A_{\eta+1} \in \mathscr{P}_{A_\eta} \quad \text{for } \eta < \kappa, \text{ and}$$
$$A_{\eta'} \subset A_\eta \quad \text{for } \eta < \eta' < \kappa.$$

For $\eta < \kappa$ there are $\bar{F}(\eta) \in \mathscr{P}^*_\omega(B_A)$ and $\bar{f}_\eta \in \lambda^{\bar{F}(\eta)}$ and $\bar{\zeta}(\eta) < \lambda$ such that

$$A_{\eta+1} = \varphi_{A_\eta}^{-1}(\{\bar{F}(\eta), \bar{f}_\eta, \bar{\zeta}(\eta)\}).$$

We assume without loss of generality, using the fact that κ is regular and $\lambda < \kappa$, that there is $\bar{\zeta} < \lambda$ such that $\bar{\zeta}(\eta) = \bar{\zeta}$ for all $\eta < \kappa$.

We set

$$V_\eta = (\cap \{U_i(\bar{f}_\eta(i)) : i \in \bar{F}(\eta)\}) \cap (\cap \{U_i(\bar{\zeta}) : i \in \bar{F}(\eta)\}) \quad \text{for } \eta < \kappa.$$

Since V_η is the intersection of a finite number of open subsets of X, the set V_η is open in X. Further for $\eta < \kappa$ we have, using the fact that $\bar{F}(\eta)$ is a finite subset of B_{A_η} and that $\{U_i : i \in B_{A_\eta}\}$ has the finite intersection property, that

$$V_\eta \supset \cap\{U_i : i \in \bar{F}(\eta)\} \neq \varnothing.$$

We claim finally that

$$V_\eta \cap V_{\eta'} = \varnothing \quad \text{for } \eta < \eta' < \kappa.$$

Indeed since $\bar{F}(\eta') \neq \varnothing$ there is $\xi \in \bar{F}(\eta')$, and from

$$\xi \in \bar{F}(\eta') \subset A_{\eta+1} = \varphi_{A_\eta}^{-1}(\{(\bar{F}(\eta), \bar{f}_\eta, \bar{\zeta})\})$$

we have

$$\varphi_{A_\eta}(\xi) = (\bar{F}(\eta), \bar{f}_\eta, \bar{\zeta})$$

and hence

$$\bigcap_{i \in F(\eta)} U_i(\bar{f}_\eta(i)) \cap U_\xi(\bar{\zeta}) = \varnothing;$$

then from $\cap_{i \in F(\eta)} U_i(\bar{f}_\eta(i)) \supset V_\eta$ and $U_\xi(\bar{\zeta}) \supset V_{\eta'}$ we have $V_\eta \cap V_{\eta'} = \varnothing$, as claimed.

It follows that $S(X) > \kappa$. This contradiction completes the proof.

5.13 *Corollary*. Let $\alpha \geq \omega$ and let X be a compact Hausdorff space such that $S(X) \leq \alpha^+$. Then $X_{(\alpha^+)}$ has calibre $(2^\alpha)^+$.

Proof. We have remarked in A.5(b) of the Appendix that $\alpha^+ \ll (2^\alpha)^+$. The statement then follows from Theorem 5.12 with λ, κ and α replaced by α, α^+ and $(2^\alpha)^+$ respectively.

We now turn to the case of singular cardinals. The first lemma generalizes Lemma 5.1.

5.14 *Lemma*. Let $\omega \leq \lambda < \kappa \ll \alpha$ with α and κ regular and let X be a compact Hausdorff space such that $S(X) \leq \kappa$. Let $\{U_\xi : \xi < \alpha\}$ be a set of non-empty, open subsets of $X_{(\lambda^+)}$ such that for $\xi < \alpha$ there is a set $\{U_\xi(\zeta) : \zeta < \lambda\}$ of open subsets of X such that

$$U_\xi = \bigcap_{\zeta < \lambda} U_\xi(\zeta) = \bigcap_{\zeta < \lambda} \mathrm{cl}_X U_\xi(\zeta),$$

$\mathrm{cl}_X U_\xi(\zeta)$ is compact for $\zeta < \lambda$, and

$\{U_\xi(\zeta) : \zeta < \lambda\}$ is closed under finite intersections.

Then the set $\{U_\xi : \xi < \alpha\}$ has a representative.

Proof. We suppose the statement fails. For $A \subset \alpha$ with $|A| = \alpha$ we choose $\xi_A \in A$. There are

a cardinal $\beta_A < \alpha$, and

a set $\{W_\upsilon^A : \upsilon < \beta_A\}$ of non-empty, open (in $X_{(\lambda^+)}$) subsets of U_{ξ_A}, such that
if $B \subset A$ and $\{U_\xi \cap W_\upsilon^A : \xi \in B\}$ has the finite intersection property for all $\upsilon < \beta_A$, then $|B| < \alpha$.

We assume without loss of generality, using Lemma 2.23, that for $\upsilon < \beta_A$ there is a set $\{W_\upsilon^A(\zeta') : \zeta' < \lambda\}$ of (non-empty) open subsets of X such that

$$W_\upsilon^A = \bigcap_{\zeta' < \lambda} W_\upsilon^A(\zeta') = \bigcap_{\zeta' < \lambda} \mathrm{cl}_X W_\upsilon^A(\zeta'), \text{ and}$$

$\mathrm{cl}_X W_\upsilon^A(\zeta')$ is compact for $\zeta' < \lambda$, and

$\{W_\upsilon^A(\zeta') : < \lambda\}$ is closed under finite intersections.

For $A \subset \alpha$ with $|A| = \alpha$ we set

$$\mathscr{B}_A = \{B \subset A : \{U_\xi \cap W_\upsilon^A : \xi \in B\} \text{ has the finite intersection property for all } \upsilon < \beta_A\}.$$

Since the set $\mathscr{B}_A$ partially ordered by inclusion is inductive, there is a maximal element B_A of $\mathscr{B}_A$. We have $|B_A| < \alpha$, and since $\{\xi_A\} \in \mathscr{B}_A$ we have $B_A \neq \varnothing$.

For $\xi \in A \backslash B_A$ it follows from the maximality of B_A that there are a finite, non-empty subset $F(\xi)$ of B_A and $\upsilon(\xi) < \beta_A$ such that

$$\left(\bigcap_{i \in F(\xi)} U_i\right) \cap W_{\upsilon(\xi)}^A \cap U_\xi = \varnothing.$$

From the conditions on the families $\{U_\xi(\zeta) : \zeta < \lambda\}$ and $\{W_\upsilon^A(\zeta') : \zeta' < \lambda\}$ it follows that there are $f_\xi \in \lambda^{F(\xi)}$ and $\zeta(\xi), \zeta'(\xi) < \lambda$ such that

$$\left(\bigcap_{i \in F(\xi)} U_i(f_\xi(i))\right) \cap W_{\upsilon(\xi)}^A(\zeta'(\xi)) \cap U_\xi(\zeta(\xi)) = \varnothing.$$

We define

$$\varphi_A : A \backslash B_A \to \mathscr{P}_\omega^*(B_A) \times \cup\{\lambda^F : F \in \mathscr{P}_\omega^*(B_A)\} \times \beta_A \times \lambda \times \lambda$$

by the rule

$$\varphi_A(\xi) = (F(\xi), f_\xi, \upsilon(\xi), \zeta'(\xi), \zeta(\xi)),$$

and we set

$$\mathscr{P}_A = \{B_A\} \cup \{\varphi_A^{-1}(\{(F, f, \upsilon, \zeta', \zeta)\}) : F \in \mathscr{P}_\omega^*(B_A), f \in \lambda^F, \upsilon < \beta_A, \zeta', \zeta < \lambda\}.$$

Then $\mathscr{P}_A$ is a partition of A, and since $|B_A| < \alpha$, $\beta_A < \alpha$ and $\lambda < \alpha$

we have $|\mathscr{P}_A| < \alpha$. It follows from (the case $\beta = \alpha$ of) Lemma 1.1 that there is a family $\{A_\eta : \eta < \kappa\}$ of subsets of α such that

$$|A_\eta| = \alpha \quad \text{for } \eta < \kappa,$$
$$A_{\eta+1} \in \mathscr{P}_{A_\eta} \quad \text{for } \eta < \kappa, \text{ and}$$
$$A_{\eta'} \subset A_\eta \quad \text{for } \eta < \eta' < \kappa.$$

For $\eta < \kappa$ there are $\bar{F}(\eta) \in \mathscr{P}^*_\omega(B_{A_\eta})$, $\bar{f}_\eta \in \lambda^{\bar{F}(\eta)}$, $\bar{v}(\eta) < \beta_{A_\eta}$, $\bar{\zeta}'(\eta) < \lambda$ and $\bar{\zeta}(\eta) < \lambda$ such that

$$A_{\eta+1} = \varphi^{-1}_{A_\eta}(\{(\bar{F}(\eta), \bar{f}_\eta, \bar{v}(\eta), \bar{\zeta}'(\eta), \bar{\zeta}(\eta))\}).$$

We assume without loss of generality, using the fact that κ is regular and $\lambda < \kappa$, that there are $\bar{\zeta}', \bar{\zeta} < \lambda$ such that $\bar{\zeta}'(\eta) = \bar{\zeta}'$ and $\bar{\zeta}(\eta) = \bar{\zeta}$ for all $\eta < \kappa$.

We set

$$V_\eta = (\cap\{U_i(\bar{f}_\eta(i)) : i \in \bar{F}(\eta)\}) \cap W^{A_\eta}_{\bar{v}(\eta)}(\bar{\zeta}') \cap (\cap\{U_i(\bar{\zeta}) : i \in \bar{F}(\eta)\}).$$

Since V_η is the intersection of a finite number of open subsets of X, the set V_η is open in X. Further for $\eta < \kappa$ we have, using the facts that $\bar{F}(\eta)$ is a finite subset of B_{A_η} and $\{U_i \cap W^A_{\bar{v}(\eta)} : i \in B_{A_\eta}\}$ has the finite intersection property and

$$V_n \supset (\bigcap_{i \in \bar{F}(\eta)} U_i) \cap W^{A_n}_{\bar{v}(\eta)},$$

that $V_\eta \neq \varnothing$. We claim finally that

$$V_\eta \cap V_{\eta'} = \varnothing \quad \text{for } \eta < \eta' < \kappa.$$

Indeed since $\bar{F}(\eta') \neq \varnothing$ there is $\xi \in \bar{F}(\eta')$, and from

$$\xi \in \bar{F}(\eta') \subset A_{\eta+1} = \varphi^{-1}_{A_\eta}(\{(\bar{F}(\eta), \bar{f}_\eta, \bar{v}(\eta), \bar{\zeta}', \bar{\zeta})\})$$

we have

$$\varphi_{A_\eta}(\xi) = (\bar{F}(\eta), \bar{f}_\eta, \bar{v}(\eta), \bar{\zeta}', \bar{\zeta})$$

and hence

$$\bigcap_{i \in \bar{F}(\eta)} U_i(\bar{f}_\eta(i)) \cap W^{A_\eta}_{v(\eta)}(\bar{\zeta}') \cap U_\xi(\bar{\zeta}) = \varnothing;$$

then from $\cap_{i \in \bar{F}(\eta)} U_i(\bar{f}_\eta(i)) \cap W^{A_\eta}_{\bar{v}(\eta)}(\bar{\zeta}') \supset V_\eta$ and $U_\xi(\bar{\zeta}) \supset V_{\eta'}$ we have $V_\eta \cap V_{\eta'} = \varnothing$, as claimed.

It follows that $S(X) > \kappa$. This contradiction completes the proof.

5.15 *Theorem.* Let $\omega \leq \lambda < \kappa \ll \alpha$ with κ regular and α singular and let X be a compact, Hausdorff space such that $S(X) \leq \kappa$ and $X_{(\lambda^+)}$ has calibre $\mathrm{cf}(\alpha)$. If there is a set $\{\alpha_\sigma : \sigma < \mathrm{cf}(\alpha)\}$ such that

$$\begin{aligned}
&\alpha_\alpha \text{ is a regular cardinal for } \sigma < \mathrm{cf}(\alpha),\\
&\kappa \ll \alpha_\sigma \quad \text{for } \sigma < \mathrm{cf}(\alpha),\\
&\mathrm{cf}(\alpha) < \alpha_{\sigma'} < \alpha_\sigma < \alpha \quad \text{for } \sigma' < \sigma < \mathrm{cf}(\alpha), \text{ and}\\
&\alpha = \sum_{\sigma < \mathrm{cf}(\alpha)} \alpha_\sigma,
\end{aligned}$$

then $X_{(\lambda^+)}$ has calibre α.

The proof is similar to that of Theorem 5.4; we here use Lemma 5.14 in place of Lemma 5.1. (We note that in order to show that for every set $\{U_\xi : \xi < \alpha\}$ of non-empty, open subsets of $X_{(\lambda^+)}$ there is $A \subset \alpha$ with $|A| = \alpha$ such that $\{U_\xi : \xi \in A\}$ has the finite intersection property, it is by Lemma 2.23 sufficient to consider the case that for $\xi < \alpha$ the set U_ξ has the form $U_\xi = \cap_{\zeta < \lambda} U_\xi(\zeta)$ with $U_\xi(\zeta)$ as in the statement of Lemma 5.14.)

5.16 *Theorem.* Let $\omega \leq \lambda < \kappa \ll \alpha$ and $\kappa \ll \mathrm{cf}(\alpha)$ with κ regular and let X be a compact, Hausdorff space such that $S(X) \leq \kappa$. Then $X_{(\lambda^+)}$ has calibre α.

Proof. It follows from Theorem 5.10 (applied to $\mathrm{cf}(\alpha)$) that $X_{(\lambda^+)}$ has calibre $\mathrm{cf}(\alpha)$. Since $\kappa \ll \alpha$ and $\mathrm{cf}(\kappa) \leq \kappa < \mathrm{cf}(\alpha)$ there is (for α singular) by Theorem A.7 of the Appendix a set $\{\alpha_\sigma : \sigma < \mathrm{cf}(\alpha)\}$ satisfying the conditions of Theorem 5.15. Hence $X_{(\lambda^+)}$ has calibre α.

We recall from Chapter 3 that for $\kappa \geq \omega$ and $\{X_i : i \in I\}$ a set of spaces, the product $\prod_{i \in I} X_i$ with the κ-box topology is denoted $(\prod_{i \in I} X_i)_\kappa$ or $(X_I)_\kappa$. It is not required that $(X_I)_\kappa$ is a P_κ-space.

5.17 *Theorem.* Let $\omega \leq \lambda < \kappa \ll \alpha$ with α and κ regular, and let $\{X_i : i \in I\}$ be a set of spaces such that $S(X_i) \leq \kappa$ for $i \in I$. Then $(X_I)_{\lambda^+}$ has precalibre α.

Proof. From the statement and theorem following (the proof of) Theorem 3.6 it is enough to consider the case $I = \lambda$. If the lemma is false in this case then there is a set $\{U^\xi : \xi < \alpha\}$ of non-empty basic open subsets of $(\prod_{\zeta < \lambda} X_\zeta)_{\lambda^+}$ such that if $B \subset \alpha$ and $\{U^\xi : \xi \in B\}$ has the finite intersection property, then $|B| < \alpha$. We write $U^\xi = \prod_{\zeta < \lambda} U^\xi_\zeta$ with U^ξ_ζ a (non-empty) open subset of X_ζ for $\xi < \alpha, \zeta < \lambda$.

For $A \subset \alpha$ with $|A| = \alpha$ we set

$$\mathscr{B}_A = \{B \subset A : \{U^\xi : \xi \in B\} \text{ has the finite intersection property}\}.$$

Since the set $\mathscr{B}_A$ partially ordered by inclusion is inductive, there is a maximal element B_A of $\mathscr{B}_A$. We have $|B_A| < \alpha$, and since $\{\xi\} \in \mathscr{B}_A$ for all $\xi \in A$ we have $B_A \neq \emptyset$.

For $\xi \in A \backslash B_A$ it follows from the maximality of B_A that there is a finite, non-empty subset $F(\xi)$ of B_A such that

$$(\bigcap_{i \in F(\xi)} U^i) \cap U^\xi = \emptyset;$$

it follows from Lemma 3.1 that there is $\zeta(\xi) < \lambda$ such that

$$(\bigcap_{i \in F(\xi)} U^i_{\zeta(\xi)}) \cap U^\xi_{\zeta(\xi)} = \emptyset.$$

We define

$$\varphi_A : A \backslash B_A \to \mathscr{P}^*_\omega(B_A) \times \lambda$$

by the rule

$$\varphi_A(\xi) = (F(\xi), \zeta(\xi)),$$

and we set

$$\mathscr{P}_A = \{B_A\} \cup \{\varphi_A^{-1}(\{(F, \zeta)\}) : F \in \mathscr{P}^*_\omega(B_A), \zeta < \lambda\}.$$

Then $\mathscr{P}_A$ is a partition of A, with $|\mathscr{P}_A| < \alpha$. It follows from (the case $\beta = \alpha$ of) Lemma 1.1 that there is a family $\{A_\eta : \eta < \kappa\}$ of subsets of α such that

$$|A_\eta| = \alpha \quad \text{for } \eta < \kappa,$$
$$A_{\eta+1} \in \mathscr{P}_{A_\eta} \quad \text{for } \eta < \kappa, \text{ and}$$
$$A_{\eta'} \subset A_\eta \quad \text{for } \eta < \eta' < \kappa.$$

For $\eta < \kappa$ there are $\bar{F}(\eta) \in \mathscr{P}^*_\omega(B_{A_\eta})$ and $\bar{\zeta}(\eta) < \lambda$ such that

$$A_{\eta+1} = \varphi_{A_\eta}^{-1}(\{(\bar{F}(\eta), \bar{\zeta}(\eta))\}).$$

We assume without loss of generality, using the fact that κ is regular and $\lambda < \kappa$, that there is $\bar{\zeta} < \lambda$ such that $\bar{\zeta}(\eta) = \bar{\zeta}$ for all $\eta < \kappa$.

We set

$$V_\eta = \bigcap_{i \in \bar{F}(\eta)} U^i_{\bar{\zeta}}.$$

Since V_η is the intersection of a finite number of open subsets of $X_{\bar{\zeta}}$, the set V_η is open in $X_{\bar{\zeta}}$. Further for $\eta < \kappa$ we have, using the fact that $\bar{F}(\eta)$ is a finite subset of B_{A_η} and that $\{U^i : i \in B_{A_\eta}\}$ has the finite intersection property, that $\bigcap_{i\in F(\eta)} U^i \neq \emptyset$ and hence $V_\eta \neq \emptyset$. We claim finally that

$$V_\eta \cap V_{\eta'} = \emptyset \quad \text{for } \eta < \eta' < \kappa.$$

Indeed since $\bar{F}(\eta') \neq \emptyset$ there is $\xi \in \bar{F}(\eta')$, and from

$$\xi \in \bar{F}(\eta') \subset A_{\eta+1} = \varphi_{A_\eta}^{-1}(\{(\bar{F}(\eta), \bar{\zeta})\})$$

we have

$$\varphi_{A_\eta}(\xi) = (\bar{F}(\eta), \bar{\zeta})$$

and hence

$$(\bigcap_{i\in \bar{F}(\eta)} U^i_{\bar{\zeta}}) \cap U^\xi_{\bar{\zeta}} = \emptyset;$$

then from $\bigcap_{i\in \bar{F}(\eta)} U^i_{\bar{\zeta}} = V_\eta$ and $U^\xi_{\bar{\zeta}} \supset V_{\eta'}$ we have $V_\eta \cap V_{\eta'} = \emptyset$, as claimed.

It follows that $S(X_{\bar{\zeta}}) > \kappa$. This contradiction completes the proof.

5.18 *Corollary*. Let $\alpha \geq \omega$ and let $\{X_i : i \in I\}$ be a set of spaces such that $S(X_i) \leq \alpha^+$ for $i \in I$. Then $(\prod_{i\in I} X_i)_{\alpha^+}$ has precalibre $(2^\alpha)^+$.

Proof. We have remarked in A.5(b) of the Appendix that $\alpha^+ \ll (2^\alpha)^+$. The statement then follows from Theorem 5.17 with λ, κ and α replaced by α, α^+ and $(2^\alpha)^+$, respectively.

5.19 *Lemma*. Let $\omega \leq \lambda < \kappa \ll \alpha$ with α and κ regular and let $\{X_\zeta : \zeta < \lambda\}$ be a set of spaces such that $S(X_\zeta) \leq \kappa$ for $\zeta < \lambda$. Then every set $\{U^\xi : \xi < \alpha\}$ of non-empty, open subsets of $(\prod_{\zeta<\lambda} X_\zeta)_{\lambda^+}$ has a representative.

Proof. If the statement fails then for $A \subset \alpha$ with $|A| = \alpha$ there are $\xi_A \in A$, a cardinal $\beta_A < \alpha$, and a set $\{W^A_\upsilon : \upsilon < \beta_A\}$ of non-empty (basic) open subsets of U^{ξ_A} such that

if $B \subset A$ and $\{U^\xi \cap W^A_\upsilon : \xi \in B\}$ has the finite intersection property for all $\upsilon < \beta_A$, then $|B| < \alpha$.

We write

$$U^\xi = \prod_{\zeta<\lambda} U^\xi(\zeta) \quad \text{for } \xi < \alpha, \text{ and}$$

$$W^A_\upsilon = \prod_{\zeta<\lambda} W^A_\upsilon(\zeta) \quad \text{for } \upsilon < \beta_A.$$

For $A \subset \alpha$ with $|A| = \alpha$ we set

$$\mathscr{B}_A = \{B \subset A : \{U^\xi \cap W^A_\upsilon : \xi \in B\} \text{ has the finite intersection property for all } \upsilon < \beta_A\}.$$

Since the set $\mathscr{B}_A$ partially ordered by inclusion is inductive, there is a maximal element B_A of $\mathscr{B}_A$. We have $|B_A| < \alpha$, and since $\{\xi_A\} \in \mathscr{B}_A$ we have $B_A \neq \varnothing$.

For $\xi \in A \backslash B_A$ it follows from the maximality of B_A that there are a a finite, non-empty subset $F(\xi)$ of B_A and $\upsilon(\xi) < \beta_A$ such that

$$\Big(\bigcap_{i \in F(\xi)} U^i\Big) \cap W^A_{\upsilon(\xi)} \cap U^\xi = \varnothing;$$

hence there is $\zeta(\xi) < \lambda$ such that

$$\Big(\bigcap_{i \in F(\xi)} U^i(\zeta(\xi))\Big) \cap W^A_{\upsilon(\xi)}(\zeta(\xi)) \cap U^\xi(\zeta(\xi)) = \varnothing.$$

We define

$$\varphi_A : A \backslash B_A \to \mathscr{P}^*_\omega(B_A) \times \beta_A \times \lambda$$

by the rule

$$\varphi_A(\xi) = (F(\xi), \upsilon(\xi), \zeta(\xi)),$$

and we set

$$\mathscr{P}_A = \{B_A\} \cup \{\varphi_A^{-1}(\{(F, \upsilon, \zeta)\}) : F \in \mathscr{P}^*_\omega(B_A), \upsilon < \beta_A, \zeta < \lambda\}.$$

The proof concludes along lines by now familiar to the reader: there is a set $\{A_\eta : \eta < \kappa\}$ of subsets of α as in the conclusion of (the case $\beta = \alpha$ of) Lemma 1.1, for $\eta < \kappa$ there are $\bar{F}(\eta) \in \mathscr{P}^*_\omega(B_{A_\eta})$, $\bar{\upsilon}(\eta) < B_{A_\eta}$ and $\bar{\zeta}(\eta) < \lambda$ such that

$$A_{\eta+1} = \varphi_{A\eta}^{-1}(\{(\bar{F}(\eta), \bar{\upsilon}(\eta), \bar{\zeta}(\eta))\});$$

we assume without loss of generality that there is $\bar{\zeta} < \lambda$ such that $\bar{\zeta}(\eta) = \bar{\zeta}$ for all $\eta < \kappa$, we set

$$V_\eta = \Big(\bigcap_{i \in \bar{F}(\eta)} U^i(\bar{\zeta})\Big) \cap W^{A_\eta}_{\bar{\upsilon}(\eta)}(\bar{\zeta}),$$

and we show as in earlier proofs that the (indexed) family $\{V_\eta : \eta < \kappa\}$ is a cellular family in $X_{\bar{\zeta}}$. This contradicts the assumption $S(X_{\bar{\zeta}}) \leq \kappa$.

5.20 *Theorem.* Let $\omega \leq \lambda < \kappa \ll \alpha$ with κ regular and α singular and let $\{X_i : i \in I\}$ be a set of spaces such that $S(X_i) \leq \kappa$ for $i \in I$ and $(X_I)_{\lambda^+}$ has precalibre cf(α). If there is a set $\{\alpha_\sigma : \sigma < \text{cf}(\alpha)\}$ such

that

$$\alpha_\sigma \text{ is a regular cardinal for } \sigma < \mathrm{cf}(\alpha),$$
$$\kappa \ll \alpha_\sigma \quad \text{for } \sigma < \mathrm{cf}(\alpha),$$
$$\mathrm{cf}(\alpha) < \alpha_{\sigma'} < \alpha_\sigma < \alpha \quad \text{for } \sigma' < \sigma < \mathrm{cf}(\alpha), \text{ and}$$
$$\alpha = \sum_{\sigma < \mathrm{cf}(\alpha)} \alpha_\sigma,$$

then $(X_I)_{\lambda^+}$ has precalibre α.

From the theorem following (the proof of) Theorem 3.6 it is enough to consider the case $I = \lambda$. The proof is now similar to that of Theorem 5.5; we use Lemma 5.19 in place of Lemma 5.1.

5.21 *Theorem.* Let $\omega \leq \lambda < \kappa \ll \alpha$ and $\kappa \ll \mathrm{cf}(\alpha)$ with κ regular, and let $\{X_i : i \in I\}$ be a set of spaces such that $S(X_i) \leq \kappa$ for $i \in I$. Then $(X_I)_{\lambda^+}$ has precalibre α.

Proof. It follows from Theorem 5.17 (with α replaced by $\mathrm{cf}(\alpha)$) that $(X_I)_{\lambda^+}$ has precalibre $\mathrm{cf}(\alpha)$. There is (for α singular), by Theorem A.7 of the Appendix, a set $\{\alpha_\sigma : \sigma < \mathrm{cf}(\alpha)\}$ satisfying the conditions of Theorem 5.20. Hence $(X_I)_{\lambda^+}$ has precalibre α.

Calibres of compact spaces: the 'exceptional' case

We recall from Appendix A that for $\alpha \geq \omega, \kappa > 2$ the cardinal numbers $\sqrt[\underline{\kappa}]{\alpha}$, $\sqrt[\kappa]{\alpha}$ are defined by the relations

$$\sqrt[\kappa]{\alpha} = \min\{\beta : \beta^\kappa \geq \alpha\},$$
$$\sqrt[\underline{\kappa}]{\alpha} = \min\{\sqrt[\lambda]{\alpha} : 1 < \lambda < \kappa\}.$$

5.22 *Lemma.* Let α and κ be regular cardinals with $2 < \sqrt[\underline{\kappa}]{\alpha} < \alpha$, let X be a space with $S(X) \leq \kappa$, and let $\{U_\xi : \xi < \alpha\}$ be a set of non-empty, open subsets of X. Then there is $A \subset \alpha$ with $|A| = \alpha$ such that: if $C \subset A$ and $|C| \leq \kappa^{\underline{\kappa}}$ and $\{U_\zeta : \zeta \in C\}$ has the finite intersection property, then
$|\{\xi \in A : \{U_\zeta : \zeta \in C \cup \{\xi\}\}$ has the finite intersection property$\}| = \alpha$.

Proof. We suppose the statement fails. For $A \subset \alpha$ with $|A| = \alpha$ there are $B_A \subset A$ with $|B_A| < \alpha$ and $C_A \subset A$ with $|C_A| \leq \kappa^{\underline{\kappa}}$ such that
$\{U_\zeta : \zeta \in C_A\}$ has the finite intersection property, and
for $\xi \in A \backslash (B_A \cup C_A)$ there is a finite, non-empty subset $F(\xi)$ of C_A such that $(\bigcap_{\zeta \in F(\xi)} U_\zeta) \cap U_\xi = \varnothing$.

We define

$$\varphi_A : A\backslash(B_A \cup C_A) \to \mathscr{P}^*_\omega(C_A)$$

by the rule

$$\varphi_A(\xi) = F(\xi),$$

and we set

$$\mathscr{P}_A = \{B_A \cup C_A\} \cup \{\varphi_A^{-1}(\{F\}) : F \in \mathscr{P}^*_\omega(C_A)\}.$$

Then $\mathscr{P}_A$ is a partition of α such that $|\mathscr{P}_A| \leq \kappa^{\underline{\kappa}}$. We set

$$\beta = (\kappa^{\underline{\kappa}})^+.$$

It follows from Theorem A.11 of the Appendix that $\kappa \ll \beta$ and $\beta \leq \alpha$, and from Lemma 1.1 it then follows that there is a family $\{A_\eta : \eta < \kappa\}$ of subsets of α such that

$$|A_\eta| = \alpha \quad \text{for } \eta < \kappa,$$

$$A_{\eta+1} \in \mathscr{P}_{A_\eta} \quad \text{for } \eta < \kappa, \text{ and}$$

$$A_{\eta'} \subset A_\eta \quad \text{for } \eta < \eta' < \kappa.$$

For $\eta < \kappa$ there is $\bar{F}(\eta) \in \mathscr{P}^*_\omega(C_{A_\eta})$ such that

$$A_{\eta+1} = \varphi_{A_\eta}^{-1}(\{\bar{F}(\eta)\}).$$

For $\eta < \kappa$ we set

$$V_\eta = \bigcap_{i \in \bar{F}(\eta)} U_i,$$

and we note that since $\bar{F}(\eta)$ is a finite subset of C_{A_η} and $\{U_i : i \in C_{A_\eta}\}$ has the finite intersection property, the open set V_η is non-empty. We claim finally that

$$V_\eta \cap V_{\eta'} = \varnothing \quad \text{for } \eta < \eta' < \kappa.$$

Indeed since $\bar{F}(\eta') \neq \varnothing$ there is $\xi \in \bar{F}(\eta')$, and from

$$\xi \in \bar{F}(\eta') \subset A_{\eta+1} = \varphi_{A_\eta}^{-1}(\{\bar{F}(\eta)\})$$

we have

$$\varphi_{A_\eta}(\xi) = \bar{F}(\eta)$$

and hence

$$V_\eta \cap V_{\eta'} \subset (\bigcap_{i \in \bar{F}(\eta)} U_i) \cap U_\xi = \varnothing.$$

It follows that $S(X) > \kappa$. This contradiction completes the proof.

We have seen in Theorem 2.1(h) for $\alpha \geq \omega$ that a space with calibre $(\alpha, 2)$ has calibre (α, n) for $0 < n < \omega$. We note now that in fact such a space has calibre (α, ω, n) for $0 < n < \omega$.

5.23 *Lemma.* Let $\alpha \geq \omega$ and let X be a space such that $S(X) \leq \alpha$.

(a) If α is regular and $\{U_\xi : \xi < \alpha\}$ is a set of non-empty, open subsets of X then there is $\zeta < \alpha$ such that

$$|\{\xi < \alpha : U_\zeta \cap U_\xi \neq \varnothing\}| = \alpha;$$

(b) X has precalibre (α, ω).

Proof. (a) We assume the result fails and we define $\{A(\eta): \eta < \alpha\} \subset \mathscr{P}(\alpha)$ and $\{\xi_\eta : \eta < \alpha\} \subset \alpha$ such that

(i) $A(0) = \alpha, \xi_0 \in A(0)$;

(ii) $A(\eta) = \{\xi < \alpha : U_\xi \cap U_{\xi_{\eta'}} = \varnothing$ for all $\eta' < \eta\}$ for $\eta < \alpha$; and

(iii) $\xi_\eta \in A(\eta)$ for $\eta < \alpha$.

We proceed by transfinite recursion. We define $A(0)$ and ξ_0 by (i), and $A(\eta)$ and ξ_η for $0 < \eta < \alpha$ by (ii) and (iii). We note that $|A(\eta)| = \alpha$ for $\eta < \alpha$, since α is regular and

$$|\{\xi < \alpha : \text{there is } \eta' < \eta \text{ such that } U_\xi \cap U_{\xi_{\eta'}} \neq \varnothing\}| < \alpha.$$

It is clear that $U_{\xi_{\eta'}} \cap U_{\xi_\eta} = \varnothing$ for $\eta' < \eta < \alpha$, a contradiction.

(b) Suppose first that α is regular and let $\{U_\xi : \xi < \alpha\} \subset \mathscr{T}^*(X)$; we show there is $A \subset \alpha$ with $|A| = \omega$ such that $\{U_\xi : \xi \in A\}$ has the finite intersection property. From part (a) there are $\xi_0 < \alpha$ and $A_0 \subset \alpha$ with $|A_0| = \alpha$ such that

$$\{U_\xi \cap U_{\xi_0} : \xi \in A_0\} \subset \mathscr{T}^*(X).$$

It is then simple using part (a) to define by recursion (for $0 < n < \omega$) an ordinal number $\xi_n \in A_{n-1}$ and $A_n \subset A_{n-1}$ with $|A_n| = \alpha$ such that $\xi_n > \xi_{n-1}$ and

$$\{U_\xi \cap U_{\xi_0} \cap U_{\xi_1} \cap \ldots \cap U_{\xi_n} : \xi \in A_n\} \subset \mathscr{T}^*(X).$$

We set $A = \{\xi_n : n < \omega\}$; it is clear that A is as required.

If α is singular we assume without loss of generality (using 2.18–2.20) that X is a compact, Hausdorff space, we assume further that X is an infinite space (since otherwise $|\mathscr{T}^*(X)| < \omega$ and the statement to be proved is obvious), we set $\beta = S(X)$, we recall from Theorem 2.14 that β is a regular cardinal, and we note from the preceding paragraph (with β replacing α) that X has precalibre (β, ω); since $\alpha \geq \beta$ it then follows that X has precalibre (α, ω), as required.

The proof is complete.

It follows from Corollary 2.24 and Lemma 5.23 that if α is an infinite cardinal and X is a locally compact, Hausdorff space such that $S(X) \leq \alpha$, then X has calibre (α, ω). We note now in passing that in this statement the condition that X is locally compact cannot be omitted.

5.24 *Theorem.* For $\alpha > \omega$ there is a space X such that X has calibre $(\alpha, 2)$ and X does not have calibre (α, ω).

Proof. Set $Y = \{0, 1\}^\alpha$ and

$$X = \{x \in Y : |\{\xi < \alpha : x_\xi \neq 0\}| < \omega\}.$$

For $J \in \mathscr{P}^*_\omega(\alpha)$ the (finite) space $\prod_{\xi \in J} \{0, 1\}_\xi$ has calibre $(\omega^+, 2)$; hence Y has calibre $(\omega^+, 2)$ by Theorem 3.6(a). Since X is dense in Y, the space X has calibre $(\omega^+, 2)$; thus X has calibre $(\alpha, 2)$ by Theorem 2.1(a).

To prove that X does not have calibre (α, ω) it is enough to define $\{U_\xi : \xi < \alpha\} \subset \mathscr{T}^*(X)$ such that $\cap_{\xi \in A} U_\xi = \varnothing$ for all $A \in [\alpha]^\omega$. For this we set

$$U_\xi = \{x \in X : x_\xi = 1\} \quad \text{for } \xi < \alpha.$$

5.25 *Theorem.* Let α and κ be (infinite) regular cardinals such that $\kappa \leq \alpha$, with $\kappa < \alpha$ in case α is a strong limit cardinal, and let X be a space such that $S(X) \leq \kappa$. Then X has precalibre $(\alpha, \omega \cdot \sqrt[\kappa]{\alpha})$.

Proof. We consider three cases.

Case 1. $\sqrt[\kappa]{\alpha} = 2$. Then $\omega \cdot \sqrt[\kappa]{\alpha} = \omega$ and the required conclusion follows from Lemma 5.23(b).

Case 2. $\sqrt[\kappa]{\alpha} = \alpha$. Since α is a strong limit cardinal and $\kappa < \alpha$ we have $\kappa \ll \alpha$. It follows from Corollary 5.2 that X has precalibre α, as required.

Case 3. $2 < \sqrt[\kappa]{\alpha} < \alpha$. We set $\beta = \sqrt[\kappa]{\alpha}$, we choose λ such that $\lambda < \kappa$ and $\beta = \sqrt[\lambda]{\alpha}$, and we note from Theorem A.9 of the Appendix

that

$$\mathrm{cf}(\beta) = \mathrm{cf}(\sqrt[\lambda]{\alpha}) \leq \lambda < \kappa.$$

We claim next that $\kappa \ll \beta$. If there are $\gamma < \beta$ and $\mu < \kappa$ such that $\gamma^\mu \geq \beta$, then from $\beta^\lambda \geq \alpha$ we have $\gamma^{\mu \cdot \lambda} \geq \alpha$ and hence (since $\mu \cdot \lambda < \kappa$)

$$\beta = \sqrt[\kappa]{\alpha} \leq \gamma < \beta,$$

a contradiction. Since $2 < \beta$ the relation $\beta < \kappa$ fails (otherwise $2^\beta < \beta$) and hence $\kappa \leq \beta$; since the three conditions

$$\mathrm{cf}(\beta) < \kappa \leq \beta, \kappa \text{ is regular}, \kappa = \beta$$

are incompatible, we have $\kappa < \beta$. The proof that $\kappa \ll \beta$ is complete.

It now follows from Theorem A.7 of the Appendix (applied to the singular cardinal β) that there is a set $\{\beta_\sigma : \sigma < \mathrm{cf}(\beta)\}$ such that

β_σ is a regular cardinal for $\sigma < \mathrm{cf}(\beta)$,
$\kappa \ll \beta_0$ for $\sigma < \mathrm{cf}(\beta)$,
$\mathrm{cf}(\beta) < \beta_{\sigma'} < \beta_\sigma < \beta$ for $\sigma' < \sigma < \mathrm{cf}(\beta)$, and
$\beta = \sum_{\sigma < \mathrm{cf}(\beta)} \beta_\sigma$.

Now let $\{U_\xi : \xi < \alpha\}$ be a set of non-empty, open subsets of X, and let A be the subset of α (with $|A| = \alpha$) given for this family by Lemma 5.22. We define $\{B_\sigma : \sigma < \mathrm{cf}(\beta)\}, \{\bar{\xi}(\sigma) : \sigma < \mathrm{cf}(\beta)\}$ such that

(i) $B_\sigma \subset A$ and $|B_\sigma| = \beta_\sigma$ for $\sigma < \mathrm{cf}(\beta)$,
(ii) $\bar{\xi}(\sigma)$ is a representative of $\{U_\xi : \xi \in B_\sigma\}$, and
(iii) $\{U_{\bar{\xi}(\sigma)} : \sigma < \mathrm{cf}(\beta)\}$ has the finite intersection property.

We proceed by recursion. We set $B_0 = \beta_0$ and, using Lemma 5.1 (applied to B_0 in place of α), we choose a representative $\bar{\xi}(0)$ of $\{U_\xi : \xi \in B_0\}$.

Now let $\sigma < \mathrm{cf}(\beta)$ and suppose for $\sigma' < \sigma$ that $B_{\sigma'}, \bar{\xi}(\sigma')$ have been defined so that

$B_{\sigma'} \subset A$ and $|B_{\sigma'}| = \beta_{\sigma'}$,
$\bar{\xi}(\sigma')$ is a representative of $\{U_\xi : \xi \in B_{\sigma'}\}$, and
$\{U_{\bar{\xi}(\sigma')} : \sigma' < \sigma\}$ has the finite intersection property.

We set

$C = \{\bar{\xi}(\sigma') : \sigma' < \sigma\}$, and
$D = \{\xi \in A : \{U_\zeta : \zeta \in C \cup \{\xi\}\}$ has the finite intersection property$\}$.

Since

$$|C| \leq |\sigma| < \mathrm{cf}(\beta) < \kappa \leq \kappa^\kappa,$$

it follows from the characteristic property of the set A (as given in Lemma 5.22) that $|D| = \alpha$. We choose $B_\sigma \subset D$ so that $|B_\sigma| = \beta_\sigma$ and, using 5.1 (applied to B_σ in place of α), we choose a representative $\bar{\xi}(\sigma)$ of $\{U_\xi : \xi \in B_\sigma\}$. We note that since

$$\bar{\xi}(\sigma) \in B_\sigma \subset D,$$

the family $\{U_{\bar{\xi}(\sigma')} : \sigma' \leq \sigma\}$ has the finite intersection property.

The definition of the sets $\{B_\sigma : \sigma < \mathrm{cf}(\beta)\}, \{\bar{\xi}(\sigma) : \sigma < \mathrm{cf}(\beta)\}$ is complete. It is clear that conditions (i), (ii) and (iii) are satisfied.

We set $B = \cup_{\sigma < \mathrm{cf}(\beta)} B_\sigma$ and we note from Lemma 5.4, with α, $\mathrm{cf}(\alpha), \alpha_\sigma, \{U_\xi : \xi < \alpha\}$ and $\{U_\xi : \xi < \alpha_\sigma\}$ replaced by $\beta, \mathrm{cf}(\beta), \beta_\sigma$, $\{U_\xi : \xi \in B\}$ and $\{U_\xi : \xi \in B_\sigma\}$, respectively, that there is a set S (this is the set A in the statement of Lemma 5.4) such that $S \subset B$, $|S| = \beta$, and

$$\{U_\xi : \xi \in S\} \text{ has the finite intersection property.}$$

The proof is complete.

5.26 *Corollary*. Assume the generalized continuum hypothesis. Let α and κ be infinite cardinals and let X be a compact, Hausdorff space such that $S(X) = \kappa$. Then

(a) if $\mathrm{cf}(\alpha) < \kappa$ then X does not have calibre α;

(b) if $\mathrm{cf}(\alpha) > \kappa$, and if either $\mathrm{cf}(\alpha)$ is a limit cardinal or $\mathrm{cf}(\alpha) = \beta^+$ with $\mathrm{cf}(\beta) \geq \kappa$, then X has calibre α;

(c) if $\mathrm{cf}(\alpha) > \kappa$ and $\mathrm{cf}(\alpha) = \beta^+$ with $\mathrm{cf}(\beta) < \kappa$, then X has calibre (β^+, β) (and hence X has calibre (α, β)).

Proof. (a) It is clear that X does not have calibre $\mathrm{cf}(\alpha)$; the required statement follows from Theorem 2.3(a).

(b) It follows from the assumptions (using Theorem A.8(b) of the Appendix in case $\mathrm{cf}(\alpha) = \beta^+$ with $\mathrm{cf}(\beta) \geq \kappa$) that $\kappa \ll \mathrm{cf}(\alpha)$. If α is regular then $\kappa \ll \alpha$ and if α is singular then α is a (strong) limit cardinal and from $\kappa < \alpha$ it follows again that $\kappa \ll \alpha$. That X has calibre α now follows from Theorem 5.6.

(c) It follows from Theorem 5.25 (with $\mathrm{cf}(\alpha)$ replacing α) that X

has calibre $(\beta^+, \omega\cdot\sqrt[\kappa]{\mathrm{cf}(\alpha)})$. For $\gamma < \beta$ and $\lambda < \kappa$ we have

$$\gamma^\lambda \leq 2^\gamma \cdot 2^\lambda \leq \beta\cdot\kappa < \mathrm{cf}(\alpha)$$

and hence $\sqrt[\kappa]{\mathrm{cf}(\alpha)} \geq \beta$; hence X has calibre (β^+, β).

Examples on calibres

The example given below in 5.28 proves, together with 6.14 for $\kappa = \omega^+$, with Kunen's example (7.9 and the remark following), with the generalized examples of Gaifman and of Argyros (6.19, 6.23 and the Notes for Chapter 6), and with the Laver–Galvin example of 7.13, that Corollary 5.26 cannot be improved in any substantive way; see the diagrams of 7.19. The example of 5.28 will be studied further in 6.29 below; a modification, providing answers to questions related to strictly positive measures, appears in 6.25 and 6.26.

5.27 *Lemma.* Let $\alpha \geq \omega$ and let X be a set with topologies $\mathscr{T}_1$ and $\mathscr{T}_2$ such that there is a base for $\mathscr{T}_1$ consisting of $\mathscr{T}_2$-compact sets and such that

$$|X| = d(X, \mathscr{T}_1) = \alpha.$$

Then $\langle X, \mathscr{T}_1\rangle$ does not have precalibre α.

Proof. Let $\langle x_\xi : \xi < \alpha\rangle$ be a well-ordering of the elements of X and for $\xi < \alpha$ let V_ξ be a $\mathscr{T}_2$-compact, $\mathscr{T}_1$-open subset of X such that

$$V_\xi \cap \{x_\zeta : \zeta \leq \xi\} = \varnothing.$$

If $\langle X, \mathscr{T}_1\rangle$ has precalibre α there is $A \subset \alpha$ with $|A| = \alpha$ such that $\{V_\xi : \xi \in A\}$ has the finite intersection property, and since V_ξ is $\mathscr{T}_2$-compact for $\xi \in A$ there is

$$x = x_{\bar\zeta} \in \bigcap_{\xi\in A} V_\xi.$$

There is $\xi \in A$ with $\bar\zeta \leq \xi < \alpha$ and we have $x \notin V_\xi$, a contradiction.

Definitions of several terms used in the following theorem (tree, height, etc.) are given in Appendix A.

5.28 *Theorem* (Argyros). Assume the generalized continuum hypothesis. Let β be an (infinite) singular cardinal. There is a completely regular, Hausdorff space X such that

$$S(X) = (\mathrm{cf}(\beta))^+, \quad \text{and}$$
$$X \text{ does not have precalibre } \beta^+.$$

Proof. There is a set $\{\beta_\sigma : \sigma < \mathrm{cf}(\beta)\}$ such that

$$\beta_\sigma \text{ is a regular cardinal for } \sigma < \mathrm{cf}(\beta),$$
$$\omega \leq \beta_\sigma < \beta_\sigma^{++} \leq \beta_{\sigma+1} < \beta \quad \text{for } \sigma < \mathrm{cf}(\beta),$$
$$\prod_{\sigma<\tau} \beta_\sigma \leq \beta_\tau \quad \text{for limit ordinals } \tau \text{ with } 0 < \tau < \mathrm{cf}(\beta), \text{ and}$$
$$\sum_{\sigma < \mathrm{cf}(\beta)} \beta_\sigma = \beta.$$

Let $\{B_\sigma : \sigma < \mathrm{cf}(\beta)\}$ be a family of subsets of β such that

$$|B_\sigma| = \beta_\sigma \quad \text{for } \sigma < \mathrm{cf}(\beta),$$
$$B_\sigma \cap \beta_{\sigma'} = \varnothing \quad \text{for } \sigma < \sigma' < \mathrm{cf}(\beta), \quad \text{and}$$
$$\bigcup_{\sigma < \mathrm{cf}(\beta)} B_\sigma = \beta,$$

and for $\sigma < \mathrm{cf}(\beta)$ let $\{B_{\sigma+1,i} : i < \beta_\sigma\}$ be a family of subsets of $B_{\sigma+1}$ such that

$$|B_{\sigma+1,i}| = \beta_{\sigma+1}, \quad \text{and}$$
$$B_{\sigma+1,i} \cap B_{\sigma+1,i'} = \varnothing \quad \text{for } i < i' < \beta_\sigma.$$

We define a tree $\langle T, \preccurlyeq \rangle$ of height $\mathrm{cf}(\beta)$ with

$$T = \bigcup_{\sigma < \mathrm{cf}(\beta)} T_\sigma \subset [\beta]^2$$

We set $T_0 = [\beta_0]^2$.

Assume inductively that T_σ has been defined, with $T_\sigma \subset [B_\sigma]^2$ and $|T_\sigma| = \beta_\sigma$, and let $T_\sigma = \{s_i : i < \beta_\sigma\}$ be a one-to-one well-ordering of T_σ. We set

$$T_{\sigma+1} = \cup\{[B_{\sigma+1,i}]^2 : i < \beta_\sigma\}$$

and for $s = s_i \in T_\sigma$ and $t \in T_{\sigma+1}$ we define $s \prec t$ if and only if $t \in [B_{\sigma+1,i}]^2$.

Let τ be a limit ordinal with $0 < \tau < \mathrm{cf}(\beta)$ and assume that $\{T_\sigma : \sigma < \tau\}$ have been defined, together with the ordering $\preccurlyeq$ on $\cup_{\sigma<\tau} T_\sigma$, so that $\langle \cup_{\sigma<\tau} T_\sigma, \preccurlyeq \rangle$ is a tree such that

$$T_\sigma = \{s : \{t : t \prec s\} \quad \text{has order-type } \sigma\}.$$

Let Σ_τ be the set of branches of length τ of this tree, and set $\beta_\tau^* = |\Sigma_\tau|$.

It is clear that

$$\beta_\tau^* \leq \prod_{\sigma<\tau} \beta_\sigma \leq \beta_\tau.$$

Let $\{\Sigma_i : i < \beta_\tau^*\}$ be a one-to-one well-ordering of $\mathbf{\Sigma}_\tau$, and let $\{B_{\tau,i} : i < \beta_\tau^*\}$ be a family of subsets of B_τ such that

$$|B_{\tau,i}| = \beta_\tau \quad \text{for } i < \beta_\tau^*, \quad \text{and}$$
$$B_{\tau,i} \cap B_{\tau,i'} = \varnothing \quad \text{for } i < i' < \beta_\tau^*.$$

We set

$$T_\tau = \cup\{[B_{\tau,i}]^2 : i < \beta_\tau^*\},$$

and for $s \in \cup_{\sigma<\tau} T_\sigma$ and $t \in T_\tau$ we define $s \prec t$ if and only if there is $i < \beta_\tau^*$ such that $s \in \Sigma_i$ and $t \in [B_{\tau,i}]^2$.

The definition of the tree $\langle T, \preccurlyeq \rangle$ is complete. We denote by $\mathbf{\Sigma}$ the set of all branches of T.

For $\sigma < \mathrm{cf}(\beta)$ we set

$$A(\sigma) = \cup\{s \in T : s \in \bigcup_{\sigma \leq \sigma' < \mathrm{cf}(\beta)} T_{\sigma'}\}, \quad \text{and}$$
$$B(\sigma) = \beta \backslash A(\sigma).$$

For $s \in T$ we denote by K_s the set (of cardinality 2) of non-constant functions in $\{0, 1\}^s$; that is, if $s = \{\eta, \zeta\}$ then

$$K_s = \{\{\langle \eta, 1\rangle, \langle \zeta, 0\rangle\}, \{\langle \eta, 0\rangle, \langle \zeta, 1\rangle\}\}.$$

For $\Sigma \in \mathbf{\Sigma}$ we set

$$A(\Sigma) = \cup\{s : s \in \Sigma\}, \quad \text{and}$$
$$V_\Sigma = \left(\prod_{s\in\Sigma} K_s\right) \times \{0, 1\}^{\beta\backslash A(\Sigma)}$$

We now define the space $\langle X, \mathscr{T} \rangle$. As set, X is the set $\{0, 1\}^\beta$; the topology $\mathscr{T}$ is defined by the subbase that contains

(i) all subsets of $\{0, 1\}^\beta$ of the form $\pi_J^{-1}(S) \times \{0, 1\}^{\beta\backslash J}$ with $|J| < \mathrm{cf}(\beta)$ and with S compact in the usual product topology of $\{0, 1\}^J$; and

(ii) all sets of the form V_Σ for $\Sigma \in \mathbf{\Sigma}$.

The base $\mathscr{B}$ determined by this subbase consists of all sets of the form $U \cap (\cap_{j=1}^m V_{\Sigma_j})$ with $U = \pi_J^{-1}(S) \times \{0, 1\}^{\beta\backslash J}$ as above and with $\{\Sigma_j : 1 \leq k \leq m\} \subset \mathbf{\Sigma}$. The elements of $\mathscr{B}$ are closed in the usual topology of $\{0, 1\}^\beta$, hence in its extension $\mathscr{T}$; thus $\langle X, \mathscr{T} \rangle$ is a completely regular, Hausdorff space.

We say that the basic set $V = U \cap (\cap_{j=1}^m V_{\Sigma_j})$ is *determined by* $(U; \Sigma_1, \ldots, \Sigma_m)$, and we say that V is *separated at level* σ if

(i) $(\Sigma_{j_1} \cup T_\sigma) \cap (\Sigma_{j_2} \cup T_\sigma) = \emptyset$ for $1 \le j_1 < j_2 \le m$, and

(ii) there is $J \subset \beta$ with $U = \pi_J^{-1}(\pi_J[U])$ such that $J \cap A(\sigma) = \emptyset$.

If $V = U \cap (\cap_{j=1}^m V_{\Sigma_j}) \in \mathscr{B}$ and V is separated at level σ we set

$$V(0, \sigma) = \pi_{B(\sigma)}[V] \quad \text{and}$$

$$V(1, \sigma) = \bigcap_{j=1}^{m} \left(\left(\prod_{s \in \Sigma_j, s \subset A(\sigma)} K_s \right) \times \{0, 1\}^{A(\sigma) \setminus A(\Sigma_j)} \right).$$

We note that $V = V(0, \sigma) \times V(1, \sigma)$. We note also for $|J| < \mathrm{cf}(\alpha)$ that $\{0, 1\}^J$ is discrete in the $\mathrm{cf}(\alpha)$-box topology; thus $V(0, \sigma)$ is open-and-closed in the $\mathrm{cf}(\alpha)$-box topology of $(\{0, 1\}^{B(\alpha)})_{\mathrm{cf}(\alpha)}$.

For $\sigma < \mathrm{cf}(\beta)$ we set

$$U(\sigma) = \{x \in X : x_\eta = 0 \quad \text{for } \eta < \sigma, x_\sigma = 1\}.$$

Then $\{U(\sigma) : \sigma < \mathrm{cf}(\beta)\}$ is a faithfully indexed cellular family in the $\mathrm{cf}(\beta)$-box topology of $\{0, 1\}^\beta$, hence in the (larger) topology $\mathscr{T}$ of X; thus $S(X) > \mathrm{cf}(\beta)$.

To complete the proof that $S(X) = (\mathrm{cf}(\beta))^+$ we show that in fact X has property $\mathrm{K}_{(\mathrm{cf}(\beta))^+, 2}$. Let $\{V_\eta : \eta < (\mathrm{cf}(\beta))^+\} \subset \mathscr{B}$; we assume without loss of generality that there is a (fixed) $\sigma < \mathrm{cf}(\beta)$ such that each of the sets V_η is separated at level σ. For $J \subset B(\sigma)$ with $|J| < \mathrm{cf}(\beta)$ we have from the generalized continuum hypothesis that $|\{0, 1\}^J| \le \mathrm{cf}(\beta)$ and hence $\{0,1\}^J$ has calibre $(\mathrm{cf}(\beta))^+$. Further from Theorem A.8(c) of Appendix A (with α replaced by $\mathrm{cf}(\beta)$) we have from the generalized continuum hypothesis that $\mathrm{cf}(\beta) \ll (\mathrm{cf}(\beta))^+$. It then follows from Theorem 3.6(a), with α, β and γ replaced by $(\mathrm{cf}(\beta))^+$ and with κ, I, X_i and Y replaced by $\mathrm{cf}(\beta)$, $B(\sigma)$, $\{0,1\}$ and $\{0,1\}^{B(\sigma)}$ respectively, that $\{0, 1\}^{B(\sigma)}$ has calibre $(\mathrm{cf}(\beta))^+$. Thus there is $A \subset (\mathrm{cf}(\beta))^+$ with $|A| = (\mathrm{cf}(\beta))^+$ such that

$$\bigcap_{\eta \in A} V_\eta(0, \sigma) \neq \emptyset.$$

If $\eta_1, \eta_2 \in A$ with $\eta_1 \neq \eta_2$ then every branch of T that determines V_{η_1} intersects at most one of the branches of T that determines V_{η_2} at level greater than or equal to σ; hence

$$V_{\eta_1}(1, \sigma) \cap V_{\eta_2}(1, \sigma) \neq \emptyset.$$

It follows that every two elements of the family $\{V_\eta : \eta \in A\}$ have non-empty intersection, as required.

It remains to show that X does not have precalibre β^+. Since the elements of the base $\mathscr{B}$ of X are compact in the usual product topology of $\{0, 1\}^\beta$, and since

$$|X| = 2^\beta = \beta^+$$

it is enough, according to Lemma 5.27, to show that $d(X) = \beta^+$.

Let $\{x_\xi : \xi < \beta\} \subset X$. We define a branch

$$\Sigma = \{s(\sigma) : \sigma < \mathrm{cf}(\beta)\} \in \mathbf{\Sigma}$$

such that if $\xi \in B_\sigma \subset \beta$ then $x_\xi \notin K_{s(\sigma)} \times \{0,1\}^{\beta \setminus s(\sigma+1)}$.

We let $s(0) \in T_0$.

If $\sigma < \mathrm{cf}(\beta)$ and $s(\sigma)$ has been defined there is $i < \beta_\sigma$ such that $s(\sigma) = s_i$ in the ordering of T_σ. For $\zeta \in B_{\sigma+1,i}$ we set

$$A_\zeta = \{\xi \in B_\sigma : x_\xi(\zeta) = 1\}$$

and we note that since $A_\zeta \in \mathscr{P}(B_\sigma)$ and

$$|\mathscr{P}(B_\sigma)| = 2^{\beta_\sigma} = \beta_\sigma^+ < \beta_{\sigma+1} = |B_{\sigma+1,i}|$$

there are distinct elements $\zeta_0, \zeta_1 \in B_{\sigma+1,i}$ such that $A_{\zeta_0} = A_{\zeta_1}$. We note that

$$x_\xi(\zeta_0) = x_\xi(\zeta_1) \quad \text{for all } \xi \in B_\sigma,$$

and we set $s(\sigma+1) = \{\zeta_0, \zeta_1\}$.

For non-zero limit ordinals $\tau < \mathrm{cf}(\beta)$ we choose for $s(\tau)$ any element t of T_τ such that

$$s(\sigma) \prec t \quad \text{for all } \sigma < \tau.$$

The definition of the branch $\Sigma = \{s(\sigma) : \sigma < \mathrm{cf}(\beta)\}$ is complete. It is clear that

$$B_\sigma \cap K_{s(\sigma)} \times \{0, 1\}^{\beta \setminus s(\sigma+1)} = \varnothing,$$

and hence that

$$V_\Sigma \cap \{x_\xi : \xi < \beta\} = \varnothing.$$

It follows that $\{x_\xi : \xi < \beta\}$ is not dense in X.

The proof is complete.

5.29 *Corollary.* Assume the generalized continuum hypothesis. Let κ and β be cardinals with κ regular and with $\mathrm{cf}(\beta) < \kappa < \beta$. Then there is an extremally disconnected, compact Hausdorff space Ω such that

$$S(\Omega) = \kappa, \quad \text{and}$$
$$\Omega \text{ does not have calibre } \beta^+.$$

Proof. From Theorem 5.28 there is a space X_0 such that $S(X_0) = (\text{cf}(\beta))^+$ and X_0 does not have precalibre β^+, and from Theorem 3.30 there is a (completely regular, Hausdorff) space X_1 such that $S(X_1) = \kappa$. Clearly the disjoint union $X = X_0 \cup X_1$ does not have precalibre β^+ and satisfies $S(X) = \kappa$; we set $\Omega = G(X)$ (the Gleason space of X) and we note from Corollaries 2.18 and 2.22 that Ω is as required.

5.30 *Problem* (Argyros). Can one define in ZFC (without the continuum hypothesis or other special set-theoretic assumptions) a compact c.c.c. space that does not have calibre 2^ω?

In 6.18 below we note (an improvement on) an observation of Erdős: Assuming $\omega^+ = 2^\omega$ there is a compact Hausdorff space with a strictly positive measure (hence, a c.c.c. space) that does not have calibre ω^+.

Notes for Chapter 5

The material of 5.1 – 5.3, and Theorem 5.6 for regular cardinals α, is due to Argyros (1977a, b); 5.4 and 5.5, together with Theorem 5.6 for singular cardinals α, are due to Tsarpalias (1978a, b). Corollaries 5.8 – 5.10, which are consequences of Theorem 5.6, are from Argyros & Tsarpalias (1978a, b, 1981).

The results in the section concerning P_{λ^+}-spaces and λ^+-box products (5.12–5.21) were noted by Negrepontis. Corollary 5.13 strengthens a result of Juhász (1971, Theorem 2.11), while Corollary 5.18 strengthens a result of Kurepa (1962) and Hedrlín (1966).

For $\kappa \geq \omega$ and X a space we set

$$\mathscr{G}_\kappa(X) = \{\cap \mathscr{U} : \mathscr{U} \subset \mathscr{T}(X), |\mathscr{U}| < \kappa\}.$$

We note a consequence of Theorem 5.12.

Theorem. Let $\omega \leq \lambda < \kappa \ll \alpha$ with α and κ regular, let X be a compact Hausdorff space such that $S(X) \leq \kappa$, let $\mathscr{F}$ be a set of continuous functions from X to a space Y such that $|Y| < \alpha$ and $\{p\} \in \mathscr{G}_{\lambda^+}(Y)$ for all $p \in Y$, and let $\Phi: \mathscr{F} \to Y$ with $\Phi(f) \in f[X]$ for $f \in \mathscr{F}$. If $|\mathscr{F}| \geq \alpha$ then there are $\mathscr{F}' \subset \mathscr{F}$ and $p \in Y$ such that

$$|\mathscr{F}'| = \alpha,$$
$$\Phi(f) = p \quad \text{for } f \in \mathscr{F}', \text{ and}$$
$$\cap\{f^{-1}(\{p\}): f \in \mathscr{F}'\} \neq \emptyset.$$

Proof. Since $|Y| < \alpha$ and α is regular, we assume without loss of generality that there is $p \in Y$ such that

$$\Phi(f) = p \quad \text{for } f \in \mathscr{F}.$$

Since $f^{-1}(\{p\}) \in \mathscr{G}_{\lambda^+}(X) \subset \mathscr{T}(X_{(\lambda^+)})$ for $f \in \mathscr{F}$ and $X_{(\lambda^+)}$ has calibre α (by Theorem 5.12), there is $\mathscr{F}'$ as required.

Taking $\lambda = \omega$, $\kappa = \omega^+$, $\alpha = (2^\omega)^+$, $Y = \mathbb{R}$ and $\Phi = \sup$ we have the following statement: If G is a compact topological group and $\mathscr{F}$ a set of continuous functions from G to $\mathbb{R}$ with $|\mathscr{F}| > 2^\omega$, then there are $\mathscr{F}' \subset \mathscr{F}$, $\bar{x} \in G$ and $p \in \mathbb{R}$ such that

$$|\mathscr{F}'| = (2^\omega)^+ \quad \text{and}$$
$$f(\bar{x}) = \sup\{f(x): x \in G\} = p \quad \text{for } f \in \mathscr{F}'.$$

(That $S(G) = \omega^+$ follows from the fact that the Haar measure μ of G satisfies $\mu(G) = 1$ and $\mu(U) > 0$ for every non-empty, open subset U of G.) This property of compact groups (with $|\mathscr{F}'| = 2$) was observed several years ago by Juhász.

Lemma 5.22, Theorem 5.25 and Corollary 5.26 are due to Argyros & Tsarpalias (1978b, 1981)

Lemma 5.23 is noted by Rosenthal (1970a); the proof given here is due to Tsarpalias.

The results from functional analysis concerning the isomorphic embeddability of l^1_α into the Banach spaces $C(X)$, discussed briefly in the introduction to this section, are given in Argyros (1977a, b), Tsarpalias (1978a, b), and Argyros & Tsarpalias (1978b). Analogous but more delicate results in functional analysis are available, using (in place of compact Hausdorff spaces X) P_{λ^+}-spaces and λ^+-box products (as in 5.12–5.21 above).

Theorem 5.28, which establishes (assuming the generalized continuum hypothesis) a conjecture of Shelah (1980), is due to Argyros (1981c).

On the existence of independent families

The methods used in this chapter to determine calibres of compact spaces are closely connected with methods used to treat the following question. Let X be a totally disconnected, compact Hausdorff space of weight α; under what conditions is there a continuous function from X onto $\{0, 1\}^\alpha$ (or, less strictly, onto $\{0, 1\}^\beta$ for some cardinal $\beta \leq \alpha$ as close to α as possible)?

A family $\{\langle A_\xi, B_\xi \rangle : \xi < \beta\}$ of ordered pairs of subsets of X, with $A_\xi \cap B_\xi = \varnothing$ for $\xi < \beta$, is called an *independent family* (*of length* β) if for every finite subset F of ω and every function $\varepsilon: F \to \{-1, +1\}$, we have

$$\bigcap_{\xi \in F} \varepsilon_\xi A_\xi \neq \varnothing$$

(where $(+1)A_\xi = A_\xi, (-1)A_\xi = B_\xi$ for $\xi < \beta$).

It is clear that the existence of a continuous function from X onto $\{0, 1\}^\beta$ is equivalent to the existence of an independent family $\{\langle A_\xi, B_\xi \rangle : \xi < \beta\}$ of length β such that A_ξ, B_ξ are closed in X for $\xi < \beta$.

We describe without proofs the current status of questions concerning the existence of continuous functions from X onto $\{0, 1\}^{wX}$. The conditions which arise are, in general, remarkably similar to those encountered in the study of calibers of compact spaces. There is, however, a striking exception: while the occurrence of the universal property of calibre has nothing to do with the completeness of the space (in the sense that a compact space X has calibre α if and only if its Gleason space $G(X)$ has calibre α), the occurrence of the (non-universal) property of the existence of a continuous function from X onto $\{0, 1\}^\alpha$ is influenced in a positive sense by the degree of completeness of X (cf. the theorem of Balcar & Franěk mentioned below in these Notes).

Theorem. Let α and κ be regular uncountable cardinals. If there is a compact, Hausdorff space Y such that $S(Y) = \kappa$ and $w(Y) \geq \alpha$, and if α is not a calibre of Y, then there is a space X such that

$S(X) = \kappa$ and $w(X) = \alpha$, and

there is no continuous function from X onto $\{0, 1\}^\alpha$.

(This result is due to Argyros (1981b, Theorem 1.9).)

Corollary. Assume the generalized continuum hypothesis, and let α and κ be regular uncountable cardinals.

(a) If there is β such that $\alpha = \kappa = \beta^+$, then there is a totally disconnected, compact Hausdorff space X such that

$S(X) = w(X) = \kappa$, and

there is no continuous function from X onto $\{0, 1\}^\kappa$.

(b) Assume in addition that V = L, that $\alpha = \kappa$, and that the cardinal α is strongly inaccessible but not weakly compact. Then there is a totally disconnected, compact Hausdorff space X such that

$S(X) = w(X) = \kappa$, and

there is no continuous function from X onto $\{0, 1\}^\kappa$.

(c) If there is a cardinal number β such that $\text{cf}(\beta) < \kappa < \beta < \beta^+ = \alpha$, then there is a totally disconnected, compact Hausdorff space X such that

$S(X) = \kappa$ and $w(X) = \alpha$, and

there is no continuous function from X onto $\{0, 1\}^\alpha$.

(For (a) we use the generalized Gaifman example (6.23, Notes for Chapter 6), or the generalized Argyros example (6.26, Notes for Chapter 6), or the example of Kunen (7.9), or the example of Laver & Galvin (7.13), or for $\kappa = \omega^+$ the example of Erdős (6.18). For (b) we use the example of Jensen (Notes for Chapter 7). For (c) we use Argyros' example (5.28), or for $\kappa = \omega^+$ the example of Argyros & Tsarpalias (6.18).

Theorem. Let α and κ be regular cardinals with $\kappa \ll \alpha$, and let X be a totally disconnected, compact Hausdorff space such that $S(X) = \kappa$. Then there is a continuous function from X onto $\{0, 1\}^\alpha$.

(This result has been proved by Argyros & Tsarpalias (1978a).)

Corollary. Assume the generalized continuum hypothesis. If either $\kappa \leq \text{cf}(\beta) \leq \beta < \beta^+ = \alpha$ or α is a regular limit cardinal such that $\alpha > \kappa$, and if X is a totally disconnected, compact Hausdorff space with $S(X) = \kappa$ and $w(X) = \alpha$, then there is a continuous function from X onto $\{0, 1\}^\alpha$.

Theorem. If κ is a weakly compact cardinal and X is a totally disconnected, compact Hausdorff space such that $S(X) = w(X) = \kappa$, then there is a continuous function from X onto $\{0, 1\}^\kappa$.

(This result follows directly from the methods of Argyros & Tsarpalias (1978a, 1981); it has also been noted by Shelah (1980).)

We turn now to results concerning singular cardinals.

Theorem. Assume the generalized continuum hypothesis, and let α be an (infinite) singular cardinal. If X is a totally disconnected, compact Hausdorff space such that

$$w(X) \geq \alpha \quad \text{and } \operatorname{cf}(\alpha) \text{ is a calibre of } X,$$

then there is a continuous function from X onto $\{0, 1\}^\alpha$.

(This result is due to Tsarpalias (1978); cf. also Argyros & Tsarpalias (1978a, 1981).)

Theorem. Let α be an (infinite) singular cardinal. If there is a totally disconnected, compact Hausdorff space Y such that $S(Y) = \kappa$ and $w(Y) \geq \alpha$, and if $\operatorname{cf}(\alpha)$ is not a calibre of Y, then there is a totally disconnected, compact Hausdorff space X such that

$S(X) = \kappa$ and $w(X) = \alpha$, and

there is no continuous function from X onto $\{0, 1\}^\alpha$.

(This result, as well as several of the theorems of Argyros & Tsarpalias (concerning precalibres, calibres of compact spaces, and independent families) described in Chapter 5 and above in these Notes, has been obtained by Shelah (1980, Section 4.7).)

Theorem. Let α be an infinite cardinal. If X is an extremally disconnected, compact Hausdorff space with $w(X) = \alpha$, then there is a continuous function from X onto $\{0,1\}^\alpha$.

(This result is due to Balcar & Franěk (1981).)

Definition. Let X be a space and $x \in X$. The *pseudoweight* of X at x, denoted $\pi(X, x)$, is the least cardinal number α for which there is a set $\mathscr{B}$ of non-empty, open subsets of X such that

$|\mathscr{B}| = \alpha$, and

if U is open in X and $x \in U$ then there is $B \in \mathscr{B}$ such that $B \subset U$.

Theorem. Let α be an infinite cardinal and X a totally disconnected, compact Hausdorff space. There is a continuous function from X onto $\{0, 1\}^{\alpha}$ if and only if there is a non-empty, closed subset F of X such that $\pi(F, x) \geq \alpha$ for all $x \in F$.

(This result is due to Šapirovskiĭ (1975); see also Juhász (1980, Section 3.18) for a complete account.)

The results described above provide, at least assuming the generalized continuum hypothesis, a nearly complete answer to the question of the existence of continuous functions from totally disconnected, compact Hausdorff spaces onto spaces of the form $\{0, 1\}^{\alpha}$. There remains to be considered the question of the highest degree of completeness possible for a space X (as above) that does not map continuously onto $\{0, 1\}^{w(X)}$. More specifically, we formulate a (standard) definition and pose a question.

Definition. Let β be an infinite cardinal and X a space with a base of open-and-closed subsets. The space X is a *β-extremally disconnected* if for every set U that is the union of fewer than β open-and-closed subsets of X, the set $\mathrm{cl}_X U$ is open in X.

Problem. Let α and κ be regular, uncountable cardinals for which there is a totally disconnected, compact Hausdorff space X such that $S(X) = \kappa$, $w(X) \geq \alpha$, and α is not a calibre of X. What is the least cardinal β for which there is no β-extremally disconnected, compact Hausdorff space X such that

$S(X) = \kappa$ and $w(X) = \alpha$, and

there is no continuous function from X onto $\{0, 1\}^{\alpha}$?

(We note that it follows from the theorem of Balcar & Franěk that $\beta \leq \alpha$.)

The various results and questions above are easily modified to yield analogues concerning the existence of continuous functions from compact Hausdorff spaces onto spaces of the form $[0, 1]^{\alpha}$.

On the isomorphic theory of Banach spaces

Some of the results cited above have significant applications to certain central questions in the isomorphic theory of Banach spaces.

We now describe some of these applications. We note that most of the results concerning the isomorphic embeddability of the spaces $l^1(\Gamma)$ into Banach spaces of the form $C(X)$ depend on a criterion of Rosenthal (1974b) involving the existence in X of an independent family of length $|\Gamma|$.)

Theorem. Let α and κ be (infinite) regular cardinals with $\kappa \ll \alpha$ and let X be a compact, Hausdorff space such that $S(X) \leq \kappa$. If B is a closed linear subspace of $C(X)$ such that $d(B) \geq \alpha$, then there is an isomorphic embedding of $l^1(\Gamma)$ into B.

(This follows from the methods of Argyros (1977a, b), Tsarpalias (1978a, b), and Argyros & Tsarpalias (1978b). Analogous but more delicate results, using (in place of compact Hausdorff spaces X) P_{λ^+}-spaces and λ^+-box products as in 5.12–5.21 above, have been given by N. Kalamidas (1981), Th. Zachariades (1981), and Negrepontis (1980).)

The theorem just cited, together with the results of Rosenthal (1970a, p. 230) connecting the Souslin number of a compact space X with the density character of the weakly compact subsets of the Banach space $C(X)$, suggests the following question; here for a Banach space B we denote by $\Sigma(B)$ the cardinal number

$$\Sigma(B) = \min\{\lambda\colon \text{no weakly compact } K \subset B \text{ satisfies } d(K) \geq \lambda\}.$$

Problem (Argyros & Negrepontis). Let B be a Banach space. Set $\alpha = \dim B$ and $\kappa = \Sigma(B)$ and suppose $\kappa \ll \alpha$ and $\kappa \ll \mathrm{cf}(\alpha)$. Does it follow that there is an embedding of $l^1(\alpha)$ into B?

It is clear from the foregoing results that for a totally disconnected, compact Hausdorff space X, these three properties are closely related: X has calibre α; there is a continuous function from X onto $\{0, 1\}^\alpha$; and there is an isomorphic embedding of $l^1(\alpha)$ into every closed, linear subspace B of $C(X)$ such that $d(B) \geq \alpha$. The following theorem limits our attempts to prove general theorems stronger than those mentioned above.

Theorem. Assume the continuum hypothesis. There is a totally disconnected, compact Hausdorff c.c.c. space X such that X does not have calibre ω^+, there is no continuous function from X onto

$\{0, 1\}^{\omega^+}$, and there is an isomorphic embedding of $l^1(\omega^+)$ into $C(X)$ – indeed for every closed, linear non-separable subspace B of $C(X)$ such that B is generated by characteristic functions of open-and-closed subsets of X, there is an isomorphic embedding of $l^1(\omega^+)$ into B.

(This example, due to Argyros & Kalamidas, appears in Kalamidas (Doctoral dissertation, 1981); a similar result has been obtained by R. Haydon (unpublished).)

A Banach space B is *injective* if for every Banach space E and every closed, linear subspace F of E, every bounded linear transformation $T:F \to B$ extends to a bounded linear transformation $S:E \to B$. If in addition it is required that $\|S\| = \|T\|$, then B is called a $\mathscr{P}_1$-*space*.

It is known that a Banach space B is a $\mathscr{P}_1$-space if and only if B is isomorphic to $C(X)$ for some extremally disconnected, compact Hausdorff space X (cf. Goodner 1950, Kelley 1954, Nachbin 1950).

No characterization of injective Banach spaces is known, and in fact there is no known example of an injective Banach space that is not a $\mathscr{P}_1$-space; Rosenthal (1970b) has remarked, however, that if there is an Ulam-measurable cardinal then there is a compact F-space X such that $C(X)$ is an injective Banach space and is not a $\mathscr{P}_1$-space.

It is conjectured by Rosenthal (1970a) that if B is an injective Banach space with $d(B) = \alpha$, then there is an isomorphic embedding of $l^1(\alpha)$ into B. The theorem of Balcar and Franěk mentioned above settles this conjecture affirmatively for $\mathscr{P}_1$-spaces B; additional interesting partial results are given by Argyros (1981a).

A more detailed survey of some of the applications of independent families to the isomorphic theory of Banach spaces is given by Negrepontis (1980). Other recent surveys in the isomorphic theory of Banach spaces, dealing in part with techniques involving infinitary combinatorics, are given by Rosenthal (1975, 1978), Pełczynski (1979), and Haydon (1980).

6

Strictly Positive Measures

We study in this chapter the condition of the existence of a strictly positive measure and its relation to the countable chain condition and to related conditions of calibre. A result of Kelley (Theorem 6.4) gives a condition (Kelley's property (**)) on the set of non-empty, open subsets of a compact, Hausdorff space which is necessary and sufficient for the existence of a strictly positive measure; we show by simple examples, given by Argyros, van Douwen, Koumoullis and Sapounakis, that not every completely regular, Hausdorff (non-compact) space satisfying Kelley's property has a strictly positive measure (Theorems 6.35 and 6.37).

The principal chain conditions satisfied by spaces with a strictly positive measure are the properties $K_{\alpha,n}$ for every cardinal α with $\mathrm{cf}(\alpha) > \omega$ and every natural number $n \geq 2$ (this is Theorem 6.15 and Corollary 6.17, of Argyros & Kalamidas). On the other hand there is a compact Hausdorff space with a strictly positive (regular, Borel) measure which, assuming the continuum hypothesis, does not have calibre ω^+. This example, noted by Erdős, is the Stone space of the Lebesgue measure algebra on $[0, 1]$. The generalization given here in Corollary 6.19 provides another example (supplementary to the space of Theorem 5.28) on the class of calibres of compact spaces.

We give three compact c.c.c. spaces which have no strictly positive measure and which, assuming (when necessary) the continuum hypothesis, have various differing calibre properties. These are as follows.

(a) A compact Hausdorff space with the countable chain condition, indeed with condition (*) defined below (a condition implied by Kelley's condition), but with no strictly positive measure. This example is due to Gaifman (Theorem 6.23). We show in addition that this space satisfies condition K_n for $2 \leq n < \omega$ and, assuming the continuum hypothesis, that it does not have calibre ω^+.

(b) For every natural number $n \geq 2$, a compact Hausdorff space with condition (*), without a strictly positive measure, satisfying condition K_n, and, assuming the continuum hypothesis, not satisfying condition K_{n+1} (Argyros, Theorem 6.26).

(c) A compact Hausdorff space with calibre ω^+ and not satisfying the (ω, ω)-chain condition (hence with no strictly positive measure). This example, due to Galvin and Hajnal, graph-theoretic and combinatorial in nature, is given in Theorem 6.32.

Once it is recognized that a compact c.c.c. space need not have a strictly positive measure it is reasonable to ask if there is a cardinal number α with this property: For every compact c.c.c. space X there is a family $\{\mu_i : i < \alpha\}$ of regular, Borel measures on X such that for all $U \in \mathscr{T}^*(X)$ there is $i < \alpha$ satisfying $\mu_i(U) > 0$. An example of Argyros (essentially the example of 5.28 above) answers this question in the negative. A different example with similar properties can be obtained by modification of the space of Galvin and Hajnal defined in 6.32.

Characterization of spaces with a strictly positive measure

Several terms which appear in the following definition are defined and discussed in Appendix C.

Definition. A space X *has a strictly positive measure* if there are a pseudobase $\mathscr{B}$ for the topology of X and a probability measure μ, defined on the σ-field generated by $\mathscr{B}$, such that

$$\mu(B) > 0 \text{ for all non-empty } B \in \mathscr{B}.$$

Definition. A space X has *property* (*) if the set $\mathscr{T}^*(X)$ of non-empty, open subsets of X can be written in the form

$$\mathscr{T}^*(X) = \bigcup_{n<\omega} \mathscr{T}_n$$

where, for $n < \omega$, no $n+2$ elements of the family $\mathscr{T}_n$ are pairwise disjoint.

We note that a space with a strictly positive measure μ has property (*). Indeed, if $\mathscr{B}$ is a pseudobase for the topology of X such that

$\mu(B) > 0$ for every non-empty $B \in \mathscr{B}$ and for $n < \omega$ we set

$\mathscr{T}_n = \{U \in \mathscr{T}^*(X)$: there is $B \in \mathscr{B}$ with $B \subset U$ and $\mu(B) \geq 1/(n+1)\}$, then the relation

$$\mathscr{T}^*(X) = \bigcup_{n<\omega} \mathscr{T}_n$$

expresses $\mathscr{T}^*(X)$ in the required form.

Definition. Let X be a space.

(a) If $\mathscr{F} = \{U_1, \dots, U_N\}$ is a non-empty, finite family of non-empty, open subsets of X, indexed (not necessarily faithfully) by $\{1, \dots, N\}$, then cal $\mathscr{F}$ denotes the largest integer k such that there is $S \subset \{1, \dots, N\}$ with $|S| = k$ such that $\bigcap_{i \in S} U_i \neq \emptyset$.

(b) If $\mathscr{S} \subset \mathscr{P}^*(X)$, then

$$\kappa(\mathscr{S}) = \inf\{(\text{cal } \mathscr{F})/N : 1 \leq N < \omega, \mathscr{F} = \{U_1, \dots, U_N\} \subset \mathscr{S}\}.$$

(c) X has *property* (**) if the set $\mathscr{T}^*(X)$ of non-empty, open subsets of X can be written in the form

$$\mathscr{T}^*(X) = \bigcup_{n<\omega} \mathscr{T}_n$$

with $\kappa(\mathscr{T}_n) > 0$ for $n < \omega$.

Intuitively, property (*) is a finite version of a chain-type condition, while property (**) is a corresponding finite version of a calibre-type condition. The relation between the two, given in Lemma 6.2(b), is as expected.

6.1 *Lemma.* Let $(X, \mathscr{S}, \mu)$ be a probability measure space and $\mathscr{A}$ a family of (μ-measurable) subsets of X such that $\mu(A) > 0$ for $A \in \mathscr{A}$, and let $n < \omega$ and

$$\mathscr{U} \subset \{U \in \mathscr{P}^*(X): \text{there is } A \in \mathscr{A}, A \subset U, \mu(A) \geq 1/(n+1)\}.$$

Then $\kappa(\mathscr{U}) \geq 1/(n+1)$.

Proof. Let $\mathscr{F} = \{U_i : 1 \leq i \leq N\} \subset \mathscr{U}$ with $N < \omega$ and choose $A_i \in \mathscr{A}$ with $A_i \subset U_i$ and $\mu(A_i) \geq 1/(n+1)$ for $1 \leq i \leq N$.
For $S \subset \{1, 2, \dots, N\}$ we set $S' = \{1, 2, \dots, N\} \backslash S$ and we set

$$V_S = (\bigcap_{i \in S} A_i) \cap (\bigcap_{i \in S'} (X \backslash A_i)).$$

We note that if $S, T \subset \{1, 2, \dots, N\}$ and $S \neq T$ then $V_S \cap V_T = \emptyset$.

We note further that

$$A_i = \cup\{V_S : i \in S\}$$

(indeed if $x \in A_i$ and $S(x) = \{j : x \in A_j\}$ then $i \in S(x)$ and $x \in V_{S(x)}$; and $V_S \subset A_i$ when $i \in S$). It follows that

$$\mu(A_i) = \sum\{\mu(V_S) : i \in S\} \quad \text{for } 1 \leq i \leq N,$$

and hence

$$N \cdot 1/(n+1) \leq \sum_{i=1}^{N} \mu(A_i) = \sum_{i=1}^{N} [\sum\{\mu(V_S) : i \in S\}];$$

for $V_S \neq \varnothing$ the summand $\mu(V_S)$ appears on the right-hand side with multiplicity $|S|$, and

$$|S| \leq \text{cal}\,\mathscr{F} \text{ when } V_S \neq \varnothing.$$

Thus we have

$$\begin{aligned} N \cdot 1/(n+1) &\leq (\text{cal}\,\mathscr{F}) \cdot \sum\{\mu(V_S) : S \subset \{1, 2, \ldots, N\}\} \\ &\leq (\text{cal}\,\mathscr{F}) \cdot 1 = \text{cal}\,\mathscr{F} \end{aligned}$$

and hence

$$(\text{cal}\,\mathscr{F})/N \geq 1/(n+1).$$

6.2 *Lemma.* Let X and Y be spaces.

(a) If X has a strictly positive measure then X has property (**).

(b) If X has property (**) then X has property (*).

(c) If $X \leq Y$ and Y has property (*), then X has property (*); in particular, X has property (*) if and only if its Gleason space ($G(X)$ has property (*).

(d) If $X \leq Y$ and Y has property (**) then X has property (**); in particular, X has property (**) if and only if its Gleason space $G(X)$ has property (**).

Proof. (a) Let μ be a strictly positive measure on X. Let $\mathscr{B}$ be a pseudobase for X such that $\mu(B) > 0$ for every non-empty $B \in \mathscr{B}$, and for $n < \omega$ set

$$\mathscr{T}_n = \{U \in \mathscr{T}^*(X) : \text{there is } B \in \mathscr{B} \text{ with } B \subset U \text{ and } \mu(B) \geq 1/(n+1)\}.$$

Then $\mathscr{T}^*(X) = \cup_{n<\omega} \mathscr{T}_n$, and from Lemma 6.1 (with $\mathscr{A}$ and $\mathscr{U}$ replaced by $\mathscr{B}$ and $\mathscr{T}_n$, respectively) it follows that $\kappa(\mathscr{T}_n) \geq 1/(n+1)$ for $n < \omega$.

(b) Let $\mathcal{T}^*(X) = \cup_{n<\omega}\mathcal{U}_n$ with $\kappa(\mathcal{U}_n) > 0$ for $n < \omega$, and let $k(n)$ satisfy

$$\kappa(\mathcal{U}_n) > 1/(k(n)+2) \quad \text{and}$$
$$k(n) \neq k(n') \quad \text{for } n < n' < \omega.$$

We claim that no $k(n)+2$ elements of $\mathcal{U}_n$ are pairwise disjoint. Indeed if $\mathcal{F} \subset \mathcal{U}_n$ and $|\mathcal{F}| = k(n)+2$ and the elements of $\mathcal{F}$ are pairwise disjoint then $\operatorname{cal}\mathcal{F} = 1$ and we have

$$1/(k(n)+2) < \kappa(\mathcal{U}_n) \leq \operatorname{cal}\mathcal{F}/(k(n)+2) = 1/(k(n)+2),$$

a contradiction.

Finally we set

$$\mathcal{T}_{k(n)} = \mathcal{U}_n \quad \text{for } n < \omega, \quad \text{and}$$
$$\mathcal{T}_m = \varnothing \quad \text{if there is no } n < \omega \text{ with } m = k(n).$$

It is clear that

$$\mathcal{T}^*(X) = \bigcup_{n<\omega} \mathcal{T}_n, \quad \text{and}$$

no $n+2$ elements of the family $\mathcal{T}_n$ are pairwise disjoint.

Hence X has property (*).

The proof is complete.

(c) Since Y has property (*), the set $\mathcal{T}^*(Y)$ can be written in the form

$$\mathcal{T}^*(Y) = \bigcup_{n<\omega} \mathcal{T}_n$$

where, for $n < \omega$, no $n+2$ elements of the family $\mathcal{T}_n$ are pairwise disjoint. Since $X \leq Y$ there is $\varphi: \mathcal{T}^*(X) \to \mathcal{T}^*(Y)$ such that if $N < \omega$ and $\{U_k : k < N\} \subset \mathcal{T}^*(X)$ and $\cap_{k<N} U_k = \varnothing$, then $\cap_{k<N} \varphi(U_k) = \varnothing$.

We set

$$\mathcal{S}_n = \{U \in \mathcal{T}^*(X) : \varphi(U) \in \mathcal{T}_n\} \quad \text{for } n < \omega,$$

we note that $\mathcal{T}^*(X) = \cup_{n<\omega}\mathcal{S}_n$, and we note further that if $\mathcal{W}$ is a set of pairwise disjoint elements of $\mathcal{S}_n$ with $|\mathcal{W}| = n+2$, then $\{\varphi(U) : U \in \mathcal{W}\}$ is a set of pairwise disjoint elements of $\mathcal{T}_n$ with

$$|\{\varphi(U) : U \in \mathcal{W}\}| = n+2.$$

The second assertion of (c) now follows from the fact that $X \equiv G(X)$ (Theorem 2.20).

(d) Since Y has property (**) the set $\mathscr{T}^*(Y)$ can be written in the form

$$\mathscr{T}^*(Y) = \bigcup_{n<\omega} \mathscr{T}_n$$

with $\kappa(\mathscr{T}_n) > 0$ for $n < \omega$. Since $X \leq Y$ there is $\varphi: \mathscr{T}^*(X) \to \mathscr{T}^*(Y)$ such that

if $N < \omega$ and $\{U_k : k < N\} \subset \mathscr{T}^*(X)$ and $\cap_{k<N} U_k = \varnothing$, then $\cap_{k<N} \varphi(U_k) = \varnothing$.

We set

$$\mathscr{S}_n = \{U \in \mathscr{T}^*(X): \varphi(U) \in \mathscr{T}_n\} \quad \text{for } n < \omega$$

and we note that $\mathscr{T}^*(X) = \cup_{n<\omega} \mathscr{S}_n$. Now for $N < \omega$ and

$$\mathscr{F} = \{U_1, \ldots, U_N\} \subset \mathscr{S}_n$$

we set

$$\mathscr{G} = \{\varphi(U_1), \ldots, \varphi(U_N)\}$$

and we note from the characteristic property of the function φ that

$$\text{cal}\,\mathscr{G} \leq \text{cal}\,\mathscr{F}$$

and hence

$$(\text{cal}\,\mathscr{G})/N \leq (\text{cal}\,\mathscr{F})/N;$$

since $\mathscr{G} \subset \mathscr{T}_n$ we have

$$\kappa(\mathscr{S}_n) \geq \kappa(\mathscr{T}_n) > 0 \quad \text{for } n < \omega.$$

The second assertion of (d) now follows from the fact that $X \equiv G(X)$ (Theorem 2.20).

6.3 *Lemma.* Let X be a compact, Hausdorff space and let $\mathscr{S}$ be a non-empty family of non-empty, open subsets of X. Then there is a regular, Borel probability measure μ on X such that

$$\inf\{\mu(\text{cl}_X U): U \in \mathscr{S}\} \geq \kappa(\mathscr{S}).$$

Proof. We denote as usual by $C(X)$ the set of real-valued, continuous functions on X with norm

$$\|f\| = \sup\{|f(x)|: x \in X\} \quad \text{for } f \in C(X);$$

and we denote by F the set of elements f of $C(X)$ such that there are

$$n < \omega \quad \text{with } n \geq 1,$$
$$\{U_1, \ldots, U_n\} \subset \mathscr{S},$$
$$\{t_1, \ldots, t_n\} \subset [0, 1] \quad \text{with } \sum_{i=1}^{n} t_i = 1,$$
$$\{f_1, \ldots, f_n\} \subset C(X) \quad \text{with } 0 \leq f_i(x) \leq 1$$

for $x \in X$, $1 \leq i \leq n$ and with $f_i | U_i \equiv 1$,

such that

$$f = t_1 f_1 + \ldots + t_n f_n.$$

We claim first that if $f \in F$ then $\| f \| \geq \kappa(\mathscr{S})$.

Let $f \in F$ with

$$f = t_1 f_1 + \ldots + t_n f_n$$

as above, let $\varepsilon > 0$, let $\{p_i : 1 \leq i \leq n\}$ be a set of positive integers with

$$\sum_{i=1}^{n} p_i = N, \quad \text{and}$$

$$\left\| \sum_{i=1}^{n} p_i f_i / N \right\| < \| f \| + \varepsilon,$$

and let $\mathscr{F}$ be the family, indexed by N, whose elements are the sets $U_i (1 \leq i \leq n)$ repeated with multiplicity p_i (i.e.,

$$\mathscr{F} = \{\underbrace{U_1, \ldots, U_1}_{p_1 \text{ times}}, \ldots, \underbrace{U_n, \ldots, U_n}_{p_n \text{ times}}\}).$$

Since $\kappa(\mathscr{S}) \leq (\text{cal } \mathscr{F})/N$ and $\text{cal } \mathscr{F} \leq \| \sum_{i=1}^{n} p_i f_i \|$, we have

$$\kappa(\mathscr{S}) \leq \left\| \sum_{i=1}^{n} p_i f_i \right\| \Big/ N < N(\| f \| + \varepsilon)/N = \| f \| + \varepsilon;$$

it follows that $\kappa(\mathscr{S}) \leq \| f \|$.

The proof of the claim is complete.

Consider now the cone K in $C(X)$ defined by

$$K = \{\alpha(f + g) + \beta h : \alpha, \beta \geq 0, f \in F, g, h \in C(X), \| g \| < \kappa(\mathscr{S}), h \geq 0\}.$$

It is clear from the claim just proved that for $k \in K$ there is $x \in X$ such that $k(x) \geq 0$; hence K does not contain (the function with constant value equal to) -1 and there is by the Hahn–Banach theorem (cf. Section C. 11 of the Appendix) a bounded linear functional Λ on

$C(X)$ such that

$$-1 = \Lambda(-1) < \Lambda(k) \quad \text{for } k \in K.$$

From the fact that K is closed under multiplication by positive reals it then follows that

$$\Lambda(k) \geq 0 \quad \text{for } k \in K.$$

Now for $0 < \varepsilon < \kappa(\mathscr{S})$ we have

$$\|\varepsilon - \kappa(\mathscr{S})\| = \|\kappa(\mathscr{S}) - \varepsilon\| < \kappa(\mathscr{S})$$

and hence

$$f + \varepsilon - \kappa(\mathscr{S}) \in K \quad \text{for } f \in F,$$

so that

$$\Lambda(f) + \Lambda(\varepsilon - \kappa(\mathscr{S})) = \Lambda(f + \varepsilon - \kappa(\mathscr{S})) \geq 0 \quad \text{for } f \in F$$

and hence

$$\Lambda(f) \geq \Lambda(\kappa(\mathscr{S}) - \varepsilon) = \kappa(\mathscr{S}) - \varepsilon \quad \text{for } f \in F;$$

if follows that

$$\Lambda(f) \geq \kappa(\mathscr{S}) \quad \text{for } f \in F.$$

There is by the Riesz representation theorem (cf. Section C. 10 of the Appendix) a regular, Borel probability measure μ on X such that

$$\Lambda(f) = \int_X f \, d\mu \quad \text{for } f \in C(X).$$

We have

$$\mu(\mathrm{cl}_X U) = \inf \{\Lambda(f) : f \in C(X), f \geq 0, f|U \equiv 1\}$$

and hence

$$\mu(\mathrm{cl}_X U) \geq \kappa(\mathscr{S}) \quad \text{for } U \in \mathscr{S},$$

as required.

As a result we have the following theorem (in particular, the rather surprising implication (a) $\Rightarrow$ (b) below).

6.4 *Theorem.* Let X be a compact Hausdorff space. The following are equivalent:

(a) X has a strictly positive measure;

(b) X has a regular Borel strictly positive measure;

(c) X satisfies condition (**); and

(d) the Gleason space $G(X)$ of X has a strictly positive measure.

Proof. (a)$\Rightarrow$(c) This is Lemma 6.2(a).

(c)$\Rightarrow$(b) Let $\mathscr{T}^*(X) = \cup_{n<\omega}\mathscr{T}_n$ with $\kappa(\mathscr{T}_n) > 0$ for $n < \omega$. There is by Lemma 6.2 a regular, Borel probability measure μ_n on X such that

$$\inf\{\mu_n(\text{cl } U): U \in \mathscr{T}_n\} \geq \kappa(\mathscr{T}_n) > 0 \quad \text{for } n < \omega.$$

We set

$$\mu = \sum_{n<\omega} \mu_n/2^n;$$

then μ is a regular, Borel probability measure on X, and it is clear that μ is strictly positive on X.

(b)$\Rightarrow$(a) This is trivial.

(a) $\Leftrightarrow$(d) We have noted in Lemma 6.2(d) that X satisfies condition (**) if and only if $G(X)$ satisfies condition (**). The equivalence (a)$\Leftrightarrow$(d) then follows from the equivalence (a)$\Leftrightarrow$(c) of this theorem, already proved.

We give two consequences of Theorem 6.4.

6.5 *Corollary.* If a space X has a strictly positive measure, then its Gleason space $G(X)$ has a strictly positive measure.

Proof. It follows from Lemma 6.2 that X (and hence $G(X)$) have property (**). It then follows from Theorem 6.4((c)$\Rightarrow$(a)) (with X replaced by the compact Hausdorff space $G(X)$) that $G(X)$ has a strictly positive measure.

6.6 *Corollary.* For a completely regular, Hausdorff space X, the following statements are equivalent.

(a) The Gleason space $G(X)$ has a strictly positive measure;

(b) the Stone–Čech compactification βX has a strictly positive measure;

(c) some Hausdorff compactification of X has a strictly positive measure;

(d) every Hausdorff compactification of X has a strictly positive measure.

Proof. Let K be either the Gleason space of X or a Hausdorff compactification of X. From either Theorem 2.20 or Theorem 2.16(b) we have $K \equiv X$, so from Lemma 6.2(d) it follows that X has property (**) if and only if K has property (**). Then from the equivalence (a)$\Leftrightarrow$(c) of Theorem 6.4 (applied to K in place of X) it follows that each of the conditions (a), (b), (c) and (d) is equivalent to the condition that X has property (**).

We note that the converse of Corollary 6.5 for compact, Hausdorff spaces is given by Theorem 6.4; and the condition that a (completely regular, Hausdorff) space X has a strictly positive measure implies the various (equivalent) conditions of Corollary 6.6. We show in 6.35–6.37 below, however, that there are completely regular, Hausdorff spaces which satisfy the conditions of Corollary 6.6 and which have no strictly positive measure.

Spaces with a strictly positive measure

6.7 *Theorem.* Let $\{X_i : i \in I\}$ be a set of spaces, each with a strictly positive measure, and let $X = \prod_{i \in I} X_i$. Then X has a strictly positive measure.

Proof. For $i \in I$ there are a pseudobase $\mathscr{B}_i$ for the topology of X_i, and a probability measure μ_i defined on the σ-algebra $\mathscr{S}_i$ generated by $\mathscr{B}_i$, such that $\mu_i(B) > 0$ for all non-empty $B \in \mathscr{B}_i$. Let $\mathscr{B}$ be the set of finite intersections of elements of the family

$$\{\pi_i^{-1}(B) : B \in \mathscr{B}_i, i \in I\}$$

and let $\mathscr{S}$ be the σ-algebra on X generated by $\mathscr{B}$; then $\mathscr{B}$ is a pseudobase for the topology of X, and $\mathscr{S}$ is equal to the σ-algebra generated by

$$\{\pi_i^{-1}(S) : S \in \mathscr{S}_i, i \in I\}.$$

It is now clear from Theorem C.2 of the Appendix that the (product) measure μ defined on $\mathscr{S}$ is a strictly positive measure for X.

6.8 *Corollary.* Every product of separable spaces has a strictly positive measure.

Proof. It is enough to show that every separable space has a strictly positive measure. Let $\{x_n : n < \omega\}$ be dense in the space X and define

$$\mu:\mathscr{P}(X)\rightarrow[0,1]$$

by $\mu(A)=\sum\{1/2^n:n<\omega, x_n\in A\}$. Then μ (or more accurately: its restriction to the σ-algebra of Borel sets of X) is a strictly positive Borel measure on X.

The following example shows that not every compact Hausdorff space with a strictly positive measure is equivalent (in the sense of the relation $\equiv$ of Chapter 2) to a product of a set of separable spaces.

The measure algebra $\mathscr{B}=\mathscr{M}_\lambda/\mathscr{N}_\lambda$ of a measure space $(X,\mathscr{S},\lambda)$ is defined in (the statement of) Theorem C.6 of the Appendix.

6.9 *Theorem.* There is an extremally disconnected, compact, Hausdorff space Ω such that

Ω has a strictly positive regular Borel measure,

Ω is not separable,

$w(\Omega)=2^\omega$, and

Ω is not homeomorphic to the Gleason space of a product of separable Hausdorff spaces.

Proof. Let λ denote (completed) Lebesgue measure on the interval $[0,1]$ and let Ω be the Stone space of the measure algebra of $([0,1], \mathscr{M},\lambda)$ (with $\mathscr{M}$ the σ-algebra of Lebesgue measurable subsets of $[0,1]$). It follows from Theorem C.6 of the Appendix that Ω is an extremally disconnected, compact Hausdorff space with a strictly positive regular Borel measure and (since λ is atomless) that Ω is not separable.

For $A\in\mathscr{M}$ and $n<\omega$ there are (since λ is regular) closed $K_n\subset A$ and open $U_n\supset A$ such that $\lambda(U_n\backslash K_n)<1/n$; since the Borel set S defined by

$$S=(\bigcap_{n<\omega}U_n)\backslash(\bigcup_{n<\omega}K_n)$$

differs from A by a λ-null set, the sets A and S determine the same element of the measure algebra. It follows that the quotient Boolean algebra $\mathscr{A}=\mathscr{M}/\mathscr{N}$ satisfies $|\mathscr{A}|=2^\omega$, and from Theorem B.14 of the Appendix we have

$$w(\Omega)=w(S(\mathscr{A}))=|\mathscr{A}|=2^\omega.$$

Suppose now that Ω is homeomorphic to the Gleason space of a product $X=\prod_{i\in I}X_i$ with X_i a separable Hausdorff space. We assume without loss of generality that $|X_i| > 1$ for $i \in I$. It then follows that for $i \in I$ there is a regular-open subset $U_i \subset X_i$ with $U_i \neq X_i$, and the function $i \to \pi_i^{-1}(U_i)$ is one-to-one from I into the set $\mathscr{R}(X_I)$ of regular-open subsets of X_I. Now from

$$|I| \leq |\mathscr{R}(X_I)| = w(S(\mathscr{R}(X_I))) = w(G(X_I)) = w(\Omega) = 2^\omega$$

we have $|I| \leq 2^\omega$ and hence X_I is separable (cf. Theorem B.2). It follows from Theorem B.18 that $G(X_I)$ and hence Ω are separable, contrary to what was proved above.

The proof is complete.

6.10 *Remark.* Let S be the Gleason space of the product $\Omega \times X$, where Ω is the space of Theorem 6.9 and X is the Gleason space of $[0, 1]$. It can be shown that S has a strictly positive regular Borel measure, that S does not have a strictly positive normal Borel measure (cf. Theorem C.7 of the Appendix), that S is not separable and that $w(S) = 2^\omega$.

6.11 *Problem* (H. Rosenthal). Is there an extremally disconnected, compact Hausdorff space X with $w(X) \leq 2^\omega$ such that X has a strictly positive measure but $C(X)$ is not isomorphic as a Banach space to $C(\Omega)$ for any space Ω with a strictly positive normal measure? We note in this connection that Argyros & Haydon (1981) have found such an example if the requirement $w(X) \leq 2^\omega$ is relaxed to $w(X) \leq (2^\omega)^+$. Indeed, in this case we may take for X the Gleason space of $\{0, 1\}^{((2^\omega)^+)}$.

The properties $\mathbf{K}_{\alpha,n}$

We recall from Chapter 2 that for $\alpha \geq \omega$ and $2 \leq n < \omega$, a space is said to have property $K_{\alpha,n}$ if it has calibre (α,α,n). (Thus a space X has property $K_{\alpha,n}$ if and only if for every set $\{U_\xi : \xi < \alpha\}$ of non-empty, open subsets of X there is $A \subset \alpha$ with $|A| = \alpha$ such that if $B \subset A$ and $|B| = n$ then $\cap_{\xi \in B} U_\xi \neq \emptyset$.)

A space has property K_n if it has property $K_{\omega^+,n}$. Property K_2 is property (K).

We show in Corollary 6.17 below that a space with a strictly positive measure has property $K_{\alpha,n}$ for all $\alpha \geq \omega$ with $\text{cf}(\alpha) > \omega$ and for all n with $2 \leq n < \omega$.

6.12 *Lemma.* Let $1 \le n < \omega$, $1 \le p < \omega$ and $1 \le N < \omega$ with $p \le 2^N$, and let $\{\theta_i : 1 \le i \le p\}$ be a set of positive real numbers. Then

(i) $(2p)^{n-1}\left(\sum_{i=1}^{p} \theta_i^n\right) > \left(\sum_{i=1}^{p} \theta_i\right)^n$, and

(ii) $\sum_{i=1}^{p} (1/2^N)\theta_i^n > (1/2^{n-1})\left(\sum_{i=1}^{p} (1/2^N)\theta_i\right)^n$.

Proof. (i) We proceed by induction on n. The statement is true for $n = 1$. We assume the statement true for $n = k < \omega$. We note for $i, j \in \{1, 2, \ldots, p\}$ that

$$\begin{aligned} \theta_i^{k+1} + \theta_j^{k+1} &\ge \theta_i^{k+1} \ge \theta_i \theta_j^k \quad \text{if } \theta_i \ge \theta_j \\ &\ge \theta_j^{k+1} \ge \theta_i \theta_j^k \quad \text{if } \theta_j \ge \theta_i \end{aligned}$$

and hence $\theta_i^{k+1} + \theta_j^{k+1} \ge \theta_i \theta_j^k$. It follows that

$$\begin{aligned} p\theta_i^{k+1} + \sum_{\substack{j=1 \\ j \ne i}}^{p} \theta_j^{k+1} &= \theta_i^{k+1} + \sum_{\substack{j=1 \\ j \ne i}}^{p} (\theta_i^{k+1} + \theta_j^{k+1}) \\ &\ge \theta_i^{k+1} + \sum_{\substack{j=1 \\ j \ne i}}^{p} \theta_i \theta_j^k = \theta_i \left[\sum_{j=1}^{p} \theta_j^k\right] \end{aligned}$$

and hence

$$\begin{aligned} (2p)^k\left(\sum_{i=1}^{p} \theta_i^{k+1}\right) &> (2p)^{k-1}(2p-1)\left(\sum_{i=1}^{p} \theta_i^{k+1}\right) \\ &= (2p)^{k-1}\left(\sum_{i=1}^{p}\left[p\theta_i^{k+1} + \sum_{\substack{j=1 \\ j \ne i}}^{p} \theta_j^{k+1}\right]\right) \\ &\ge (2p)^{k-1}\left(\sum_{i=1}^{p} \theta_i\left[\sum_{j=1}^{p} \theta_j^k\right]\right) \\ &= \left(\sum_{i=1}^{p} \theta_i\right)(2p)^{k-1}\left(\sum_{j=1}^{p} \theta_j^k\right) \\ &> \left(\sum_{i=1}^{p} \theta_i\right)\left(\sum_{i=1}^{p} \theta_i\right)^k = \left(\sum_{i=1}^{p} \theta_i\right)^{k+1}; \end{aligned}$$

the statement is proved for $n = k + 1$.

(ii) From (i) and the assumption $p \leq 2^N$ we have

$$\sum_{i=1}^{p} (1/2^N)\theta_i^n = (1/2^N)(1/(2p))^{n-1}(2p)^{n-1} \sum_{i=1}^{p} \theta_i^n$$
$$> (1/2^N)(1/2^{n-1})(1/2^N)^{n-1}\left(\sum_{i=1}^{p} \theta_i\right)^n$$
$$= (1/2^{n-1})\left(\sum_{i=1}^{p} (1/2^N)\theta_i\right)^n.$$

6.13 *Lemma.* Let α be a cardinal with $\mathrm{cf}(\alpha) > \omega$, μ the Haar probability measure on $\{0, 1\}^I$, $\delta > 0$ and $\{U_\xi : \xi < \alpha\}$ a set of open-and-closed subsets of $\{0, 1\}^I$ such that $\mu(U_\xi) \geq \delta$ for $\xi < \alpha$. Then there is $A \subset \alpha$ with $|A| = \alpha$ such that $\mu(\cap_{\xi \in F} U_\xi) \geq \delta^n/4^{n-1}$ for all $F \subset A$ with $|F| \leq n$.

Proof. Let $S(\xi)$ be a finite subset of I such that U_ξ depends on $S(\xi)$. We distinguish two cases.

Case 1. α is a regular cardinal. We show (for use in Case 2 below) that in fact $\mu(\cap_{\xi \in F} U_\xi) \geq \delta^n/2^{n-1}$ for $F \subset A$ with $|F| \leq n$. By the Erdős–Rado theorem for quasi-disjoint families (Theorem 1.4) there are $C \subset \alpha$ with $|C| = \alpha$ and a set J such that

$$S(\xi) \cap S(\xi') = J \quad \text{for } \xi, \xi' \in C, \xi \neq \xi'.$$

We assume without loss of generality that $J \neq \varnothing$ and we set $|J| = N$. We set

$$H_\xi^x = \{y \in \{0, 1\}^{S(\xi) \backslash J} : \langle x, y \rangle \in \pi_{S(\xi)}[U_\xi]\}$$
$$\text{for } x \in \{0, 1\}^J, \xi \in C, \quad \text{and}$$
$$T_\xi = \{x \in \{0, 1\}^J : H_\xi^x \neq \varnothing\} \quad \text{for } \xi \in C.$$

We note that

$$\pi_{S(\xi)}[U_\xi] = \cup\{\{x\} \times H_\xi^x : x \in \{0, 1\}^J\}$$
$$= \cup\{\{x\} \times H_\xi^x : x \in T_\xi\} \quad \text{for } \xi \in C.$$

It then follows, denoting by μ_ξ the Haar probability measure on $\{0, 1\}^{S(\xi) \backslash J}$, that

$$\text{(1)} \qquad \delta \leq \mu(U_\xi) = (1/2^N)\left(\sum\{\mu_\xi(H_\xi^x) : x \in T_\xi\}\right) \quad \text{for } \xi \in C.$$

Since α is an infinite cardinal there are $B \subset C$ with $|B| = \alpha$ and a set $T = \{x_i : 1 \le i \le p\} \subset \{0, 1\}^J$ such that

$$T_\xi = T \quad \text{for } \xi \in B;$$

we note that $|T| = p \le 2^N$. Finally since $\mathrm{cf}(\alpha) > \omega$ there are $A \subset B$ with $|A| = \alpha$ and a set $\{\theta_i : 1 \le i \le p\}$ of positive rational numbers such that

$$(2) \qquad \mu_\xi(H_\xi^{x_i}) = \theta_i \text{ for } 1 \le i \le p,\ \xi \in A.$$

Now let $F \subset A$ with $|F| = n$ and set $S = \cup_{\xi \in F} S(\xi)$. Since

$$\bigcap_{\xi \in F} U_\xi = \left(\bigcup_{i=1}^{p} \left(\{x_i\} \times \prod_{\xi \in F} H_\xi^{x_i} \right) \right) \times \{0, 1\}^{I \setminus S}$$

we have from (1), (2) and Lemma 6.10 that

$$\begin{aligned} \mu\left(\bigcap_{\xi \in F} U_\xi \right) &= \sum_{i=1}^{p} (1/2^N) \left(\prod_{\xi \in F} \mu_\xi(H_\xi^{x_i}) \right) = \sum_{i=1}^{p} (1/2^N) \theta_i^n \\ &\ge (1/2^{n-1}) \left(\sum_{i=1}^{p} (1/2^N) \theta_i \right)^n \ge \delta^n / 2^{n-1}, \end{aligned}$$

as required.

Case 2. α is a singular cardinal. There is by Theorem A.7 of the Appendix a set $\{\alpha_\sigma : \sigma < \mathrm{cf}(\alpha)\}$ of regular cardinals such that

$$\begin{aligned} &\omega < \alpha_\sigma < \alpha_{\sigma'} < \alpha \quad \text{for } \sigma < \sigma' < \mathrm{cf}(\alpha), \text{ and} \\ &\alpha = \sum_{\sigma < \mathrm{cf}(\alpha)} \alpha_\sigma. \end{aligned}$$

By the extension to singular cardinals of the Erdős–Rado theorem (Theorem 1.9), there are Γ, J, $\{D_\sigma : \sigma \in \Gamma\}$ and $\{J_\sigma : \sigma \in \Gamma\}$ such that

$$\begin{aligned} &\Gamma \subset \mathrm{cf}(\alpha) \quad \text{and } |\Gamma| = \mathrm{cf}(\alpha), \\ &D_\sigma \subset \alpha \quad \text{and } |D_\sigma| = \alpha_\sigma \quad \text{for } \sigma \in \Gamma, \\ &D_\sigma \cap D_{\sigma'} = \varnothing \quad \text{for } \sigma', \sigma \in \Gamma, \sigma \ne \sigma', \quad \text{and} \\ &S(\xi) \cap S(\xi') = J_\sigma \quad \text{for } \xi, \xi' \in D_\sigma, \xi \ne \xi' \\ &\qquad\qquad\quad = J \quad \text{for } \xi \in D_\sigma, \xi' \in D_{\sigma'}, \sigma, \sigma' \in \Gamma, \sigma \ne \sigma'. \end{aligned}$$

We assume without loss of generality that $J \ne \varnothing$ and we set $|J| = N$.

We set $D = \cup_{\sigma < \mathrm{cf}(\alpha)} D_\sigma$, as before we set

$$H_\xi^x = \{y \in \{0, 1\}^{S(\xi)\setminus J} : \langle x, y \rangle \in \pi_{S(\xi)}[U_\xi]\}$$
$$\text{for } x \in \{0, 1\}^J, \xi \in D, \quad \text{and}$$
$$T_\xi = \{x \in \{0, 1\}^J : H_\xi^x \neq \emptyset\} \quad \text{for } \xi \in D,$$

and we note that

$$\pi_{S(\xi)}[U_\xi] = \cup \{\{x\} \times H_\xi^x : x \in T_\xi\} \quad \text{for } \xi \in D.$$

For $\sigma \in \Gamma$ there are $C_\sigma \subset D_\sigma$ with $|C_\sigma| = \alpha_\sigma$ and $T(\sigma) \subset \{0, 1\}^J$ such that

$$T_\xi = T(\sigma) = \{x_i(\sigma) : 1 \leq i \leq p(\sigma)\} \quad \text{for } \xi \in C_\sigma,$$

and there are $B_\sigma \subset C_\sigma$ with $|B_\sigma| = \alpha_\sigma$ and set $\{\theta_i(\sigma) : 1 \leq i \leq p(\sigma)\}$ of positive rational numbers such that, denoting by μ_ξ the Haar probability measure on $\{0, 1\}^{S(\xi)\setminus J}$, we have

$$\mu_\xi(H_\xi^{x_i(\sigma)}) = \theta_i(\sigma) \quad \text{for } 1 \leq i \leq p(\sigma), \xi \in B_\sigma.$$

Since $\mathrm{cf}(\alpha) > \omega$ there are $\Delta \subset \Gamma$ with $|\Delta| = \mathrm{cf}(\alpha)$ and a set $T = \{x_i : 1 \leq i \leq p\} \subset \{0, 1\}^J$ and a set $\{\theta_i : 1 \leq i \leq p\}$ of positive rational numbers such that

$$p(\sigma) = p \quad \text{and } T(\sigma) = T \quad \text{for } \sigma \in \Delta, \quad \text{and}$$
$$\theta_i(\sigma) = \theta_i \quad \text{for } 1 \leq i \leq p, \quad \sigma \in \Delta.$$

Since

$$\mu_\xi(H_\xi^{x_i}) = \mu(H_\xi^{x_i} \times \{0, 1\}^{(I\setminus S(\xi)) \cup J}) = \theta_i \quad \text{for } \xi \in B_\sigma$$

and α_σ is a regular, uncountable cardinal number, there is by Case 1 a set $A_\sigma \subset B_\sigma$ with $|A_\sigma| = \alpha_\sigma$ such that

$$\mu\Big(\bigcap_{\xi \in F} (H_\xi^{x_i} \times \{0, 1\}^{(I\setminus S(\xi)) \cup J})\Big) \geq \theta_i^{|F|}/2^{|F|-1} \quad \text{for } F \subset A_\sigma, |F| \leq n. \tag{3}$$

We set $A = \cup_{\sigma \in \Delta} A_\sigma$ and we note $|A| = \alpha$.

Now let $F \subset A$ with $|F| = n$, and set $S = \cup_{\xi \in F} S(\xi)$,

$\{\sigma \in \Delta : F \cap A_\sigma \neq \emptyset\} = \{\sigma_j : 1 \leq j \leq m\}$ with $1 \leq m \leq n$, and

$F_j = F \cap A_{\sigma_j}$ for $1 \leq j \leq m$.

Since

$$\bigcap_{\xi \in F} U_\xi = \bigcap_{j=1}^{m} \Big(\bigcap_{\xi \in F_j} U_\xi\Big)$$
$$= \bigcup_{i=1}^{p} \Big(\{x_i\} \times \prod_{j=1}^{m} \Big(\bigcap_{\xi \in F_j} H_\xi^{x_i} \times \{0, 1\}^{(I\setminus S(\xi)) \cup J}\Big)\Big)$$

we have from (3) and Lemma 6.12 that

$$\begin{aligned}\mu(\bigcap_{\xi\in F} U_\xi) &= \sum_{i=1}^{p} (1/2^N) \prod_{j=1}^{m} \theta_i^{|F_j|}/2^{|F_j|-1} \\ &= \sum_{i=1}^{p} (1/2^N)\theta_i^n/2^{n-m} \geq 1/2^{n-1} \sum_{i=1}^{p} (1/2^N)\theta_i^n \\ &\geq (1/2^{n-1})(1/2^{n-1})\left(\sum_{i=1}^{p} (1/2^N)\theta_i\right)^n \\ &\geq \delta^n/4^{n-1},\end{aligned}$$

as required.

6.14 *Corollary.* Let α be a cardinal with $\mathrm{cf}(\alpha) > \omega$, μ the Haar probability measure on $\{0, 1\}^I$, and $\{B_\xi : \xi < \alpha\}$ a set of Borel subsets of $\{0, 1\}^I$ such that $\mu(B_\xi) > 0$ for $\xi < \alpha$. Then for $1 \leq n < \omega$ there are $A_n \subset \alpha$ with $|A_n| = \alpha$ and $\delta_n > 0$ such that $\mu(\cap_{\xi\in F} B_\xi) \geq \delta_n$ for all $F \subset A_n$ with $|F| \leq n$.

Proof. We assume without loss of generality that there is $\delta > 0$ such that $\mu(B_\xi) > \delta$ for $\xi < \alpha$; and we set $\delta_n = (\delta/8)^n/2$.

Since the Borel measure μ is regular (cf. Appendix C), for $\xi < \alpha$ there is an open-and-closed subset U_ξ of $\{0, 1\}^I$ such that

$$\mu(U_\xi \Delta B_\xi) < (\delta/8)^n/(2n) \leq (\delta/8)^n/2 < \delta/2$$

and hence

$$\mu(U_\xi) \geq \mu(B_\xi) - \mu(U_\xi \Delta B_\xi) > \delta - \delta/2 = \delta/2.$$

It follows from Lemma 6.13 that there is $A_n \subset A$ with $|A_n| = \alpha$ such that if $F \subset A_n$ with $|F| = n$ then

$$\mu(\bigcap_{\xi\in F} U_\xi) \geq (\delta/2)^n/4^{n-1};$$

from

$$\bigcap_{\xi\in F} U_\xi \subset (\bigcap_{\xi\in F} B_\xi) \cup (\bigcup_{\xi\in F} (U_\xi \Delta B_\xi))$$

we then have

$$\begin{aligned}(\delta/8)^n < (\delta/2)^n/4^{n-1} &\leq \mu(\bigcap_{\xi\in F} U_\xi) \leq \mu(\bigcap_{\xi\in F} B_\xi) + \sum_{\xi\in F} \mu(U_\xi \Delta B_\xi) \\ &\leq \mu(\bigcap_{\xi\in F} B_\xi) + (\delta/8)^n/2\end{aligned}$$

and hence

$$\mu(\bigcap_{\xi\in F} B_\xi) \geq (\delta/8)^n/2 = \delta_n,$$

as required.

6.15 *Theorem* (Argyros and Kalamidas). Let α be a cardinal with $\mathrm{cf}(\alpha) > \omega$, $(X, \mathscr{S}, \mu)$ a probability measure space and $\{S_\xi : \xi < \alpha\} \subset \mathscr{S}$ with $\mu(S_\xi) > 0$ for $\xi < \alpha$. Then for $1 \leq n < \omega$ there are $A_n \subset A$ with $|A_n| = \alpha$ and $\delta_n > 0$ such that $\mu(\cap_{\xi \in F} S_\xi) \geq \delta_n$ for all $F \subset A_n$ with $|F| \leq n$.

Proof. Let $(\mathscr{B}, \lambda)$ be the measure algebra of $(X, \mathscr{S}, \mu)$(cf. Theorem C.6 of the Appendix) and set $\Omega = S(\mathscr{B})$. It is by C.6 enough to show that if $\{V_\xi : \xi < \alpha\} \subset \mathscr{B}$ with $\lambda(V_\xi) > 0$ for $\xi < \alpha$ then for $1 \leq n \leq \omega$ there are $A_n \subset \alpha$ with $|A_n| = \alpha$ and $\delta_n > 0$ such that $\lambda(\cap_{\xi \in F} V_\xi) \geq \delta_n$ for all $F \subset A_n$ with $|F| \leq n$.

Now let $\gamma, \delta, \{p_i : i < \gamma\}, \{\Omega_m : m < \delta\}, \{\alpha_m : m < \delta\}$ and $\{\mu_m : m < \delta\}$ be as in the statement of Maharam's structure theorem (C.8 of the Appendix). Since

$$(\bigcup_{n<\delta} \Omega_n) \cup \{p_i : i < \gamma\}$$

is dense in Ω and $\gamma + \delta \leq \omega < \mathrm{cf}(\alpha)$, either there are $A \subset \alpha$ with $|A| = \alpha$ and $i < \gamma$ such that $p_i \in \cap_{\xi \in A} V_\xi$, or there are $A \subset \alpha$ with $|A| = \alpha$ and $m < \delta$ such that $V_\xi \cap \Omega_m \neq \varnothing$ for $\xi \in A$. In the first case we set $\delta_n = \lambda(\{p_i\}) > 0$ and the proof is complete. In the second case it is enough to show that if $\{B_\xi : \xi \in A\}$ is a family of Borel subsets of $\{0, 1\}^{\alpha_m}$ with $\mu_m(B_\xi) > 0$ for $\xi \in A$, then for $1 \leq n < \omega$ there are $A_n \subset A$ with $|A_n| = \alpha$ and $\delta_n > 0$ such that $\mu_m(\cap_{\xi \in F} B_\xi) \geq \delta_n$ for all $F \subset A_n$ with $|F| \leq n$. This is the content of Corollary 6.14.

The proof is complete.

Remark. We note in passing that the proof of Theorem 6.15 does not require Maharam's structure theorem in case the (uncountable) cardinal number α is regular. Indeed for $n = 1$ the statement follows for arbitrary α from the assumption $\mathrm{cf}(\alpha) > \omega$, and then for regular $\alpha > \omega$ the proof is immediate (by induction) from the following lemma.

6.16 *Lemma.* Let α be an (infinite) regular cardinal, $(X, \mathscr{S}, \mu)$ a probability measure space, $2 \leq n < \omega$, $\{S_\xi : \xi \in A\} \subset \mathscr{S}$ with $|A| = \alpha$, $\delta > 0$ and k an integer such that $k\delta/2 > 1$. If

$$\mu(\bigcap_{\xi \in F} S_\xi) \geq \delta \quad \text{for all } F \subset A \text{ with } |F| < n$$

then there is $B \subset A$ with $|B| = \alpha$ such that

$$(1) \qquad \mu(\bigcap_{\xi \in F} S_\xi) \geq \delta/(4k) \quad \text{for all } F \subset B \text{ with } |F| \leq n.$$

Proof. Suppose the statement fails. We set $A_1 = A$ and we set

$$\mathscr{C}_1 = \{B : B \subset A_1, \{S_\xi : \xi \in B\} \text{ satisfies (1)}\}.$$

For $C \subset A_1$ with $|C| \leq n-1$ we have $C \in \mathscr{C}_1$ and hence $\mathscr{C}_1 \neq \emptyset$; further the set $\mathscr{C}_1$ partially ordered by inclusion is inductive. Hence there is a maximal element C_1 of $\mathscr{C}_1$. We have $|C_1| < \alpha$.

For $\xi \in A_1 \backslash C_1$ there is $F(\xi) \subset C_1$ with $|F(\xi)| < n$ such that

$$\mu(S_\xi \cap \bigcap_{\zeta \in F(\xi)} S_\zeta) < \delta/(4k),$$

and since α is regular and $|C_1| < \alpha$ there are $A_2 \subset A_1$ with $|A_2| = \alpha$ and a (fixed) set $\bar{F}(1) \subset C_1$ with $|\bar{F}(1)| < n$ such that, defining

$$T(1) = \bigcap_{\zeta \in \bar{F}(1)} S_\zeta,$$

we have

$$\mu(S_\xi \cap T(1)) < \delta/(4k) \quad \text{for } \xi \in A_2.$$

We repeat this argument with A_2 in place of A_1 and we find $C_2 \subset A_2$ with $|C_2| < \alpha$ and $A_3 \subset A_2$ with $|A_3| = \alpha$ and $\bar{F}(2) \subset C_2$ with $|\bar{F}(2)| < n$ such that, setting

$$T(2) = \bigcap_{\zeta \in \bar{F}(2)} S_\zeta,$$

we have

$$\mu(S_\xi \cap T(2)) < \delta/(4k) \quad \text{for } \xi \in A_3.$$

Repeating the argument we finally find $\{A_m : 1 \leq m \leq k+1\}$ with

$$A_{k+1} \subset A_k \subset \ldots \subset A_1 \quad \text{and}$$

$$|A_m| = \alpha,$$

and we find $\bar{F}(m) \subset C_m \subset A_m$ with $|\bar{F}(m)| < n$ such that, defining

$$T(m) = \bigcap_{\zeta \in \bar{F}(m)} S_\zeta,$$

we have

$$\mu(S_\xi \cap T(m)) < \delta/(4k) \quad \text{for } \xi \in A_{m+1}.$$

For $1 \leq m < j \leq k$ there is $\zeta \in \bar{F}(j) \subset A_j \subset T_m$ and hence

$$\mu(T(j) \cap T(m)) \leq \mu(S_\xi \cap T(m)) < \delta/(4k);$$

it then follows, defining $U(1) = T(1)$ and $U(j) = T(j) \setminus \cup_{m=1}^{j-1} T(m)$ for $1 < j \leq k$, that

$$\begin{aligned} &\mu(U(1)) \geq \delta \quad \text{and} \\ &\mu(U(j)) \geq \delta - (j-1)\delta/(4k) \\ &\qquad\qquad \geq \delta - (k-1)\delta/(4k) > \delta/2 \quad \text{for } 1 < j \leq k. \end{aligned}$$

Since $U(m) \cap U(j) = \varnothing$ for $1 \leq m < j \leq k$ we have

$$\mu(X) \geq \sum_{j=1}^{k} \mu(U(j)) \geq k\delta/2 > 1,$$

a contradiction.

The proof is complete.

The next result, a consequence of Theorem 6.15, is the principal result of this section.

6.17 *Corollary*. Every space with property (**) (in particular: every space with a strictly positive measure) has property $K_{\alpha,n}$ for all cardinals α with $\mathrm{cf}(\alpha) > \omega$ and for all natural numbers $n \geq 2$.

Remark. Corollary 6.17 gives (strong) chain conditions satisfied by every space with a strictly positive measure. The weakest chain condition stronger than property $K_{\alpha,n}$ for all n is precalibre α. We show next that for $\mathrm{cf}(\beta) = \omega$ and assuming $\beta^+ = 2^\beta$, a space with a strictly positive measure need not have (pre-) calibre β; in particular, assuming the continuum hypothesis, there is a space with a strictly positive measure that does not have precalibre ω^+.

6.18 *Theorem*. Let β be a strong limit cardinal with $\mathrm{cf}(\beta) = \omega$, and assume that $2^\beta = \beta^+$. Then there is an extremally disconnected compact Hausdorff space Ω such that

Ω has a strictly positive regular Borel measure (and hence satisfies c.c.c.), and

Ω does not have calibre β^+.

Proof. Let Ω be the Stone space of the measure algebra of $(\{0, 1\}^\beta, \mathscr{S}, \lambda)$, where λ is the (complete) Haar probability measure on the compact Abelian group $\{0, 1\}^\beta$ and $\mathscr{S}$ is the σ-algebra of λ-measurable subsets of $\{0, 1\}^\beta$. It follows from Theorem C.6 of the Appendix that Ω is an extremally disconnected compact Hausdorff space with a strictly positive regular Borel measure.

We claim first that if $A \subset \{0, 1\}^\beta$ with $|A| \leq \beta$ then $A \in \mathscr{S}$ and $\lambda(A) = 0$. We note that if $|A| = \beta$ then since $\mathrm{cf}(\beta) = \omega$ there is a sequence $\{A_n : n < \omega\}$ such that $A = \cup_{n<\omega} A_n$ and $|A_n| < \beta$ for $n < \omega$; hence it is enough to prove the claim in the special case that $|A| < \beta$.

Let B be the smallest closed subgroup of $\{0, 1\}^\beta$ containing A. Then $B \in \mathscr{S}$, and since β is a strong limit cardinal we have

$$|B| \leq 2^{2^{|A|}} < \beta;$$

hence the number of cosets in the quotient $\{0, 1\}^\beta$ modulo B is certainly infinite, and since the probability measure λ is translation-invariant (cf. Theorem C.8), we have $\lambda(B) = 0$. Since λ is complete, from $A \subset B$ we have $A \in \mathscr{S}$ and $\lambda(A) = 0$, as required.

We show next that there is a family $\{K_\eta : \eta < \beta^+\}$ of subsets of $\{0, 1\}^\beta$ such that

K_η is compact for $\eta < \beta^+$,

$\lambda(K_\eta) > 0$ for $\eta < \beta^+$, and

if $A \subset \beta^+$ with $|A| = \beta^+$ then $\{K_\eta : \eta \in A\}$ does not have the finite intersection property.

Since $\beta^+ = 2^\beta$ there is a faithful indexing $\{x_\xi : \xi < \beta^+\}$ of $\{0, 1\}^\beta$ by β^+. We note that the set

$$\{x_\xi : \xi < \eta\}$$

has cardinality $|\eta| \leq \beta$ for all $\eta < \beta^+$, and hence from the claim proved above we have

$$\lambda(\{x_\xi : \xi < \eta\}) = 0 \quad \text{for } \eta < \beta^+.$$

Since the (complete) Haar measure λ is regular (cf. Theorem C.3), there is for $\eta < \beta^+$ a compact subset K_η of $\{x_\xi : \xi \geq \eta\}$ such that $\lambda(K_\eta) > 0$. Let $A \subset \beta^+$ with $|A| = \beta^+$. If $\{K_\eta : \eta \in A\}$ has the finite intersection property, then by compactness there is

$$x \in \bigcap_{\eta \in A} K_\eta.$$

Then $x = x_\xi$ for some $\xi < \beta^+$, and since $|A| = \beta^+$, there is $\eta \in A$ such that $\eta > \xi$; we have $x \notin K_\eta$, a contradiction. The proof that the family $\{K_\eta : \eta < \beta^+\}$ is as required is complete.

Now for $\eta < \beta^+$ let $[K_\eta]$ denote the $\mathcal{N}_\lambda$-equivalence class of K_η (where $\mathcal{N}_\lambda = \{A \in \mathcal{S} : \lambda(A) = 0\}$), so that $K_\eta \in [K_\eta] \in \mathcal{A} = \mathcal{M}_\lambda / \mathcal{N}_\lambda$, and let U_η denote the unique open-and-closed subset of Ω that corresponds to $[K_\eta]$ via Stone's duality.

We claim finally that if $A \subset \beta$ with $|A| = \beta^+$, then $\cap_{\eta \in A} U_\eta = \varnothing$. It is enough to prove that $\{U_\eta : \eta \in A\}$ does not have the finite intersection property, i.e., that $\{[K_\eta] : \eta \in A\}$ does not have the finite intersection property. This follows from the fact that $\{K_\eta : \eta \in A\}$ does not have the finite intersection property.

The proof is complete.

6.19 *Corollary*. Assume the generalized continuum hypothesis, and let α be a cardinal number such that

$$\mathrm{cf}(\alpha) = \beta^+ \quad \text{and } \mathrm{cf}(\beta) = \omega.$$

Then there is an extremally disconnected compact Hausdorff space Ω such that

Ω has a strictly positive regular Borel measure (and hence satisfies c.c.c.), and

Ω does not have calibre α.

Proof. Since β is a strong limit cardinal and $\beta^+ = 2^\beta$, there is by Theorem 6.18 an extremally disconnected, compact Hausdorff space Ω such that

Ω has a strictly positive measure (and hence satisfies c.c.c.), and

Ω does not have calibre β^+.

That Ω does not have calibre α follows from Theorem 2.3(a).

6.20 *Remarks*. (a) Corollary 6.19 shows that, at least assuming the generalized continuum hypothesis, the result of Corollary 5.26 is best possible.

(b) For $\alpha = \omega^+$, $\beta = \omega$ the space Ω of Corollary 6.19 is a compact c.c.c. space that does not have calibre ω^+. In Theorems 7.9 and 7.13 below, again assuming $\omega^+ = 2^\omega$, we define two more such spaces. We note (from Corollary 6.17) that in fact the spaces of Corollary 6.19 have property $\mathrm{K}_{\alpha,n}$ for all natural numbers $n \geq 2$ and all cardinals α with $\mathrm{cf}(\alpha) > \omega$.

(c) It follows from Theorem 6.18 that, at least assuming the generalized continuum hypothesis, for arbitrarily large regular cardinals α there are extremally disconnected compact Hausdorff spaces with c.c.c. that fail to have calibre α.

6.21 *Theorem.* Let $\omega^+ \ll \alpha$ and $\omega^+ \ll \mathrm{cf}(\alpha)$ and let X be a compact Hausdorff space with a strictly positive regular Borel measure μ. If $\{S_\xi : \xi < \alpha\}$ is a set of μ-measurable subsets of X with $\mu(S_\xi) > 0$ for $\xi < \alpha$, then there is $A \subset \alpha$ with $|A| = \alpha$ such that $\cap_{\xi \in A} S_\xi \neq \emptyset$.

Proof. Let $\mathscr{B}$ be the σ-algebra of Borel sets in X and let Ω be the Stone space of the measure algebra of $(X, \mathscr{B}, \mu)$. It follows from Theorem C.6 of the Appendix that Ω has a strictly positive probability measure (and hence has the c.c.c.); and from Theorem 5.6 it then follows that the (compact) space Ω has calibre α.

Since μ and its Carathéodory completion (cf. Theorem C.3 of the Appendix) are regular Borel measures there is for $\xi < \alpha$ a compact set $K_\xi \subset S_\xi$ such that $\mu(K_\xi) > 0$. The μ-equivalence class $[K_\xi]$ corresponds via Stone's duality to a non-empty, open-and-closed subset of Ω, and since Ω has calibre α there is $A \subset \alpha$ with $|A| = \alpha$ such that

$$\cap\{[K_\xi] : \xi \in A\} \neq \emptyset.$$

Since $\{[K_\xi] : \xi \in A\}$ has the finite intersection property in the measure algebra, the family $\{K_\xi : \xi \in A\}$ has the finite intersection property in X and we have from compactness that

$$\bigcap_{\xi \in A} S_\xi \supset \bigcap_{\xi \in A} K_\xi \neq \emptyset,$$

as required.

If instead of a strictly positive measure we assume only the weaker (ω, ω)-chain condition (defined below), we can nevertheless establish a chain condition of calibre type, namely property (K) (cf. Lemma 6.22(b)); that the stronger properties K_n for $2 < n < \omega$ need not hold in this context will follow, at least assuming the continuum hypothesis, from Argyros' example below (Theorem 6.25).

Definition. Let κ be a cardinal and X a space. Then X satisfies the (κ, κ)-*chain condition* if the set $\mathscr{T}^*(X)$ of non-empty, open subsets of X can be written in the form

$$\mathscr{T}^*(X) = \bigcup_{i < \kappa} \mathscr{T}_i$$

where, for $i < \kappa$, no κ elements of the set $\mathscr{T}_i$ are pairwise disjoint.

6.22 *Lemma.* Let X be a space.

(a) If X has property (*) then X satisfies the (ω, ω)-chain condition.

(b) If X satisfies the (ω, ω)-chain condition then X has property (K).

Proof. (a) is obvious. We prove (b). Let $\mathscr{T}^*(X) = \cup_{n<\omega} \mathscr{T}_n$ where, for $n < \omega$, the set $\mathscr{T}_n$ has no infinite, pairwise disjoint subfamily. Let $\{U_\xi : \xi < \omega^+\} \subset \mathscr{T}^*(X)$. There are $\bar{n} < \omega$ and $A \subset \omega^+$ with $|A| = \omega^+$ such that $\{U_\xi : \xi \in A\} \subset \mathscr{T}_{\bar{n}}$. For $\{\xi, \xi'\} \in [A]^2$ we write

$$\begin{aligned} \{\xi, \xi'\} &\in P_0 \quad \text{if } U_\xi \cap U_{\xi'} \neq \varnothing \\ &\in P_1 \quad \text{if } U_\xi \cap U_{\xi'} = \varnothing. \end{aligned}$$

Since $[A]^2 = P_0 \cup P_1$ and $\omega \ll \omega^+$, it follows from Theorem 1.7 that there is $B \subset A$ such that either

$$\begin{aligned} &\text{type } B = \omega + 1 \quad \text{and } [B]^2 \subset P_1, \text{or} \\ &|B| = \omega^+ \quad \text{and } [B]^2 \subset P_0. \end{aligned}$$

Since $\mathscr{T}_{\bar{n}}$ has no infinite, pairwise disjoint subfamily the first possibility cannot occur; hence $B \subset A \subset \omega^+$ and $|B| = \omega^+$ and

$$U_\xi \cap U_{\xi'} \neq \varnothing \quad \text{for } \xi, \xi' \in B,$$

as required.

Gaifman's example

6.23 *Theorem.* There is an extremally disconnected, compact Hausdorff space X such that

(a) X has property (*) (hence, X is a c.c.c. space),

(b) X does not have property (**) (hence, X does not have a strictly positive measure),

(c) X has property K_n for $2 \leq n < \omega$, and

(d) assuming the continuum hypothesis, X does not have calibre ω^+.

Proof. It follows from Lemma 6.2 and Corollary 2.22 that it is enough to find a (compact, Hausdorff) space X with properties (a), (b), (c) and (d); for then the Gleason space of X will have all the required properties.

Let $\{T_n : 2 \leq n < \omega\}$ be an enumeration of the set of open intervals in $\mathbb{R}$ with rational endpoints, for $2 < n < \omega$ let

$$\{T_{n,k} : 1 \leq k \leq n^2\}$$

be a family of subintervals of T_n with rational endpoints such that

$$T_{n,k} \cap T_{n,k}' = \emptyset \text{ for } 1 \leq k < k' \leq n^2,$$

and set

$$X = \{p \in \{0,1\}^{\mathbb{R}} : |\{k : 1 \leq k \leq n^2, p|T_{n,k} \not\equiv 0\}| < n \quad \text{for } 2 \leq n < \omega\}.$$

For every basic open subset $U = \prod_{t \in \mathbb{R}} U_t$ of $\{0, 1\}^{\mathbb{R}}$ we denote as usual by $R(U)$ the (finite) restriction set $R(U)$ of U:

$$R(U) = \{t \in \mathbb{R} : U_t \neq \{0, 1\}\}.$$

We set

$$R^0(U) = \{t \in R(U) : \pi_t[U] = \{0\}\} \text{ and}$$
$$R^1(U) = \{t \in R(U) : \pi_t[U] = \{1\}\},$$

for $2 \leq n < \omega$ we set

$$R^1(U, n) = \{k : 1 \leq k \leq n^2, R^1(U) \cap T_{n,k} \neq \emptyset\},$$

and we note that

$$X = \{0, 1\}^{\mathbb{R}} \setminus \cup \{U : U \text{ is basic open}, |R^1(U, n)| \geq n \text{ for some } n \geq 2\}.$$

Thus X is a compact Hausdorff space.

Now for $2 \leq n < \omega$, $1 \leq m < \omega$ we set

$$\mathscr{B}_{n,m} = \{U : U \text{ is basic open in } \{0, 1\}^{\mathbb{R}}, |R(U)| \leq n/m\},$$

and we claim that if $\{U_i : 1 \leq i \leq m\} \subset \mathscr{B}_{n,m}$ with

$$U_i \cap X \neq \emptyset \quad \text{for } 1 \leq i \leq m,$$
$$\bigcap_{i=1}^{m} U_i \neq \emptyset, \quad \text{and}$$
$$R^1(U_i, k) = R^1(U_{i'}, k) \quad \text{for } 1 \leq i \leq i' \leq m, 2 \leq k \leq n,$$

then $(\cap_{i=1}^{m} U_i) \cap X \neq \emptyset$.

Indeed since $\cap_{i=1}^{m} U_i \neq \emptyset$ we have

$$\left(\bigcup_{i=1}^{m} R^0(U_i)\right) \cap \left(\bigcup_{i=1}^{m} R^1(U_i)\right) = \emptyset;$$

we define $p \in \{0, 1\}^{\mathbb{R}}$ by

$$p_t = 1 \quad \text{if } t \in \bigcup_{i=1}^{m} R^1(U_i)$$
$$= 0 \quad \text{otherwise.}$$

We have $p \in \cap_{i=1}^{m} U_i$; we verify that $p \in X$. For $2 \leq j < \omega$, set

$$I(j) = \{k : 1 \leq k \leq j^2, p|T_{j,k} \not\equiv 0\}.$$

Then for $n < j$ we have

$$|I(j)| \leq |\{t \in \mathbb{R} : p_t = 1\}| = \left|\bigcup_{i=1}^{m} R^1(U_i)\right|$$
$$\leq mn/m = n < j,$$

and for $2 < j \leq n$ we have

$$|I(j)| = \left|\bigcup_{i=1}^{m} R^1(U_i, j)\right| = |R^1(U_1, j)| < j$$

(the inequality since $U_1 \cap X \neq \varnothing$). It follows that $p \in X$, and the claim is proved.

For $2 \leq n < \omega$ and $J_k \subset \{1, 2, \dots, k^2\}$ for $2 \leq k \leq n$ we set $\mathscr{B}_{n,m}(J_2, \dots, J_n) = \{U \in \mathscr{B}_{n,m} : R^1(U, k) = J_k \text{ for } 2 \leq k \leq n\}$ for $1 \leq m < \omega$; and we define

$$s(n) = \sum_{2 \leq k \leq n^2} 2^{(k^2)}, \quad p_n = 2^{n/2} \cdot 2^{s(n)}$$

We verify statements (a), (b), (c) and (d).

(a) From (the case $m = 2$ of) the claim just proved it follows that if $\mathscr{F} \subset \mathscr{B}_{n,2}$ with

$$U \cap X \neq \varnothing \quad \text{for } U \in \mathscr{F},$$
$$U \cap V \cap X = \varnothing \quad \text{for } U, V \in \mathscr{F}, U \neq V, \text{ and}$$
$$R^1(U, k) = R^1(V, k) \text{ for } U, V \in \mathscr{F}, 2 \leq k \leq n,$$

then the elements of $\mathscr{F}$ are pairwise disjoint. Since the (product) measure μ on $\{0, 1\}^{\mathbb{R}}$ determined by the coordinate measures

$$\mu_t(\{0\}) = \mu_t(\{1\}) = 1/2$$

satisfies $\mu(\{0, 1\}^{\mathbb{R}}) = 1$ and $\mu(U) \geq (1/2)^{n/2}$ for $U \in \mathscr{B}_{n,2}$, we have $|\mathscr{F}| \leq 2^{n/2}$.

We note that if $\mathscr{F} \subset \mathscr{B}_{n,2}$ with

$$U \cap X \neq \varnothing \quad \text{for } U \in \mathscr{F} \text{ and}$$
$$U \cap V \cap X = \varnothing \quad \text{for } U, V \in \mathscr{F}, U \neq V,$$

then

$$|\mathscr{F}| \leq \sum \{|\mathscr{F} \cap \mathscr{B}_{n,2}(J_2, \dots, J_n)| : J_k \subset \{1, \dots, k^2\}, 2 \leq k \leq n\} \leq p_n.$$

Finally for $2 \le n < \omega$ set

$$\mathscr{T}_n = \{W \in \mathscr{T}^*(X): \text{there is } U \in \mathscr{B}_{n,2} \text{ such that } W \supset U \cap X \neq \varnothing\}.$$

Then

$$\mathscr{T}^*(X) = \cup_{2 \le n < \omega} \mathscr{T}_n, \quad \text{and}$$

no $p_n + 1$ elements of the family $\mathscr{T}_n$ are pairwise disjoint.

It then follows that X has property (*), as required.

(b) We show that X does not have property (**), i.e., that if

$$\mathscr{T}^*(X) = \bigcup_{n<\omega} \mathscr{S}_n$$

then there is $\bar{n} < \omega$ such that $\kappa(\mathscr{S}_{\bar{n}}) = 0$.

For $t \in \mathbb{R}$ we have $\pi_t^{-1}(\{1\}) \cap X \neq \varnothing$. We set

$$I_n = \{t \in \mathbb{R}: \pi_t^{-1}(\{1\}) \cap X \in \mathscr{S}_n\} \quad \text{for } n < \omega$$

and we note from the Baire category theorem that not all of the sets I_n are nowhere dense; that is, there are $\bar{n} < \omega$ and a non-empty interval $T \subset \mathbb{R}$ such that

$$I_{\bar{n}} \cap T \text{ is dense in } T.$$

We set

$$A = \{n < \omega: T_n \subset T\},$$

we choose

$$t_{n,k} \in I_{\bar{n}} \cap T_{n,k} \quad \text{for } n \in A, 1 \le k \le n^2,$$

and we set

$$\mathscr{F}_n = \{\pi_{t_{n,k}}^{-1}(\{1\}) \cap X : 1 \le k \le n^2\} \quad \text{for } n \in A.$$

Then $\mathscr{F}_n \subset \mathscr{S}_{\bar{n}}$ for $n \in A$; since no n elements of $\mathscr{F}_n$ have non-empty intersection we have

$$\text{cal } \mathscr{F}_n \le n - 1$$

and hence

$$\kappa(\mathscr{S}_{\bar{n}}) \le (\text{cal } \mathscr{F}_n)/|\mathscr{F}_n| \le (n-1)/n^2 < 1/n \quad \text{for } n \in A.$$

Since $|A| = \omega$ we have $\kappa(\mathscr{S}_{\bar{n}}) = 0$, as required.

(c) We fix m such that $1 \le m < \omega$ and we show that X has property K_m. Let $\{U_\xi : \xi < \omega^+\}$ be a set of basic open subsets of $\{0, 1\}^{\mathbb{R}}$ with $U_\xi \cap X \neq \varnothing$ for $\xi < \omega^+$. Since $\{0, 1\}^{\mathbb{R}}$ has calibre ω^+ (cf. Corollary 3.7(a)), we assume without loss of generality that $\cap_{\xi < \omega^+} U_\xi \neq \varnothing$.

We assume further without loss of generality, passing if necessary to a subfamily $\{U_\xi : \xi \in I\}$ with $I \subset \omega^+$ and $|I| = \omega^+$, that there are n with $2 \leq n < \omega$ and $J_2, \ldots, J_n \subset \{1, 2 \ldots, n^2\}$ such that

$$\{U_\xi : \xi < \omega^+\} \subset \mathscr{B}_{n,m}(J_2, \ldots, J_n).$$

It then follows from the claim proved above that for $A \subset \omega^+$ with $|A| = m$ we have $(\cap_{\xi \in A} U_\xi) \cap X \neq \emptyset$, as required.

(d) Assume the continuum hypothesis, and let $S = \{x_\xi : \xi < \omega^+\}$ be a faithfully indexed Lusin set of $\mathbb{R}$ (cf. Theorem B.4 of the Appendix). We consider $\{\pi_{x_\xi}^{-1}(\{1\}) \cap X : \xi < \omega^+\}$, a family of non-empty, open subsets of X. Suppose there are $A \subset \omega^+$ with $|A| = \omega^+$ and $p \in \{0,1\}^{\mathbb{R}}$ such that

$$p \in \bigcap_{\xi \in A} \pi_{x_\xi}^{-1}(\{1\}) \cap X.$$

Since S is of Category II in $\mathbb{R}$ there is n with $2 \leq n < \omega$ such that $S \cap T_n$ is dense in T_n (and hence $S \cap T_{n,k} \neq \emptyset$ for $1 \leq k \leq n^2$). It follows that

$$|\{k : 1 \leq k \leq n^2, p | T_{n,k} \not\equiv 0\}| = n^2 > n,$$

contradicting the condition $p \in X$.

The proof of the theorem is complete.

6.24 *Problem.* Is there a compact Hausdorff space X such that X has property (*), X does not have a strictly positive measure, and X has calibre ω^+?

We think that, even without special set-theoretic assumptions, it should be possible to define such a space.

The example of Argyros

6.25 *Theorem* (Argyros). There is an extremally disconnected, compact Hausdorff space X such that

(a) X has property (*) (and hence property (K));

(b) X does not have property (**) (hence, X does not have a strictly positive measure); and

(c) assuming the continuum hypothesis, X does not have property K_3.

Proof. It is enough to define a (completely regular, Hausdorff) space X satisfying (a), (b) and (c). For then, according to Theorem 2.17(d) and Lemma 6.2(c), (d), the Gleason space of X will have all the required properties.

We begin by defining a tree $\langle T, \preccurlyeq \rangle$ with

$$T = \cup_{n<\omega} T_n \subset [\omega]^2.$$

Let $\{S_{n,j} : n < \omega, 1 \leq j \leq 3^n\}$ be a family such that

$$S_{n,j} \in [\omega]^3 \quad \text{and}$$
$$S_{n,j} \cap S_{n',j'} = \varnothing \quad \text{for } \langle n, j \rangle \neq \langle n', j' \rangle,$$

set $T_n = \cup \{[S_{n,j}]^2 : 1 \leq j \leq 3^n\}$ for $n < \omega$ (so that $|T_n| = 3^{n+1}$), let

$$T_n = \{s_j : 1 \leq j \leq 3^{n+1}\}$$

be an enumeration of T_n, and set $T = \cup_{n<\omega} T_n$. The order $\preccurlyeq$ of T is defined so that the (immediate) successors of elements of T_n are in T_{n+1}, and is determined by the following rule: if $s \in T_n$ and $t \in T_{n+1}$, then $s \prec t$ if and only if there is j with $1 \leq j \leq 3^{n+1}$ such that $s = s_j$ and $t \in [S_{n+1,j}]^2$.

We set

$$A(n) = \cup \{s \in T : s \in \cup_{n \leq p < \omega} T_p\} \quad \text{and}$$
$$B(n) = \omega \backslash A(n) \quad \text{for } n < \omega.$$

For $s \in T$ we denote by K_s the set (of cardinality 2) of non-constant functions in $\{0, 1\}^s$; that is, if $s = \{k, l\}$ then

$$K_s = \{\{\langle k, 1 \rangle, \langle l, 0 \rangle\}, \{\langle k, 0 \rangle, \langle l, 1 \rangle\}\}.$$

We note that if $S \subset \omega$ with $|S| = 3$ and if there is $n < \omega$ such that

$$[S]^2 = \{s_1, s_2, s_3\} \subset T_n,$$

then the family $\{K_{s_q} \times \{0,1\}^{\omega \backslash s_q} : 1 \leq q \leq 3\}$, which has three elements, has empty intersection and any two elements of the family have non-empty intersection.

We denote by $\boldsymbol{\Sigma}$ the set of (finite or infinite) branches of T, and for $\Sigma \in \boldsymbol{\Sigma}$ we set

$$A(\Sigma) = \cup \{s : s \in \Sigma\} \quad \text{and}$$
$$V_\Sigma = (\textstyle\prod_{s \in \Sigma} K_s) \times \{0, 1\}^{\omega \backslash A(\Sigma)}.$$

We now define the space $\langle X, \mathscr{T} \rangle$. As a set, X is the Cantor set $\{0, 1\}^\omega$; the topology $\mathscr{T}$ is defined by the subbase that contains

(i) all subsets of $\{0, 1\}^\omega$ that are open-and-closed in the (usual) product topology of $\{0, 1\}^\omega$, and

(ii) all sets of the form V_Σ for $\Sigma \in \mathbf{\Sigma}$.

The base $\mathscr{B}$ determined by this subbase consists of all sets of the form $U \cap (\cap_{j=1}^m V_{\Sigma_j})$ with U open-and-closed in the usual topology and with $\{\Sigma_j : 1 \leq j \leq m\} \subset \mathbf{\Sigma}$. The elements of $\mathscr{B}$ are closed in the usual topology of $\{0, 1\}^\omega$, hence in its extension $\mathscr{T}$; thus $\langle X, \mathscr{T} \rangle$ is a completely regular, Hausdorff space.

We say that the basic set $V = U \cap (\cap_{j=1}^m V_{\Sigma_j})$ is *determined by* $(U; \Sigma_1, \ldots, \Sigma_m)$, and we say that V is *separated at level* n if

(i) $(\Sigma_{j_1} \cap T_n) \cap (\Sigma_{j_2} \cap T_n) = \varnothing$ for $1 \leq j_1 < j_2 \leq m$, and

(ii) there is a finite subset F of ω on which U depends such that $F \cap A(n) = \varnothing$.

If $V = U \cap (\cap_{j=1}^m V_{\Sigma_j}) \in \mathscr{B}$ and V is separated at level n we set

$$V(0, n) = \pi_{B(n)}[V] \quad \text{and}$$

$$V(1, n) = \bigcap_{j=1}^{m} \left(\left(\prod_{s \in \Sigma_j,\, s \subset A(n)} K_s \right) \times \{0, 1\}^{A(n) \setminus A(\Sigma_j)} \right);$$

then $V(0, n)$ is open-and-closed in the usual topology of $\{0, 1\}^{B(n)}$, and $V = V(0, n) \times V(1, n)$.

For use in the sequel we verify the following claim:

Statement (1). Let $\{V_i : i \in I\} \subset \mathscr{B}$ with V_i determined by $(U^i; \Sigma_1^i, \ldots, \Sigma_{m(i)}^i)$ and suppose there is $n < \omega$ such that each V_i is separated at level n. Then the following are equivalent:

($\tilde{a}$) $\cap_{i \in I} V_i \neq \varnothing$; and

($\tilde{b}$)1. $\cap_{i \in I} V_i(0, n) \neq \varnothing$, and

2. of any three distinct elements of the set $\{\Sigma_j^i \cap T_p : n \leq p < \omega,\ i \in I,\ 1 \leq j \leq m(i)\}$, some two have empty intersection.

It is clear that ($\tilde{a}$) $\Rightarrow$ ($\tilde{b}$)1. If ($\tilde{a}$) holds and $\{s_q : 1 \leq q \leq 3\}$ are distinct elements of the indicated set with

$$s_q = \Sigma_{j(q)}^{i(q)} \cap T_{p(q)} \quad \text{for } 1 \leq q \leq 3,$$

no two with empty intersection, then there is $p < \omega$ such that each natural number $p(q)$ is equal to p and there is $S \subset \omega$ with $|S| = 3$ such that

$$[S]^2 = \{s_q : 1 \leq q \leq 3\} \subset T_p;$$

we have

$$\bigcap_{i\in I} V_i \subset \bigcap_{q=1}^{3} V_{i(q)} \subset \bigcap_{q=1}^{3} K_{s_q} \times \{0,1\}^{\omega\setminus s_q} = \emptyset$$

a contradiction.

We prove $(\tilde{\text{b}}) \Rightarrow (\tilde{\text{a}})$. Since $\cap_{i\in I} V_i(0,n) \neq \emptyset$ by $(\tilde{\text{b}})1$, it is enough to show $\cap_{i\in I} V_i(1,n) \neq \emptyset$. We note (from the definition of T) that if $s_1, s_2 \in T$ with $s_1 \cap s_2 \neq \emptyset$ then there is $p < \omega$ such that $s_1, s_2 \in T_p$; and if $s_1, s_2, s_3 \in T$ with $s_1 \cap s_2 \neq \emptyset$ and $s_1 \cap s_3 \neq \emptyset$ then $s_2 \cap s_3 \neq \emptyset$. It then follows from $(\tilde{\text{b}})2$ that there is a sequence $\{B_r : r < \omega\}$ such that

$$\bigcup_{r<\omega} B_r = \{\Sigma^i_j \cap T_p : n \leq p < \omega, i \in I, 1 \leq j \leq m(i)\},$$

$$\begin{aligned}
&1 \leq |B_r| \leq 2 \quad \text{for } r < \omega,\\
&B_r \cap B_{r'} = \emptyset \quad \text{for } r < r' < \omega, \quad \text{and}\\
&\text{if } s \in B_r, s' \in B_{r'} \text{ with } r < r' < \omega \text{ then } s \cap s' = \emptyset.
\end{aligned}$$

For $r < \omega$ there is $x(r) \in \cap_{s\in B_r} K_s$, and there is $x \in \{0,1\}^{A(n)}$ such that $x_s = x(r)_s$ for $s \in B_r$. It is then clear that $x \in \cap_{i\in I} V(1,n)$, as required.

The proof of Statement (1) is complete.

We note a consequence of Statement (1).

Statement (2). Let $\{V_i : i \in I\} \subset \mathscr{B}$ and suppose there is $n < \omega$ such that each V_i is separated at level n. If $\cap_{i\in I} V(0,n) \neq \emptyset$, then $\cap_{i\in I} V_i \neq \emptyset$ if and only if every three elements of the set $\{V_i : i \in I\}$ have non-empty intersection.

Finally we verify that the space $\langle X, \mathscr{T} \rangle$ satisfies (a), (b) and (c).

(a) For $n < \omega$ let μ denote the usual Haar measure on $\{0,1\}^{B(n)}$ and set

$$\mathscr{T}_{n,m} = \{U \in \mathscr{T}^*(X) : \text{there is } V \in \mathscr{B} \text{ such that } V \subset U, V \text{ is separated at level } n, \text{ and } (\mu(V(0,n)) \geq 1/(m+1)\}.$$

If $\mathscr{U} \subset \mathscr{T}_{n,m}$ with $|\mathscr{U}| = m+2$ then there are $U_0, U_1 \in \mathscr{U}$ and $V_0, V_1 \in \mathscr{B}$ such that

$$\begin{aligned}
&V_0 \subset U_0, V_1 \subset U_1,\\
&V_0 \text{ and } V_1 \text{ are separated at level } n, \text{ and}\\
&V_0(0,n) \cap V_1(0,n) \neq \emptyset.
\end{aligned}$$

Since each branch of T determining V_0 intersects at most one branch of T determining V_1 at a level greater than or equal to n, the family $\{V_0, V_1\}$ satisfies (not only condition ($\tilde{\text{b}}$)1 but also) condition ($\tilde{\text{b}}$)2 of Statement (1); hence

$$U_0 \cap U_1 \supset V_0 \cap V_1 \neq \emptyset.$$

It follows that no $m+2$ elements of $\mathcal{T}_{n,m}$ are pairwise disjoint. That X has property (*) now follows easily.

(b) If X has property (**) then the set $\mathcal{T}^*(X)$ can be written in the form

$$\mathcal{T}^*(X) = \bigcup_{n<\omega} \mathcal{T}'_n$$

with $\kappa(\mathcal{T}'_n) > 0$ for $n > \omega$. We set

$$\mathcal{T}_n = \{U \in \mathcal{T}^*(X)\text{: there is } V \subset U \text{ such that } V \in \mathcal{T}'_n\},$$

and we note that

$$\mathcal{T}^*(X) = \bigcup_{n<\omega} \mathcal{T}_n,$$

$$\kappa(\mathcal{T}_n) > 0 \quad \text{for } n < \omega, \text{ and}$$

$$\text{if } U, V \in \mathcal{T}^*(X) \quad \text{with } V \subset U, \text{ then}$$

$$\min\{n : U \in \mathcal{T}_n\} \leq \min\{n : V \in \mathcal{T}_n\}.$$

We claim there are $\bar{n} < \omega$ and a finite branch $\bar{\Sigma}$ of T such that if Σ is a finite branch and $\bar{\Sigma} \subset \Sigma$ then $V_\Sigma \in \mathcal{T}_{\bar{n}}$. If the claim fails there are a sequence $\{n_k : k < \omega\}$ of natural numbers and a sequence $\{\Sigma_k : k < \omega\}$ of finite branches such that

$$n_k < n_{k+1} \text{ and } \Sigma_k \subset \Sigma_{k+1} \text{ for } k < \omega, \text{ and}$$

$$\min\{n : V_{\Sigma_k} \in \mathcal{T}_n\} = n_k \quad \text{for } k < \omega.$$

We set $\Sigma = \bigcup_{k<\omega} \Sigma_k$ and we choose $n < \omega$ such that $V_\Sigma \in \mathcal{T}_n$. Since $\Sigma_k \subset \Sigma$ we have $V_\Sigma \subset V_{\Sigma k}$ and hence $V_{\Sigma k} \in \mathcal{T}_n$ for all $k < \omega$, a contradiction The proof of the claim is complete. We choose $\bar{n}$ and $\bar{\Sigma}$ as given and we note by m the length of $\bar{\Sigma}$.

We choose $k < \omega$ such that $(2/3)^k < \kappa(\mathcal{T}_{\bar{n}})$ and we set

$$\mathcal{S} = \{V_\Sigma : \bar{\Sigma} \subset \Sigma, \Sigma \text{ is finite, length of } \Sigma \text{ is } m+k\}.$$

We note that $\mathcal{S} \subset \mathcal{T}_{\bar{n}}$ and $|\mathcal{S}| = 3^k$. We note further that if $\mathcal{U} \subset \mathcal{S}$ with $|\mathcal{U}| \geq 2^k + 1$ then there are $V_{\Sigma(1)}, V_{\Sigma(2)}, V_{\Sigma(3)} \in \mathcal{U}$ and p with $m \leq p < m+k$ such that

$$\Sigma(1) \cap T_p = \Sigma(2) \cap T_p = \Sigma(3) \cap T_p \text{ and}$$

$$\text{the sets } \Sigma(1) \cap T_{p+1}, \Sigma(2) \cap T_{p+1}, \Sigma(3) \cap T_{p+1} \text{ are distinct.}$$

No two of the sets $\Sigma(1) \cap T_{p+1}$, $\Sigma(2) \cap T_{p+1}$, $\Sigma(3) \cap T_{p+1}$ have empty intersection, so from Statement (1) we have

$$\cap \mathscr{U} \subset V_{\Sigma(1)} \cap V_{\Sigma(2)} \cap V_{\Sigma(3)} = \varnothing.$$

It follows that cal $\mathscr{S} \leq 2^k$ and hence

$$\kappa(\mathscr{T}_{\bar{n}}) \leq \text{cal}\, \mathscr{S}/3^k \leq (2/3)^k,$$

a contradiction.

(c) We note from (b) that X is not a separable space (cf. Corollary 6.7 and Lemma 6.1(a)). It follows from the continuum hypothesis that

$$|X| = d(X) = \omega^+,$$

and then from Lemma 5.27 (with $\mathscr{T}_1 = \mathscr{T}$ and with $\mathscr{T}_2$ the usual topology of $\{0, 1\}^{\omega}$) it follows that the space $\langle X, \mathscr{T} \rangle$ does not have calibre ω^+. There is $\{V_\xi : \xi < \omega^+\} \subset \mathscr{B}$ such that if $I \subset \omega^+$ with $|I| = \omega^+$ then $\cap_{\xi \in I} V_\xi = \varnothing$. There are $n < \omega$ and $A \subset \omega^+$ with $|A| = \omega^+$ such that each of the sets V_ξ with $\xi \in A$ is separated at level n. We assume without loss of generality, using the fact that the number of open-and-closed subsets of $\{0, 1\}^{B(n)}$ is ω, that there is a (fixed) open-and-closed subset $V(0, n)$ of $\{0, 1\}^{B(n)}$ such that

$$V_\xi(0, n) = V(0, n) \quad \text{for all } \xi \in A.$$

For $I \subset A$ with $|I| = \omega^+$ we then have

$$\bigcap_{\xi \in I} V_\xi(0, n) = V(0, n) \neq \varnothing,$$

and since $\cap_{\xi \in I} V_\xi = \varnothing$ it then follows from Statement (2) that some three elements of the set $\{V_\xi : \xi \in I\}$ have empty intersection. Thus X does not have property K_3.

The proof is complete.

We indicate how the method of (the proof of) Theorem 6.25 can be used to prove a more general result.

6.26 *Theorem* (Argyros). Let $2 \leq m < \omega$. There is an extremally disconnected, compact Hausdorff space X such that

(a) X has property (*) and property K_m;

(b) X does not have property (**) (hence, X does not have a strictly positive measure); and

(c) assuming the continuum hypothesis, X does not have property K_{m+1}.

Proof. (We note that Theorem 6.25 is the case $m = 2$.) As before, it is enough to find a (completely regular, Hausdorff) space X satisfying (a), (b) and (c).

We fix a natural number p and a family $\{F_i : 1 \le i \le m+1\}$ of subsets of $\{0, 1\}^p$ such that

$$\bigcap_{i=1}^{m+1} F_i = \varnothing \quad \text{and}$$
$$\bigcap_{\substack{i=1\\ i\neq j}}^{m+1} F_i \neq \varnothing \quad \text{for } 1 \le j \le m+1.$$

Let $\{S_{n,j} : n < \omega, 1 \le j \le (m+1)^n\}$ be a family such that

$$S_{n,j} \in [\omega]^p \quad \text{and}$$
$$S_{n,j} \cap S_{n',j'} = \varnothing \quad \text{for } \langle n, j\rangle \neq \langle n', j'\rangle,$$

let $\varphi_{n,j}$ be a one-to-one function from $S_{n,j}$ onto p, and set

$$F_{i,n,j} = \{t \circ \varphi_{n,j} : t \in F_i\} \quad \text{for } 1 \le i \le m+1, \text{ and}$$
$$\mathscr{F}_{n,j} = \{F_{i,n,j} : 1 \le i \le m+1\} \quad \text{for } n < \omega, 1 \le j \le (m+1)^n.$$

We note that $\mathscr{F}_{n,j}$ is a family of subsets of $\{0, 1\}^{S_{n,j}}$ such that $\cap \mathscr{F}_{n,j} = \varnothing$ and such that if $\mathscr{G} \subset \mathscr{F}_{n,j}$ with $|\mathscr{G}| \le m$ then $\cap \mathscr{G} \neq \varnothing$.

We set

$$T_n = \bigcup_{j=1}^{(m+1)^n} \mathscr{F}_{n,j} \quad \text{for } n < \omega,$$

we note that $|T_n| = (m+1)^{n+1}$, we let

$$T_n = \{s_j : 1 \le j \le (m+1)^{n+1}\}$$

be an enumeration of T_n, and we set $T = \cup_{n<\omega} T_n$. The order $\preccurlyeq$ of the tree T is defined so that the (immediate) successors of elements of T_n are in T_{n+1}, and is determined by the following rule: if $s \in T_n$ and $t \in T_{n+1}$, then $s \prec t$ if and only if there is j with $1 \le j \le (m+1)^{n+1}$ such that $s = s_j$ and $t \in \mathscr{F}_{n+1,j}$.

Again we denote by $\boldsymbol{\Sigma}$ the set of all branches of T, and for $\Sigma \in \boldsymbol{\Sigma}$ we set

$$V_\Sigma = \left(\prod_{F \in \Sigma} F\right) \times \{0, 1\}^{\omega \setminus \cup \Sigma}.$$

We now define the space $\langle X, \mathscr{T} \rangle$. As a set, X is the Cantor set $\{0, 1\}^\omega$; the topology $\mathscr{T}$ is defined by the subbase that contains

(i) all subsets of $\{0, 1\}^\omega$ that are open-and-closed in the (usual) product topology of $\{0, 1\}^\omega$, and

(ii) all sets of the form V_Σ for $\Sigma \in \boldsymbol{\Sigma}$.

The verification that the space $\langle X, \mathscr{T} \rangle$ is as required is analogous to the case $m = 2$ (Theorem 6.25), and is left to the reader.

6.27 *Problem* (Argyros and Negrepontis). Assume the generalized continuum hypothesis. Let $\omega^+ < \alpha = \beta^+$ with $\mathrm{cf}(\beta) = \omega$ and let $2 \leq m < \omega$. Is there a c.c.c. space X such that X has property $\mathrm{K}_{\alpha,m}$ and X does not have property $\mathrm{K}_{\alpha,m+1}$?

In the Notes to this chapter we outline a proof that if the requirement that X be a c.c.c. space is omitted then such a space exists; in this case the assumption $\mathrm{cf}(\beta) = \omega$ may be omitted.

6.28 *Lemma*. Let β be an infinite cardinal and X a compact Hausdorff space. The following statements are equivalent.

(a) The set $\mathscr{T}^*(X)$ can be written in the form $\mathscr{T}^*(X) = \cup_{i<\beta} \mathscr{T}_i$ with $\kappa(\mathscr{T}_i) > 0$ for $i < \beta$; and

(b) there is a set $\{\mu_i : i < \beta\}$ of regular, Borel probability measures on X such that for every $U \in \mathscr{T}^*(X)$ there is $i < \beta$ with $\mu_i(U) > 0$.

Proof. (a) $\Rightarrow$ (b). Let $\{\mathscr{T}_i : i < \beta\}$ be as in (a) and for $i < \beta$ let μ_i be a regular, Borel probability measure such that

$$\mu_i(\mathrm{cl}_X V) \geq \kappa(\mathscr{T}_i) \quad \text{for } V \in \mathscr{T}_i.$$

For $U \in \mathscr{T}^*(X)$ there is $V \in \mathscr{T}^*(X)$ such that $\mathrm{cl}_X V \subset U$, and if $V \in \mathscr{T}_i$ we have

$$\mu_i(U) \geq \mu_i(\mathrm{cl}_X V) \geq \kappa(\mathscr{T}_i) > 0.$$

(b) $\Rightarrow$ (a). Let $\{\mu_i : i < \beta\}$ be as in (b) and for $i < \beta$, $n < \omega$ set

$$\mathscr{T}_{i,n} = \{U \in \mathscr{T}^*(X) : \mu_i(U) \geq 1/(n+1)\}.$$

Then $\mathscr{T}^*(X) = \cup_{i<\beta, n<\omega} \mathscr{T}_{i,n}$, and from Lemma 6.1 (with $\mathscr{A} = \mathscr{U} = \mathscr{T}_{i,n}$) we have

$$\kappa(\mathscr{T}_{i,n}) \geq 1/(n+1) > 0,$$

as required.

6.29 *Theorem* (Argyros). Let β be a strong limit cardinal with $\mathrm{cf}(\beta) = \omega$. There is an extremally disconnected, compact Hausdorff space X such that

(a) X has property (*), and

(b) the set $\mathscr{T}^*(X)$ of non-empty, open subsets of X cannot be written in the form $\mathscr{T}^*(X) = \cup_{i<\beta} \mathscr{T}_i$ with $\kappa(\mathscr{T}_i) > 0$ for $i < \beta$ (i.e., for every set $\{\mu_i : i < \beta\}$ of regular, Borel measures on X there is a non-empty, open subset U of X such that $\mu_i(U) = 0$ for $i < \beta$).

Proof. It is enough to find a (completely regular, Hausdorff) space X satisfying (a) and the first part of (b). For then the Gleason space of X is an extremally disconnected, compact Hausdorff space satisfying (a) and the first part of (b); the equivalence of the two parts of (b) (for compact Hausdorff spaces) is given by Lemma 6.28.

There is a sequence $\{\alpha_n : n < \omega\}$ such that $\omega \leq \alpha_n < \alpha_{n'} < \beta$ for $n < n' < \omega$, and $\sum_{n<\omega} \alpha_n = \beta$. We set

$\beta_0 = (2^{\alpha_0})^+$, and

$\beta_{n+1} = \max\{(2^{\alpha_n})^+, (2^{\beta_n})^+\}$ for $0 < n < \omega$,

and we note that

β_n is a regular cardinal for $n < \omega$,

$\beta_n < \beta_{n'} < \beta$ for $n < n' < \omega$,

$\sum_{n<\omega} \beta_n = \beta$, and

$(\beta_n)^+ \ll \beta_{n'}$ for $n < n' < \omega$.

Now exactly as in the proof of Theorem 5.28 we choose a family $\{B_n : n < \omega\}$ of (pairwise disjoint) subsets of β with $|B_n| = \beta_n$ for $n < \omega$, we choose a family $\{B_{n+1,i} : i < \beta_n\}$ of (pairwise disjoint) subsets of B_{n+1} with $|B_{n+1,i}| = \beta_{n+1}$ for $n < \omega, i < \beta_n$, we define a tree $T = \cup_{n<\omega} T_n$ of height ω with

$$T_0 = [B_0]^2 \quad \text{and}$$

$$T_{n+1} = \cup\{[B_{n+1,i}]^2 : i < \beta_n\} \quad \text{for } n < \omega,$$

we denote by $\boldsymbol{\Sigma}$ the set of all branches of T, we define K_s for $s \in T$ and V_Σ for $\Sigma \in \boldsymbol{\Sigma}$ and we define a space $(X, \mathscr{T})$. (As a set, X is the set $\{0,1\}^\beta$. The topology $\mathscr{T}$ is defined by the subbase that contains

(i) all subsets of $\{0,1\}^\beta$ of the form $\pi_J^{-1}(S) \times \{0,1\}^{\beta\backslash J}$ with $|J| < \omega$ and with S compact in the usual product topology of $\{0,1\}^J$, and

(ii) all sets of the form V_Σ for $\Sigma \in \boldsymbol{\Sigma}$.

The proof that X has property (*) is entirely analogous to the proof of (a) in Theorem 6.25; we will not repeat the details here.

We prove (b). Suppose that $\mathscr{T}^*(X) = \cup_{i<\beta}\mathscr{T}_i$ with $\kappa(\mathscr{T}_i) > 0$ for $i < \beta$. We assume without loss of generality, replacing if necessary the set $\mathscr{T}_i$ by $\{U \in \mathscr{T}^*(X)$: there is $V \in \mathscr{T}_i$ such that $V \subset U\}$, that if $U, V \in \mathscr{T}^*(X)$ and $V \subset U$ then

$$\min\{i: U \in \mathscr{T}_i\} \leq \min\{i: V \in \mathscr{T}_i\}.$$

We claim that there are $\bar{n} < \omega$ and a finite branch $\bar{\Sigma}$ of T such that if Σ is a finite branch and $\bar{\Sigma} \subset \Sigma$ then there is $i < \beta_{\bar{n}}$ such that $V_\Sigma \in \mathscr{T}_i$. If the claim fails there are a sequence $\{n_k : k < \omega\}$ of natural numbers and a sequence $\{\Sigma_k : k < \omega\}$ of finite branches such that

$$n_k < n_{k+1} \text{ and } \Sigma_k \subset \Sigma_{k+1} \quad \text{for } k < \omega, \text{ and}$$

$$\min\{n : V_{\Sigma_k} \in \bigcup_{i<\beta_n} \mathscr{T}_i\} = n_k \quad \text{for } k < \omega.$$

We set $\Sigma = \cup_{k<\omega}\Sigma_k$ and we choose $n < \omega$ such that $V_\Sigma \in \cup_{i<\beta_n}\mathscr{T}_i$. Since $\Sigma_k \subset \Sigma$ we have $V_\Sigma \subset V_{\Sigma_k}$ and hence $V_{\Sigma_k} \in \cup_{i<\beta_n}\mathscr{T}_i$, a contradiction. We choose $\bar{n}$ and $\bar{\Sigma}$ as given, we set

$$S_m = \{i < \beta_{\bar{n}} : \kappa(\mathscr{T}_i) > 1/m\} \quad \text{for } 1 \leq m < \omega,$$

we note that $\beta_{\bar{n}} = \cup\{S_m : 1 \leq m < \omega\}$, and we claim that there are $\bar{m}$ with $1 \leq \bar{m} < \omega$ and a finite branch Σ' of T with $\bar{\Sigma} \subset \Sigma'$ such that if Σ is a finite branch and $\Sigma' \subset \Sigma$ then there is $i \in S_{\bar{m}}$ such that $V_\Sigma \in \mathscr{T}_i$. (The proof of this claim is analogous to the proof just given; we omit the details.) We choose such $\bar{m}$ and Σ', we choose $k < \omega$ such that $(2/3)^k < 1/\bar{m}$, and we choose a finite branch $\tilde{\Sigma}$ such that

$$\Sigma' \subset \tilde{\Sigma} \text{ and } |\tilde{\Sigma}| \geq \bar{n} + 1.$$

We denote by $\tilde{s}$ the largest element of $\tilde{\Sigma}$ and we denote by $\tilde{n}$ that integer n such that $\tilde{s} \in T_n$.

We recall from the definition of the tree T that for every $t \in T$ there is $B_t \subset \beta$ such that $[B_t]^2$ is the set of immediate successors of t. We set

$$W(0) = \{\tilde{s}\} \quad \text{and}$$
$$W(p) = \bigcup_{t \in W(p-1)} [B_t]^2 \quad \text{for } 1 \leq p \leq k.$$

Now for $0 \leq p \leq k$ we define a family $\{P_{p,i} : i < \beta_{\bar{n}}\}$ such that

$W(p) = \cup_{i<\beta_{\bar{n}}} P_{p,i}$. (We begin with $p = k$ and we proceed by 'downward induction'.)

For $t \in W(k)$ let $\Sigma(t)$ be the (finite) branch of T determined by t (i.e., $\Sigma(t) = \{s \in T : s \preccurlyeq t\}$) and let

$$t \in P_{k,i} \quad \text{if } V_{\Sigma(t)} \in \mathscr{T}_i.$$

Let $0 < p \leq k$ and assume that the sets $\{P_{p,i} : i < \beta_{\bar{n}}\}$ have been defined; we define $\{P_{p-1,i} : i < \beta_{\bar{n}}\}$. Let $t \in W(p-1)$. There is $n < \omega$ such that $t \in T_n$; we have $\bar{n} < n$. There is $B_t \subset \beta$ with $|B_t| = \beta_{n+1}$ such that the immediate successors of t are the elements of the set

$$[B_t]^2 \subset W(p) = \bigcup_{i<\beta_{\bar{n}}} P_{p,i}.$$

Since β_{n+1} and $(\beta_{\bar{n}})^+$ are regular cardinals and $(\beta_{\bar{n}})^+ \ll \beta_{n+1}$, we have from Theorem 1.5(a) the arrow relation $\beta_{n+1} \to (\beta_{\bar{n}}^+)^2_{\beta_{\bar{n}}}$ and *a fortiori* $\beta_{n+1} \to (3)^2_{\beta_{\bar{n}}}$. Hence there are $C \subset B_t$ with $|C| = 3$ and $i < \beta_{\bar{n}}$ such that $[C]^2 \subset P_{p,i}$; we let $t \in P_{p-1,i}$.

The definition of $\{P_{p,i} : i < \beta_{\bar{n}}\}$ for $p \leq k$ is complete.

There is $\bar{i} < \beta_{\bar{n}}$ such that $\tilde{s} \in P_{0,\bar{i}}$.

It follows from the definition of the families $\{P_{p,i} : p \leq k, i < \beta_{\bar{n}}\}$ that there is $S \subset T_{\bar{n}+k}$ with $|S| = 3^k$ such that

$$\begin{aligned} &\tilde{\Sigma} \subset \Sigma(t) \quad \text{for } t \in S, \\ &\text{cal}\,\{V_{\Sigma(t)} : t \in S\} \leq 2^k \quad \text{and} \\ &V_{\Sigma(t)} \in \mathscr{T}_{\bar{i}} \quad \text{for } t \in S. \end{aligned}$$

We have

$$1/\bar{m} < \kappa(\mathscr{T}_{\bar{i}}) \leq \text{cal}\,\{V_{\Sigma(t)} : t \in S\}/|S| \leq (2/3)^k < 1/\bar{m},$$

a contradiction.

The proof is complete.

6.30 *Corollary.* Let α be an infinite cardinal. There is an extremally disconnected, compact Hausdorff space X such that X has property (*) (hence, X is a c.c.c. space) but for every set $\{\mu_i : i < \alpha\}$ of regular Borel probability measures on X there is a non-empty, open subset U of X such that $\mu_i(U) = 0$ for all $i < \alpha$.

Proof. Since there is a strong limit cardinal $\beta > \alpha$ such that $\text{cf}(\beta) = \omega$, this is immediate from Theorem 6.29.

Remark. For an infinite cardinal β, we say that a space X has *measure-calibre* β if for every family $\{U_\xi : \xi < \beta\}$ of non-empty, open subsets of X there are $B \subset \beta$ with $|B| = \beta$ and a regular, Borel probability measure μ on X such that $\mu(U_\xi) > 0$ for $\xi \in B$. (This notion may be compared with the notion of compact-calibre β described above in Chapter 2 and considered below in Chapter 8.)

It is clear that a space with calibre β, or a space with the property of Lemma 6.28(b), has measure-calibre β. It can be proved (for β a strong limit cardinal with $\mathrm{cf}(\beta) = \omega$) that the space X of Theorem 6.29 does not even have measure-calibre β.

The Galvin–Hajnal example

6.31 *Lemma.* Let $\kappa \geq \omega$. There is a family $\{S_\eta : \eta < 2^\kappa\}$ such that

$$S_\eta \subset \eta \quad \text{for } \eta < 2^\kappa,$$
$$[S_\eta]^2 \subset \bigcup_{\xi < 2^\kappa} (S_\xi \times \{\xi\}) \quad \text{for } \eta < 2^\kappa,$$
$$\text{type } S_\eta \leq \kappa \quad \text{for } \eta < 2^\kappa, \text{ and}$$
$$\text{if } S \subset 2^\kappa \text{ with } [S]^2 \subset \bigcup_{\xi < 2^\kappa} (S_\xi \times \{\xi\})$$

and type $S \leq \kappa$, then there is $\eta < 2^\kappa$ such that $S = S_\eta$.

Proof. Set $\mathscr{A} = \{A \subset 2^\kappa : \text{type } A \leq \kappa\}$ and let

$$\{A_\eta : \eta < 2^\kappa\}$$

be a well-ordering of $\mathscr{A}$ such that

$$|\{\eta : \eta < 2^\kappa, A = A_\eta\}| = 2^\kappa \quad \text{for } A \in \mathscr{A}.$$

Recursively for $\eta < 2^\kappa$ we define the sets S_η. We set

$$S_0 = \varnothing.$$

If $\eta < 2^\kappa$ and S_ξ has been defined for $\xi < \eta$, we define S_η. We consider two cases.

Case 1. $A_\eta \subset \eta$ and $[A_\eta]^2 \subset \bigcup_{\xi < \eta} \{\{\zeta, \xi\} : \zeta \in S_\xi\}$. We set

$$S_\eta = A_\eta.$$

Case 2. Case 1 fails. We set

$$S_\eta = \varnothing.$$

The definition of S_η for $\eta < 2^\kappa$ is complete.

We have $S_\eta = A_\eta$ or $S_\eta = \varnothing$ for $\eta < 2^\kappa$ and hence

$$\text{type } S_\eta \leq \kappa;$$

further for $[S_\eta]^2 \neq \varnothing$ we have

$$[S_\eta]^2 = [A_\eta]^2 \subset \bigcup_{\xi<\eta} \{\{\zeta,\xi\} : \zeta \in S_\xi\} \subset \bigcup_{\xi<2^\kappa} (S_\xi \times \{\xi\}).$$

Finally if $S \subset 2^\kappa$ with type $S \leq \kappa$ and if

$$[S]^2 \subset \bigcup_{\xi<2^\kappa} (S_\xi \times \{\xi\})$$

then (since $\mathrm{cf}(2^\kappa) > \kappa$) there is $\bar{\xi} < 2^\kappa$ such that

$$[S]^2 \subset \{\{\zeta, \xi\} : \zeta \in S_\xi, \xi < \bar{\xi}\}.$$

There is $\bar{\eta}$ such that $\bar{\xi} \leq \bar{\eta} < 2^\kappa$ and $S = A_{\bar{\eta}}$ and from

$$[A_\eta]^2 = [S]^2 \subset \bigcup_{\xi<\bar{\eta}} \{\{\zeta, \xi\} : \zeta \in S_\xi\}$$

we have

$$S_{\bar{\eta}} = A_{\bar{\eta}} = S,$$

as required.

The proof is complete.

Remark. Let X and Y be spaces and let κ be an infinite cardinal. If $X \leq Y$ and Y satisfies the (κ, κ)-chain condition, then X satisfies the (κ, κ)-chain condition; in particular, X satisfies the (κ, κ)-chain condition if and only if its Gleason space $G(X)$ satisfies the (κ, κ)-chain condition.

(This remark is proved by imitating the proof of Theorem 6.2(c). We omit the details.)

6.32 *Theorem.* Let κ be an infinite cardinal. There is an extremally disconnected, compact Hausdorff space X such that

(i) X has calibre α for every regular cardinal $\alpha > \kappa$; and

(ii) X does not satisfy the (κ, κ)-chain condition.

Proof. It follows from the remark above and Corollary 2.22 that it is enough to find a space X with (i) and (ii); for then the Gleason space of X will have all the required properties.

Let X as a set be equal to $\{0, 1\}^{2^\kappa}$, set

$$V_\eta = \{x \in X : x | S_\eta \equiv 0, x_\eta = 1\} \quad \text{for } \eta < 2^\kappa,$$

and give X the topology determined by the subbase $\{V_\eta : \eta < 2^\kappa\}$.

We verify that the space X satisfies (i) and (ii).

(i) Let α be regular with $\alpha > \kappa$ and let $\{U_\xi : \xi < \alpha\}$ be a family of non-empty, basic open subsets of X; for $\xi < \alpha$ there is a finite subset $F(\xi)$ of 2^κ such that

$$U_\xi = \bigcap_{\eta \in F(\xi)} V_\eta.$$

For $\xi < \alpha$ there is $x \in U_\xi$ and since $x|F(\xi) \equiv 1$ and $x|(\cup_{\eta \in F(\xi)} S_\eta) \equiv 0$ we have

$$(\bigcup_{\eta \in F(\xi)} S_\eta) \cap F(\xi) = \varnothing.$$

For $A \subset \alpha$ we set $F(A) = \cup_{\xi \in A} F(\xi)$. It is enough to show there is $A \subset \alpha$ with $|A| = \alpha$ such that

$$(\bigcup_{\eta \in F(A)} S_\eta) \cap F(A) = \varnothing.$$

We assume without loss of generality, using the Erdős–Rado theorem for quasi-disjoint families (Theorem 1.4), that there is a set J such that

$$F(\xi) \cap F(\xi') = J \quad \text{for } \xi < \xi' < \alpha;$$

we set

$$B(\xi) = F(\xi) \backslash J \quad \text{for } \xi < \alpha$$

and we assume without loss of generality, using the fact that α is a regular, uncountable cardinal, that there is $n < \omega$ such that

$$|B(\xi)| = n \quad \text{for } \xi < \alpha.$$

If $n = 0$ then $F(\xi) = J = F(0)$ for $\xi < \alpha$ and with $A = \alpha$ we have $F(A) = F(0)$ and hence

$$(\bigcup_{\eta \in F(A)} S_\eta) \cap F(A) = (\bigcup_{\eta \in F(0)} S_\eta) \cap F(0) = \varnothing.$$

We assume $n > 0$, we let

$$B(\xi) = \{\eta_\xi^1 < \eta_\xi^2 < \ldots < \eta_\xi^n\}$$

be the listing of the elements of $B(\xi)$ in the natural order inherited from 2^κ, and we choose $B \subset \alpha$ such that

$$|B| = \alpha \quad \text{and}$$
$$\eta_\xi^k < \eta_{\xi'}^k \quad \text{for } 1 \le k \le n, \xi < \xi', \xi, \xi' \in B.$$

We set

$$g(\eta) = \{\zeta \in B : B(\zeta) \cap S_\eta \neq \varnothing\} \quad \text{for } \eta < 2^\kappa \text{, and}$$
$$f(\xi) = \cup \{g(\eta) : \eta \in B(\xi)\} \quad \text{for } \xi \in B.$$

Since type $S_\eta \leq \kappa$ for $\eta < 2^\kappa$ and

$$B(\zeta) \cap B(\zeta') = \varnothing \quad \text{for } \zeta, \zeta' \in B, \zeta \neq \zeta',$$

we have

$$\text{type } g(\eta) \leq \kappa \quad \text{for } \eta < 2^\kappa$$

and hence

$$\text{type} f(\xi) \leq \kappa \cdot n \text{ (ordinal product) for } \xi \in B.$$

Since $\xi \notin f(\xi)$ for $\xi \in B$, it follows from (the case $\rho = \kappa \cdot n$ of) Theorem 1.11 that there is $A \subset B$ such that

$$|A| = \alpha, \text{ and}$$
$$\zeta \notin f(\xi) \quad \text{for } \xi, \zeta \in A.$$

Now suppose there are $\xi, \zeta \in A$ and $\eta \in F(\xi)$ such that $S_\eta \cap F(\zeta) \neq \varnothing$. Then $\xi \neq \zeta$ and from $\eta \notin F(\zeta)$ we have $n \notin J$ and hence $\eta \in B(\xi)$; further from

$$S_\eta \cap J \subset S_\eta \cap F(\xi) = \varnothing$$

we have $S_\eta \cap B(\zeta) \neq \varnothing$ and hence

$$\zeta \in g(\eta) \subset f(\xi),$$

a contradiction. Thus $(\cup_{\eta \in F(A)} S_\eta) \cap F(A) = \varnothing$, as required.

(ii) It is enough to show that if $\mathscr{T}^*(X) = \cup_{i<\kappa} \mathscr{T}_i$ then there are $\bar{i} < \kappa$, and $A \subset 2^\kappa$ with $|A| = \kappa$, such that

$$\{V_\eta : \eta \in A\} \subset \mathscr{T}_{\bar{i}}, \quad \text{and}$$
$$[A]^2 \subset \bigcup_{\xi < 2^\kappa} (S_\xi \times \{\xi\});$$

indeed then if there is $x \in V_\eta \cap V_{\eta'}$ with $\eta, \eta' \in A$ and $\eta < \eta'$ we have from $\{\eta, \eta'\} \in S_{\eta'} \times \{\eta'\}$ that $\eta \in S_{\eta'}$ and hence $x_\eta = 0$ (since $x \in V_{\eta'}$) and $x_\eta = 1$ (since $x \in V_\eta$), a contradiction.

We fix $f : \kappa \to \kappa$ such that

$$|f^{-1}(\{i\})| = \kappa \quad \text{for } i < \kappa$$

and we define a subset $\{\eta_\xi : \xi < \kappa\}$ of 2^κ such that

$\eta_0 = 0;$

$\{\eta_{\xi'} : \xi' < \xi\} \subset S_{\eta_\xi}$ for $\xi < \kappa$; and

if $\xi < \kappa$ and $f(\xi) = i$, and if there is $\eta < 2^\kappa$ such that $V_\eta \in \mathscr{T}_i$ and $\{\eta_{\xi'} : \xi' < \xi\} \subset S_\eta$, then $V_{\eta_\xi} \in \mathscr{T}_i$.

We proceed by recursion.

We set $\eta_0 = 0$.

If $\xi < \kappa$ and $\eta_{\xi'}$ has been defined for $\xi' < \xi$, then

$$[\{\eta_{\xi'} : \xi' < \xi\}]^2 \subset \bigcup_{\eta < 2^\kappa} (S_\eta \times \{\eta\})$$

and hence there is $\eta < 2^\kappa$ such that $\{\eta_{\xi'} : \xi' < \xi\} \subset S_\eta$. We choose for η_ξ such an ordinal η; if there is such an ordinal η so that (in addition) $V_\eta \in \mathscr{T}_{f(\xi)}$, we choose η_ξ to satisfying also this condition.

The definition of the set $\{\eta_\xi : \xi < \kappa\}$ is complete. Since

$$\eta_{\xi'} < \eta_\xi \quad \text{for } \xi' < \xi < \kappa$$

we have type $\{\eta_\xi : \xi < \kappa\} = \kappa$ and hence there is $\bar{\eta} < \kappa$ such that

$$\{\eta_\xi : \xi < \kappa\} = S_{\bar{\eta}}.$$

There is $\bar{i} < \kappa$ such that $V_{\bar{\eta}} \in \mathscr{T}_{\bar{i}}$. We set

$$A = \{\eta_\xi : \xi < \kappa, f(\xi) = \bar{i}\}$$

and we note that

$$|A| = \kappa,$$

$$V_\eta \in \mathscr{T}_{\bar{i}} \quad \text{for } \eta \in A, \text{ and}$$

$$[A]^2 \subset [S_{\bar{\eta}}]^2 \subset \bigcup_{\eta < 2^\kappa} (S_\eta \times \{\eta\}),$$

as required.

The proof is complete.

6.33 *Corollary*. There is an extremally disconnected, compact Hausdorff space X such that

(i) every uncountable, regular cardinal number is a calibre for X, and

(ii) X does not have a strictly positive measure (and in fact X does not have property (*)).

Proof. This follows from Lemmas 6.2 and 6.22(a) and (the case $\kappa = \omega$ of) Theorem 6.31.

6.34 *Problem.* Is there a compact Hausdorff space satisfying the (ω, ω)-chain condition but not condition $(*)$?

Property $(**)$ for non-compact spaces

Every space with a strictly positive measure has property $(**)$, and the converse holds for compact, Hausdorff spaces (cf. Theorem 6.4). We show now that this converse fails for appropriate completely regular, Hausdorff spaces; the examples we define have precalibre α for every cardinal α such that $\mathrm{cf}(\alpha) > \omega$.

The first of these examples is σ-compact (and hence normal); the second, the Pixley–Roy space of 6.37, is not normal and is first-countable.

6.35 *Theorem.* There is a completely regular, Hausdorff space X such that

(a) X does not have a strictly positive measure;

(b) X has property $(**)$ (hence $G(X)$ and βX have strictly positive measures); and

(c) X has precalibre α for every cardinal α with $\mathrm{cf}(\alpha) > \omega$.

Proof. Using the notation of Chapter 4, we set

$$X = \Sigma_\omega \{0, 1\}^{\omega^+},$$

i.e.,

$$X = \{x \in \{0, 1\}^{\omega^+} : |\{\eta < \omega^+ : x_\eta \neq 0\}| < \omega\}.$$

We verify statements (a), (b) and (c).

(a) We suppose there is a strictly positive measure μ on X, we choose a pseudobase $\mathscr{B}$ for X such that $\mu(B) > 0$ for all $B \in \mathscr{B}$, we set

$$U_\xi = \{x \in X : x_\xi = 1\} \quad \text{for } \xi < \omega^+,$$

and for $\xi < \omega^+$ we choose $B_\xi \in \mathscr{B}$ such that $B_\xi \subset U_\xi$. There are $\varepsilon > 0$ and $A \subset \omega^+$ with $|A| = \omega^+$ such that

$$\mu(B_\xi) \geq \varepsilon \quad \text{for } \xi \in A.$$

Let $\{\xi(n):n<\omega\}$ be a subset of A faithfully indexed by ω and set

$$S_k = \bigcap_{n\geq k} B_{\xi(n)} \quad \text{for } k<\omega.$$

From the definition of the set X it follows that $\cap_{k<\omega} S_k = \varnothing$, and from Theorem C.1 of the Appendix we have $\mu(\cap_{k<\omega} S_k) \geq \varepsilon$. This contradiction completes the proof of (a).

(b) The space $\{0,1\}^{\omega^+}$, a product of separable spaces, is a Hausdorff compactification of X. The space $\{0,1\}^{\omega^+}$ has a strictly positive measure by Corollary 6.8, and the statements of (b) follow from Theorem 6.4 and Corollary 6.6.

(c) If $\mathrm{cf}(\alpha)>\omega$ then $\{0,1\}^{\omega^+}$ has calibre α (and hence precalibre α) by Theorem 3.18(a). Hence the space X, a dense subspace of $\{0,1\}^{\omega^+}$, has precalibre α (cf. Theorems 2.16(b) and 2.17(b)).

The proof is complete.

We note that it follows from the Hewitt–Marczewski–Pondiczery theorem (B.2 of the Appendix) that the space $\{0,1\}^{\omega^+}$ is itself separable. That $\{0,1\}^{\omega^+}$ has property (**) is then immediate, and statement (c) of Theorem 6.35 follows from Theorem 2.7.

That the space X of Theorem 6.35 is σ-compact is a special case of Lemma 8.9 below. For a direct argument one may set

$$X_n = \{x\in\{0,1\}^{\omega^+} : |\{\eta<\omega^+ : x_\eta \neq 0\}| \leq n\} \quad \text{for } n<\omega$$

and note that

X_n is closed in $\{0,1\}^{\omega^+}$ (hence X_n is compact), and

$$X = \bigcup_{n<\omega} X_n.$$

We define the Pixley–Roy space. For a non-empty, finite subset F of $\mathbb{R}$ and an open subset V of $\mathbb{R}$, we denote by $U(F,V)$ the set

$$U(F,V) = \{G : F\subset G\subset V, |G|<\omega\}.$$

We note that if F is not a subset of V then $U(F,V)=\varnothing$.

Definition. The elements of the Pixley–Roy space X are the non-empty, finite subsets of $\mathbb{R}$. The topology $\mathscr{T}(X)$ is the topology generated by the subbase

$$\mathscr{S}(X) = \{U(F,V) : F\in X, V\in\mathscr{T}^*(\mathbb{R})\}.$$

6.36 *Lemma.* The Pixley–Roy space X has these properties:

(a) the subbase $\mathscr{S}(X)$ is a base for (the topology of) X;

(b) the elements of $\mathscr{S}(X)$ are open-and-closed in X;

(c) X is a completely regular, Hausdorff space.

Proof. (a) It is enough to show that if $\{U(F_k, V_k) : k < n\} \subset \mathscr{S}(X)$ with $n < \omega$ then $\cap_{k<n} U(F_k, V_k) \in \mathscr{S}(X)$. Set

$$F = \bigcup_{k<n} F_k \text{ and } V = \bigcap_{k<n} V_k.$$

Then $\cap_{k<n} U(F_k, V_k) = U(F, V) \in \mathscr{S}(X)$, as required.

(b) We show that if $U(F, V) \in \mathscr{S}(X)$, then $U(F, V)$ is closed in X. If $G \in X \backslash U(F, V)$, then one of the inclusions $F \subset G$, $G \subset V$ must fail. If $F \not\subset G$ there is $p \in F \backslash G$ and then $U(G, \mathbb{R} \backslash \{p\})$ is a neighborhood of G such that

$$U(F, V) \cap U(G, \mathbb{R} \backslash \{p\}) = \varnothing,$$

and if $G \not\subset V$ then $U(G, \mathbb{R})$ is a neighborhood of G such that

$$U(F, V) \cap U(G, \mathbb{R}) = \varnothing.$$

(c) It is clear that for $F \in X$ we have

$$\{F\} = \cap \{U(F, V) : F \subset V \in \mathscr{T}(\mathbb{R})\},$$

and hence $\{F\}$ is a closed subset of X. Statement (c) now follows from (a) and (b).

6.37 *Theorem.* Let X be the Pixley–Roy space. Then

(a) X does not have a strictly positive measure;

(b) βX is separable (hence βX has a strictly positive measure and X has property (**)); and

(c) X has precalibre α for every cardinal α with $\mathrm{cf}(\alpha) > \omega$.

Proof. (a) We suppose there is a strictly positive measure μ on X, we choose a pseudobase $\mathscr{B}$ for X such that $\mu(B) > 0$ for all $B \in \mathscr{B}$, and for $F \in X$ we choose $B(F) \in \mathscr{B}$ such that

$$B(F) \subset U(F, \mathbb{R}).$$

We set

$$X_n = \{F \in X : \mu(B(F)) \geq 1/n\} \text{ for } n < \omega$$

and we note that since $X = \cup_{n<\omega} X_n$ there is $\bar{n} < \omega$ such that $|X_{\bar{n}}| \geq \omega$. Let $\{F_n : n < \omega\}$ be a subset of $X_{\bar{n}}$ faithfully indexed by ω, and set

$$S_k = \bigcup_{n \geq k} B(F_n) \text{ for } k < \omega, \text{ and}$$

$$S = \bigcap_{k<\omega} S_k.$$

From Theorem C.1 of the Appendix we have $\mu(S) \geq 1/\bar{n}$ and hence there is $F \in S$. For $k < \omega$ there is $n(k) \geq k$ such that

$$F \in B(F_{n(k)}) \subset U(F_{n(k)}, \mathbb{R})$$

and hence

$$\bigcup_{k<\omega} F_{n(k)} \subset F;$$

it follows that $|F| = \omega$, a contradiction.

(b) Let $\mathscr{V} = \{V_n : n < \omega\}$ be a countable base for the usual topology of $\mathbb{R}$ such that $\mathscr{V}$ is closed under finite unions, and for $n < \omega$ set

$$\mathscr{U}_n = \{U(F, V) : F \in X, V \in \mathscr{T}(\mathbb{R}), F \subset V_n \subset V\}.$$

It is then clear, retaining the notation

$$\mathscr{S}(X) = \{U(F, V) : F \in X, V \in \mathscr{T}^*(\mathbb{R})\},$$

that $\mathscr{U}_n \subset \mathscr{S}(X)$ for $n < \omega$. Further, since $\mathscr{V}$ is a basis for $\mathbb{R}$ closed under finite unions, for $F \in X$ and $V \in \mathscr{T}^*(\mathbb{R})$ with $F \subset V$ there is $n < \omega$ such that

$$F \subset V_n \subset V$$

and hence $U(F, V) \in \mathscr{U}_n$; it follows that

$$\mathscr{S}(X) = \bigcup_{n<\omega} \mathscr{U}_n.$$

For $n < \omega$ the set $\mathscr{U}_n$ has the finite intersection property; indeed if $m < \omega$ and

$$U(F(k), V(k)) \in \mathscr{U}_n \text{ for } k < m,$$

then with

$$F = \bigcup_{k<m} F(k), \quad V = \bigcap_{k<m} V(k)$$

we have

$$\bigcap_{k<m} U(F(k), V(k)) = U(F, V) \in \mathscr{U}_n.$$

Hence for $n < \omega$ there is

$$p_n \in \cap \{\mathrm{cl}_{\beta X} U : U \in \mathscr{U}_n\}.$$

We claim that $\{p_n : n < \omega\}$ is dense in βX.

If W is a non-empty, open subset of βX there is a non-empty open subset of βX such that

$$S \subset \mathrm{cl}_{\beta X} S \subset W,$$

there is $U(F, V) \in \mathscr{S}(X)$ such that

$$\varnothing \neq U(F, V) \subset S \cap X,$$

and there is $n < \omega$ such that $U(F, V) \in \mathscr{U}_n$; it follows that

$$p_n \in \mathrm{cl}_{\beta X} U(F, V) \subset \mathrm{cl}_{\beta X}(S \cap X) \subset W,$$

as required.

The proof that βX is separable is complete. The remaining statements of (b) follow from 6.2–6.7 above.

(c) For $\mathrm{cf}(\alpha) > \omega$, the (separable) space βX has calibre α (by Theorem 2.7); hence X has precalibre α.

The proof is complete.

We note in passing that the Pixley–Roy space is first-countable. Indeed for $F \in X$ set

$$V_n = \cup \{(x - 1/n, x + 1/n) : x \in F\};$$

then $\{U(F, V_n) : n < \omega\}$ is a countable local neighborhood basis in X at F.

Another class of examples of spaces with property (**), but with no strictly positive measure, described by Koumoullis, is discussed in the Notes to this chapter.

6.38 *Problem.* Find a topological characterization of the (completely regular, Hausdorff) spaces with a strictly positive measure.

A possible characterization is the following condition, a strengthening of (**) suggested by the examples of 6.35 and 6.37: The set $\mathscr{T}^*(X)$ can be written in the form $\mathscr{T}^*(X) = \cup_{n<\omega} \mathscr{T}_n$ where, for $n < \omega, \kappa(\mathscr{T}_n) > 0$ and

$$\text{if } \{U_k : k < \omega\} \subset \mathscr{T}_n \text{ then } \cap_{m<\omega} [\cup_{k>m} U_k] \neq \varnothing.$$

6.39 *Problem* (Negrepontis). Is there a compact c.c.c. space X, with no strictly positive measure, that is *point-homogeneous* (in the sense that for $p, q \in X$ there is a homeomorphism h of X onto X such that $h(p) = q$)?

Recall in this connection that every compact topological group has a (strictly positive) Haar probability measure.

Notes for Chapter 6

The study of strictly positive measures originated in the theory of Boolean algebras (especially in the work of Tarski) and in topological measure theory; for older references the reader should consult the text of Sikorski (1964), especially section 42. Our definition of a (space with a) strictly positive measure, which appears at first glance unorthodox, was dictated by two considerations: (a) The definition should be sufficiently general to ensure Theorem 6.7 and (b) for compact Hausdorff spaces the definition should be equivalent to the property that there exist a regular Borel probability measure assigning positive measure to every non-empty open set.

Lemmas 6.2(a), (b) and 6.3, and Theorem 6.4 ((b)$\Leftrightarrow$(c)) for the case of totally disconnected compact Hausdorff spaces, are due to Kelley (1959). The equivalence ((a)$\Leftrightarrow$ (d)) of Theorem 6.4 has also been noted by Hebert & Lacey (1968, Theorems 4.6 and 4.12).

The implication (a)$\Rightarrow$(d) of Theorem 6.4, that if a compact space X has a strictly positive measure then its Gleason space $G(X)$ does also, can be proved without any appeal to condition (**) or the relation $\equiv$. The canonical projection from $G(X)$ onto X furnishes an embedding of $C(X)$ into $C(G(X))$, and the given (strictly positive) measure on X furnishes a positive linear functional Λ on $C(X)$ such that $\Lambda(1) = 1$. There is by the Hahn–Banach theorem a positive linear extension $\bar{\Lambda}$ of Λ defined on $C(G(X))$, and by the Riesz representation theorem (C.10 of the Appendix) there is a regular Borel probability measure $\bar{\mu}$ on $G(X)$ such that

$$\bar{\Lambda}(f) = \int_{G(X)} f \, d\bar{\mu} \quad \text{for } f \in C(G(X));$$

that $\bar{\mu}$ is strictly positive on $G(X)$ follows from the fact that the set $\mathscr{R}(X)$ of regular-open subsets of X is identified, via the Stone representation of $\mathscr{R}(X)$, with a base for $G(X)$. A similar argument, using the fact that every element of $C^*(X)$ extends to an element of $C^*(\beta X)$ (Theorem B.5(a) of the Appendix), shows that if a completely regular, Hausdorff space has a strictly positive measure then its Stone–Čech compactification does also.

For $0 < \theta \leq 1$, a space X is said to have *property* $(**)_\theta$ if the set $\mathscr{T}^*(X)$ can be written in the form $\mathscr{T}^*(X) = \cup_{n<\omega} \mathscr{T}_n$ with $\kappa(\mathscr{T}_n) \geq \theta$ for $n < \omega$.

For a compact, Hausdorff space X we denote by $M(X)$ the set of finite, regular Borel measures on X with the w^*-topology (as dual of the Banach space $C(X)$), and by $M_1^+(X)$ and $S_{M(X)}$ the subset of probability measures and the unit ball, respectively. The following statements have been proved.

(a) X is separable if and only if X has property $(**)_1$.

(b) (Fakhoury 1976, Mägerl & Namioka 1980). The following conditions are equivalent.

(1) X has property $(**)_\theta$ for some $\theta, 0 < \theta < 1$;

(2) X has property $(**)_\theta$ for all $\theta, 0 < \theta < 1$;

(3) $M_1^+(X)$ is w^*-separable;

(4) $S_{M(X)}$ is w^*-separable;

(5) there is an isomorphic embedding of $C(X)$ into $l^\infty = C(\beta \mathbb{N})$; and

(6) there is an isometric embedding of $C(X)$ into l^∞.

(c) $M(X)$ is separable if and only if there is a one-to-one continuous, linear operator from $C(X)$ into l^∞.

(d) X is separable $\Rightarrow X$ has property $(**)_\theta$ for some (or all) θ, $0 < \theta < 1 \Rightarrow M(X)$ is w^*-separable $\Rightarrow X$ has property $(**)$.

(e) The space Ω of Theorem 6.9 (and of Theorem 6.18 for $\beta = \omega$), which is the space Ω for which $C(\Omega) = L[0, 1]$, is a non-separable, extremally disconnected Hausdorff space with property $(**)_\theta$ for $0 < \theta < 1$ such that, assuming $\omega^+ = 2^\omega$, Ω does not have calibre ω^+.

(f) Talagrand (1980b) has shown, assuming $\omega^+ = 2^\omega$, that there is a compact, Hausdorff space Ω such that $M(\Omega)$ is w^*-separable and $M_1^+(\Omega)$ is not w^*-separable (i.e., Ω does not have property $(**)_\theta$ for $0 < \theta < 1$).

(g) Kalamidas has shown that if $\{X_i : i \in I\}$ is a set of compact, Hausdorff spaces with property $(**)_\theta$ for $0 < \theta < 1$ and if $|I| \leq 2^\omega$, then the product space $\prod_{i \in I} X_i$ also has property $(**)_\theta$.

(h) There are of course spaces Ω with property (**) such that $M(\Omega)$ is not w^*-separable. (Indeed if $M(\Omega)$ is w^*-separable then, as is obvious from (c) above, we have $w(M(\Omega)) \leq 2^\omega$; but for every cardinal α there is Ω with property (**) such that $w(\Omega) \geq \alpha$.) It follows from Pełczynski (1958) and Rosenthal (1970b) (Theorem 4.8 and following Remarks) that with Ω the space such that $C(\Omega) = L^\infty(\{0,1\}^{\omega^+})$, the space Ω has property (**) and weight equal to 2^ω and $M(\Omega)$ is not w^*-separable.

(i) (Argyros & Negrepontis 1982).

(1) X has a strictly positive measure if and only if $M_1^+(X)$ has a strictly positive measure;

(2) if $\alpha > \omega^+$ with α regular and X has a strictly positive measure, then $M_1^+(X)$ has calibre α.

(3) assuming $\omega^+ = 2^\omega$, the space X defined by Haydon (1978) (for another purpose) has a strictly positive measure and $M_1^+(X)$ does not have calibre ω^+.

(4) for $2 \leq n < \omega$, X has property K_n if and only if $M_1^+(X)$ has property K_n.

(5) X is productively c.c.c. if and only if $M_1^+(X)$ is productively c.c.c.

(6) X^I is c.c.c. for all sets I if and only if $M_1^+(X)$ is c.c.c.

(7) $M_1^+(X)$ has calibre ω^+ if and only if for every family $\{V_\xi : \xi < \omega^+\}$ of non-empty, open subsets of X, and for $0 < \theta < 1$, there is $A \subset \omega^+$ with $|A| = \omega^+$ such that $\kappa(\{V_\xi : \xi \in A\}) \geq \theta$.

We cite two related problems. Here as above, X remains a compact, Hausdorff space.

Problem (Mägerl & Namioka, 1980). If $M(X)$ is w^*-separable, must $M(G(X))$ be w^*-separable? (Here as usual $G(X)$ denotes the Gleason space of X.)

Problem (Rosenthal, 1970a). If X is extremally disconnected and $M(X)$ is w^*-separable, must $M_1^+(X)$ be w^*-separable?

The measure-theoretic argument of Corollary 6.8 was introduced by Oxtoby (1961, statement 2.7) to show that the product of separable spaces has the countable chain condition (cf. also the case $\kappa = \omega$, $\alpha = \omega^+$ of Theorem 3.28 above).

The material in 6.12–6.17 is due to Argyros & Kalamidas (1981); for regular $\alpha > \omega$, Corollary 6.17 has been known for some years to Kunen. Lemma 6.12(i) is, of course, an immediate consequence of Jensen's elementary inequality (see for example Spivak (1967, p. 199)).

It was noted by Erdős many years ago, but not published at the time, that assuming the continuum hypothesis the Stone space Ω of the measure algebra of Lebesgue measure on $[0, 1]$ does not have calibre ω^+; the more general statement in Theorem 6.18 is due to Argyros & Tsarpalias (1981). Four proofs of Erdős' theorem are given by Tall (1974) and Kunen & Tall (1979); in the former paper it is noted (Example 7.5) that if Martin's axiom is assumed then Ω does not have calibre 2^ω.

Assuming $\omega^+ = 2^\omega$, Haydon (1978) has given a construction in functional analysis, roughly related to Erdős' example, which answers in the negative a conjecture of Pelczynski (1968). This example has proved useful on various occasions; the (projective limit) construction that yields this example apparently has considerable similarity with the following constructions (all assuming CH): the result of Kunen (1981) establishing the existence of a compact L-*space* (i.e. a non-separable, hereditarily Lindelöf space); an example of Losert (1979, Theorem 3) on the existence of a compact separable space with a probability measure μ such that μ has a uniformly distributed sequence, but μ does not have a well-distributed sequence; and the example of Talagrand (1980a) mentioned above, establishing the existence of a compact space X such that $M(X)$ is w^*-separable but $M^1_+(X)$ is not w^*-separable. Fleissner & Negrepontis (1979) have found a model of $\mathrm{ZFC} + (\omega^+ < 2^\omega)$ in which Haydon's construction can be repeated. This raises the question, probably to be answered positively, whether it is possible to find, in some situation where the continuum hypothesis fails, a compact Hausdorff space with a strictly positive measure and without calibre ω^+.

The following statement, in its essentials a special case of Theorem 6.21, is attributed by Knaster (1945, footnote 11) to Marczewski ($\equiv$ Szpilrajn) (1945): Of any ω^+ subsets of $\mathbb{R}$ with positive Lebesgue measure, some ω^+ have pairwise non-empty intersection. See in this connection the later paper of Marczewski (1947, 1.1 (iii)).

That every Stone space $\Omega = S(\mathscr{A})$ for $\mathscr{A}$ a measure algebra has property (K) (a special case of Lemma 6.22) is due to Horn & Tarski (1948) (Theorem 2.4 (i)$\Rightarrow$(iii)); it is noted by Halmos (1963, Section 24, Exercise 2) that Ω is not separable if $(X, \mathscr{S}, \lambda)$ is atomless.

Theorem 6.22(b) is due to Kalamidas.

Rosenthal (1970a) has studied the existence of a strictly positive measure on a (compact, Hausdorff) space X in relation to the Banach space $C(X)$ of real-valued, continuous functions on X and its dual space $M(X)$ of regular Borel measures on X. He has shown for example that

(a) if X is a compact, Hausdorff space with c.c.c. and $C(X)$ is a conjugate Banach space, then X has a strictly positive measure (Rosenthal 1970a, Theorem 4.1); and

(b) a compact, Hausdorff space X has a strictly positive measure if and only if there is a weakly compact subspace of $M(X)$ whose convex hull is weak*-dense in $M(X)$ (Rosenthal 1970a, Theorem 4.5(b)).

The example of 6.23, due to Gaifman (1964), was for a long time the only space known with the countable chain condition and with no strictly positive measure. Parts (c) and (d) of Theorem 6.23 are due to Kalamidas and Negrepontis, respectively. A simplified version of Gaifman's example is given by Fremlin (1980).

Theorems 6.25, 6.26 and 6.29 are due to Argyros (1981d).

Rubin & Shelah (1981) have also constructed, assuming the continuum hypothesis, for $2 \leq m < \omega$ a (compact, Hausdorff, extremally disconnected) space which has property K_m and does not have property K_{m+1}. Their construction, motivated by certain model-theoretic considerations, is related both to property (c) of Lemma 7.7 (and the resulting property (i) of Theorem 7.9) and to property (ii) of Lemma 7.12 (and the resulting property $S(X) \leq \alpha^+$ in Theorem 7.13).

The constructions of Gaifman and Argyros can be generalized as follows.

Replacing in Theorem 6.23 the space $\mathbb{R}$ by the Stone space of the α-homogeneous, α-universal Boolean algebra, whose essential properties are given by Negrepontis (1969), Negrepontis (1981) proved that for $\alpha = \alpha^{\underline{\alpha}}$ there is a (generalized Gaifman) extremally disconnected, compact Hausdorff space X such that

(a) X has property $(*(\alpha))$ – i.e., the set $\mathcal{T}^*(X)$ can be written in the form $\mathcal{T}^*(X) = \cup_{i<\alpha}\mathcal{T}_i$ where, for $i < \alpha$, no k_i elements of the family $\mathcal{T}_i$ are pairwise disjoint (here $\{k_i : i < \alpha\}$ is a set of natural numbers);

(b) X does not have property $(**(\alpha))$ – i.e., the set $\mathcal{T}^*(X)$ cannot be written in the form $\mathcal{T}^*(X) = \cup_{i<\alpha}\mathcal{T}_i$ with $\kappa(\mathcal{T}_i) > 0$ for $i < \alpha$;

(c) X has property $K_{\alpha^+,n}$ for $2 \leq n < \omega$; and

(d) assuming $\alpha^+ = 2^\alpha$, X does not have calibre α^+.

Replacing in Theorem 6.26 the set $\{0, 1\}^\omega$ by the space $(\{0, 1\}^\alpha)_\alpha$, whose essential properties are given by Hung & Negrepontis (1973, 1974), Argyros and Negrepontis proved that for $\alpha = \alpha^{\underline{\alpha}}$ and $2 \leq m < \omega$ there is a (generalized Argyros) extremally disconnected, compact Hausdorff space X such that

(a) X has property $(*(\alpha))$;

(b) X does not have property $(**(\alpha))$;

(c) X has property $K_{\alpha^+,m}$; and

(d) assuming $\alpha^+ = 2^\alpha$, X does not have property $K_{\alpha^+,m+1}$.

For $2 \leq m < \omega$, a space X is *σ-m-linked* if the set $\mathcal{T}^*(X)$ of non-empty, open subsets of X can be written in the form $\mathcal{T}^*(X) = \cup_{n<\omega}\mathcal{T}_n$ where, for $n < \omega$, every m elements of $\mathcal{T}_n$ have non-empty intersection. It has been shown by Bell (1982b), in ZFC with no special set-theoretic assumptions, that for $2 \leq m < \omega$ there is a compactification $B\omega$ of ω such that the remainder $B\omega\backslash\omega$ is σ-m-linked but not σ-$(m + 1)$-linked.

The work of Gaifman (1964) left unresolved the question, posed by Horn & Tarski (1948, p. 482), whether every c.c.c. space has property $(*)$. The example of 6.32, due to Galvin & Hajnal (1981), settles a strong (and generalized) form of this question; in addition to the properties given in the statement of Theorem 6.32, we note that the space X defined there can be proved to have density character κ^+ and to have both weight and pseudoweight equal to 2^κ. We note further that, using the free-set theorem of Hajnal (1961) for singular cardinals (a modification of Theorem 1.11 not proved in this monograph) and also the extension to singular cardinals of the Erdős–Rado theorem on quasi-disjoint families (Theorem 1.9), the Galvin–Hajnal space of Theorem 6.32 has, for example in the case $\kappa = \omega$, calibre α whenever $\mathrm{cf}(\alpha) > \omega$.

A preliminary and weak version of Theorem 6.32, based in part on the work of Erdős, Galvin & Hajnal (1975), was obtained some years ago by Hajnal (unpublished notes).

We thank Fred Galvin for providing us with historical information and notes concerning Theorem 6.32.

That the σ-compact space $\sum_{\omega}\{0, 1\}^{\omega^+}$ has the properties given in Theorem 6.35 was proved by Argyros (October, 1979).

The Pixley–Roy space of Lemma 6.36 was introduced by Pixley & Roy (1969) in connection with the study of Moore spaces. Our proofs of Lemma 6.36 and Theorem 6.37(b) follow closely work of van Douwen (1977a, b), who has defined and investigated several classes of spaces of Pixley–Roy type and computed a number of cardinal invariants associated with them (including the density character of the associated Stone–Čech compactification); independently A. Sapounakis observed (August, 1979) that the Pixley–Roy space does not have a strictly positive measure and its Stone–Čech compactification does.

G. Koumoullis has observed (November, 1979) that every non-separable Banach space X, with its weak topology (induced by its dual space X^*), satisfies property (**) but does not have a strictly positive measure.

Indeed, X with this topology satisfies property (**) because it is homeomorphic to a dense subspace of $\mathbb{R}^{(X^*)}$ (cf. 2.16(b) and (6.2(d)). If X has a strictly positive measure μ, then the unit ball S of X^* with the weak* topology satisfies the following conditions: S is homeomorphic to a subspace of $\mathbb{R}^X$ (in the product topology); S is compact (and convex); and if $f, g \in S$ and $f = g$ μ-almost everywhere, then $f = g$. According to a theorem of A. I. Tulcea (1974), it now follows that S with the weak* topology is metrizable. Since X is a subspace of $C(S)$ with the uniform norm and $C(S)$ is separable, it follows that X itself is separable, a contradiction.

The following remarks point to a connection between the spaces with a strictly positive measure and the geometric theory of Banach spaces.

A norm $\| \ \|$ on a linear space B is said to be *strictly convex* if $x, y \in B$ with $x \neq y$ and $\|x\| = \|y\| = 1$ imply $\|(x+y)/2\| < 1$; a Banach space is *strictly convexifiable* if it has an equivalent norm that is strictly convex.

It is a difficult problem in the geometry of Banach spaces to determine those compact spaces X for which the Banach space $C(X)$ is strictly convexifiable (cf. Dashiell & Lindenstrauss, 1973). We outline now a proof that $C(X)$ is strictly convexifiable whenever X has a strictly positive measure. First, every space of the form $L^{\infty}(\mu)$, for μ a finite measure, is the dual of $L^{1}(\mu)$, a weakly compactly generated space; by the theorem of Amir & Lindenstrauss (1968), there are a set Γ and a one-to-one bounded linear operator $T: L^{\infty}(\mu) \rightarrow c_0(\Gamma)$. It then follows from Day's theorem (1955) and a remark of Klee (1953) that $L^{\infty}(\mu)$ is strictly convexifiable. Now if X has a strictly positive measure μ, then $C(X)$ is a subspace of $L^{\infty}(\mu)$ in a natural way and is therefore itself strictly convexifiable. (A more direct proof, also noted by R. R. Phelps, is as follows. Let μ be a strictly positive measure on X, identify $C(X)$ in the natural way with a subspace of $L^{2}(X, \mu)$, let $\| \ \|_u$ and $\| \ \|_2$ denote respectively the uniform and the L^2 norms on $C(X)$, and define $\|f\| = \|f\|_u + \|f\|_2$ for $f \in C(X)$. Then $\| \ \|$ is a compatible norm for $C(X)$, and $\| \ \|$ is strictly convex because $\| \ \|_2$ is strictly convex. A simple extension of this argument shows that a compact space has a strictly positive measure if and only if there is a one-to-one positive continuous linear function from $C(X)$ into a space of the form $L^{2}(Y, \nu)$ with ν a finite measure on Y.)

It can be verified for the spaces X of Argyros given in Theorem 6.26 that $C(X)$ is not strictly convexifiable; Argyros' spaces are to our knowledge the only examples known of c.c.c. spaces X for which $C(X)$ is not strictly convexifiable. These remarks suggest this question, posed by Argyros and Negrepontis: Does every compact c.c.c. space X such that $C(X)$ has an equivalent strictly convex norm have a strictly positive measure? We indicate in the Notes for Chapter 7, using results on Corson-compact spaces, that this question, assuming the continuum hypothesis, has a negative answer.

7

Between Property (K) *and the Countable Chain Condition*

We have seen in Chapter 6 that the c.c.c. property (for compact spaces) branches into two stronger and logically independent properties: the property of calibre ω^+ and the existence of a strictly positive measure. These two properties have in fact a common denominator, stronger than c.c.c.: Knaster's property (K) (indeed, property K_n for $2 \le n < \omega$). Between the c.c.c. and property (K) there is a significant qualitative difference: the c.c.c. cannot be proved in ZFC to be productive, but property (K) is productive (Theorem 2.2(a)).

In this chapter we are concerned with differentiating property (K) from c.c.c., and with properties that lie between the two. The principal results, both assuming the continuum hypothesis and using combinatorial methods, are these:

(a) There is a c.c.c. space X such that $X \times X$ is not a c.c.c. space (Theorem 7.13, due to R. Laver and F. Galvin); and

(b) there is a productively c.c.c. space that does not have property (K) (Theorem 7.9, due to K. Kunen).

In the Notes to this chapter we describe the effect of Martin's axiom on the countable chain properties.

Kunen's example

7.1 *Lemma.* Let α be an infinite cardinal and X a space such that $S(X) \le \mathrm{cf}(\alpha)$. If $\{V_\xi : \xi < \alpha\}$ is a set of non-empty open subsets of X such that $V_{\xi'} \subset V_\xi$ for $\xi < \xi' < \alpha$, then $\{\mathrm{cl}\, V_\xi : \xi < \alpha\}$ stabilizes.

Proof. We set $W_\xi = \mathrm{int\ cl}\ V_\xi$ for $\xi < \alpha$. Since $V_\xi \subset W_\xi$, we have

$$\mathrm{cl}\, V_\xi \subset \mathrm{cl}\, W_\xi = \mathrm{cl\,int\,cl}\, V_\xi \subset \mathrm{cl\,cl}\, V_\xi = \mathrm{cl}\, V_\xi$$

and hence

$$\mathrm{cl}\, V_\xi = \mathrm{cl}\, W_\xi \quad \text{for } \xi < \alpha;$$

thus it is enough to show that $\{\mathrm{cl}\, W_\xi : \xi < \alpha\}$ stabilizes.

If $\{\mathrm{cl}\, W_\xi : \xi < \alpha\}$ does not stabilize, there is a set $\{\xi(\sigma) : \sigma < \mathrm{cf}(\alpha)\}$ such that

$$\xi(\sigma') < \xi(\sigma) < \alpha \text{ for } \sigma' < \sigma < \mathrm{cf}(\alpha) \text{ and } \mathrm{cl}\, W_{\xi(\sigma)} \backslash \mathrm{cl}\, W_{\xi(\sigma+1)} \neq \varnothing.$$

It follows that $W_{\xi(\sigma)} \backslash \mathrm{cl}\, W_{\xi(\sigma+1)} \neq \varnothing$, since otherwise from

$$W_{\xi(\sigma)} \subset \mathrm{cl}\, W_{\xi(\sigma+1)} \subset \mathrm{cl}\, W_{\xi(\sigma)}$$

we have the contradiction $\mathrm{cl}\, W_{\xi(\sigma)} = \mathrm{cl}\, W_{\xi(\sigma+1)}$. We note finally that for $\sigma' < \sigma < \mathrm{cf}(\alpha)$ we have $\xi(\sigma'+1) \leq \xi(\sigma)$ and hence

$$\begin{aligned}(W_{\xi(\sigma')} \backslash \mathrm{cl}\, W_{\xi(\sigma'+1)}) \cap (W_{\xi(\sigma)} \backslash \mathrm{cl}\, W_{\xi(\sigma+1)}) &\subset W_{\xi(\sigma)} \backslash \mathrm{cl}\, W_{\xi(\sigma'+1)} \\ &\subset W_{\xi(\sigma'+1)} \backslash \mathrm{cl}\, W_{\xi(\sigma'+1)} = \varnothing;\end{aligned}$$

thus $\{W_{\xi(\sigma)} \backslash \mathrm{cl}\, W_{\xi(\sigma+1)}) : \sigma < \mathrm{cf}(\alpha)\}$ is a (faithfully indexed) cellular family in X, contradicting the assumption $S(X) \leq \mathrm{cf}(\alpha)$.

We recall from Chapter 2 that for α an infinite cardinal and ρ an ordinal such that $\rho \leq \alpha$, a space X has precalibre (α, ρ) if for every set $\{U_\xi : \xi < \alpha\}$ of non-empty, open subsets of X there is $A \subset \alpha$ with type $A = \rho$ such that $\{U_\xi : \xi \in A\}$ has the finite intersection property.

7.2 *Theorem.* Let X be a c.c.c. space and ρ an ordinal such that $\rho < \omega^+$. Then X has precalibre (ω^+, ρ).

Proof. It follows from Corollary 2.22 that it is enough to show that the Gleason space $G(X)$ of X has precalibre (ω^+, ρ); we assume without loss of generality in what follows that X is the (compact, Hausdorff) space $G(X)$.

Let $\{U_\eta : \eta < \omega^+\}$ be a set of non-empty, open subsets of X, and set

$$V_\xi = \bigcup_{\xi \leq \eta < \omega^+} U_\eta \quad \text{for } \xi < \omega^+.$$

Since $V_{\xi'} \subset V_\xi$ for $\xi < \xi' < \omega^+$ and X is a c.c.c. space, it follows from (the case $\alpha = \omega^+$ of) Lemma 7.1 that there is $\bar{\xi} < \omega^+$ such that

$$\mathrm{cl}\, V_\xi = \mathrm{cl}\, V_{\bar{\xi}} \quad \text{for } \bar{\xi} \leq \xi < \omega^+,$$

i.e., such that V_ξ is dense in $\mathrm{cl}\, V_{\bar{\xi}}$ for $\bar{\xi} \leq \xi < \omega^+$.

We claim that for $\bar{\xi} < \xi < \omega^+$ there is $f(\xi) < \omega^+$ such that

$$\cup\{U_\eta : \xi \leq \eta < f(\xi)\}$$

is dense in cl $V_{\bar{\xi}}$. We set

$$\mathscr{W} = \{W \in \mathscr{T}^*(X): \text{there is } \eta \geq \xi \text{ with } W \subset U_\eta\},$$

and we choose a maximal (faithfully indexed) cellular subfamily $\{W_i : i \in I\}$ of $\mathscr{W}$. We have $|I| \leq \omega$ (since X is a c.c.c. space), and from the maximality condition it follows that $\cup_{i \in I} W_i$ is dense in $V_{\bar{\xi}}$ and hence in cl $V_{\bar{\xi}}$. For $i \in I$ there is $\eta(i)$ such that $\xi \leq \eta(i) < \omega^+$ and $W_i \subset U_{\eta(i)}$; we set

$$f(\xi) = \sup\{\eta(i) : i \in I\} + 1.$$

The claim is proved.

We define $g : \omega^+ \to \omega^+$ by the rule

$$g(0) = \bar{\xi},$$
$$g(\xi + 1) = f(g(\xi)) \quad \text{for } \xi < \omega^+, \text{and}$$
$$g(\xi) = \sup\{g(\xi') : \xi' < \xi\}$$

for non-zero limit ordinals $\xi < \omega^+$, and for $\xi < \rho$ we set $X_\xi = \cup\{U_\eta : g(\xi) \leq \eta < g(\xi + 1)\}$. Then $\{X_\xi : \xi < \rho\}$ is a set of dense, open subspaces of the compact, Hausdorff space cl $V_{\bar{\xi}}$. Since ρ is a countable ordinal, it follows from the Baire category theorem that there is $p \in \cap_{\xi < \rho} X_\xi$. For $\xi < \rho$ there is $\eta(\xi)$ such that

$$g(\xi) \leq \eta(\xi) < g(\xi + 1) \quad \text{and}$$
$$p \in U_{\eta(\xi)}.$$

Since the function $\xi \to \eta(\xi)$ is an ordered-set isomorphism from ρ into ω^+ we have

$$\text{type}\{\eta(\xi) : \xi < \rho\} = \rho,$$

and since $p \in \cap_{\xi < \rho} U_{\eta(\xi)}$ the set $\{U_{\eta(\xi)} : \xi < \rho\}$ has the finite intersection property.

The proof is complete.

7.3 *Problem.* Let $\kappa > \omega^+$. Is there a c.c.c. space X such that X does not have precalibre (κ, ω^+)?

Definition. A space is *productively c.c.c.* if its product with every c.c.c. space is a c.c.c. space.

Definition. A space X has *property* P if for every set $\{U_\xi : \xi < \omega^+\}$ of non-empty, open subsets of X there are $A \subset \omega^+$ with $|A| = \omega^+$ and an ordinal number ρ with $2 \le \rho < \omega^+$ such that:

if $B \subset A$ and type $B = \rho$ then there are $\xi, \xi' \in B$ with $\xi \neq \xi'$ such that $U_\xi \cap U_{\xi'} \neq \emptyset$.

We remark that if X and Y are spaces and $X \equiv Y$, then X has property P if and only if Y has property P. (The proof is similar to the proof of Corollary 2.18(c); we omit the details.) It follows from Theorem 2.20 that a space X has property P if and only if its Gleason space $G(X)$ has property P.

7.4 *Theorem.* Let X be a space.

(a) If X has property (K) then X has property P.

(b) If X has property P then X is productively c.c.c.

Proof. (a) This statement is immediate from the definitions; one may take for ρ any ordinal such that $2 \le \rho < \omega^+$.

(b) Let X have property P, let Y be a c.c.c. space and let $\{U_\xi \times V_\xi : \xi < \omega^+\}$ be a set of non-empty, open subsets of $X \times Y$. We show there are $\xi, \xi' < \omega^+$ with $\xi \neq \xi'$ such that

$$(U_\xi \times V_\xi) \cap (U_{\xi'} \times V_{\xi'}) \neq \emptyset.$$

There are $A \subset \omega^+$ with $|A| = \omega^+$ and an ordinal ρ with $2 \le \rho < \omega^+$ such that if $B \subset A$ with type $B = \rho$ then there are $\xi, \xi' \in B$ with $\xi \neq \xi'$ such that $U_\xi \cap U_{\xi'} \neq \emptyset$. It follows from Theorem 7.2 that Y has precalibre (ω^+, ρ); we choose $\bar{B} \subset A$ with type $\bar{B} = \rho$ such that $\{V_\xi : \xi \in \bar{B}\}$ has the finite intersection property.

There are $\bar{\xi}, \bar{\xi}' \in \bar{B}$ with $\bar{\xi} \neq \bar{\xi}'$ such that $U_{\bar{\xi}} \cap U_{\bar{\xi}'} \neq \emptyset$. It is clear that

$$(U_{\bar{\xi}} \times V_{\bar{\xi}}) \cap (U_{\bar{\xi}'} \times V_{\bar{\xi}'}) \neq \emptyset,$$

as required.

7.5 *Corollary.* Let X be a space such that either X has calibre ω^+ or X has a strictly positive measure. Then X is productively c.c.c.

Proof. It is clear that if X has calibre ω^+ then X has property (K) (cf. Theorem 2.1(a)); it follows from Lemmas 6.2 and 6.22 (or from Corollary 6.17) that if X has a strictly positive measure then X has property (K). Thus the required statement follows from Theorem 7.4.

7.6 *Problems.* (a) (Galvin) Assume the continuum hypothesis. Is there a productively c.c.c. space that does not satisfy property P?

(b) (Argyros) Is there a (c.c.c.) space X that can serve as a 'test space' for productively c.c.c. spaces in the following sense: if Y is a space and $X \times Y$ is a c.c.c.space, then Y is productively c.c.c.

It is conjectured that (a) will be answered in the affirmative and (b) in the negative.

7.7 *Lemma.* Assume the continuum hypothesis. There is a family $\{S_\eta : \eta < \omega^+\}$ of subsets of ω^+ such that

(a) $S_\eta \subset \eta$ for $\eta < \omega^+$;

(b) $|S_\eta \cap S_\zeta| < \omega$ for $\eta < \zeta < \omega^+$; and

(c) for all $X \subset \omega^+$, either

(i) there is finite $F \subset \omega^+$ such that $X \subset \cup_{\eta \in F} S_\eta$, or

(ii) there is $\eta < \omega^+$ such that $X \cap S_\zeta \neq \varnothing$ for $\eta < \zeta < \omega^+$.

Proof. From the continuum hypothesis we have $|\{A \subset \omega^+ : |A| = \omega\}| = \omega^+$. Let $\{A_\eta : \eta < \omega^+\}$ be a well-ordering of this set of countable subsets of ω^+. We define $\{S_\eta : \eta < \omega^+\}$ satisfying conditions (a), (b) and (d) defined as follows:

(d) if $\xi < \eta < \omega^+$ with $A \subset \eta$, and if there is no finite $F \subset \eta$ with $A_\xi \subset \cup_{\zeta \in F} S_\zeta$, then $A_\xi \cap S_\eta \neq \varnothing$.

We proceed by recursion.

We set $S_0 = \varnothing$.

If $\eta < \omega^+$, and if S_ζ has been defined for $\zeta < \eta$, we enumerate $\{S_\zeta : \zeta < \eta\}$ as $\{S^{(n)} : n < \omega\}$ and we enumerate

$$\{A_\xi : A_\xi \subset \eta, \xi < \eta, \text{ there is no finite } F \subset \eta \text{ with } A_\xi \subset \cup_{\zeta \in F} S_\zeta\}$$

as $\{A^{(n)} : n < \omega\}$; we note that

$$A^{(n)} \not\subset \bigcup_{k \le n} S^{(k)} \quad \text{for } n < \omega,$$

we choose

$$s_n \in A^{(n)} \setminus \bigcup_{k \le n} S^{(k)} \quad \text{for } n < \omega,$$

and we set

$$S_\eta = \{s_n : n < \omega\}.$$

The definition of the family $\{S_\eta : \eta < \omega^+\}$ is complete. It is clear that conditions (a) and (d) are satisfied; we note further that if $\zeta < \eta$ and $S_\zeta = S^{(k)}$ (in the enumeration used to define S_η) then

$$S_\eta \cap S_\zeta = S_\eta \cap S^{(k)} \subset \{s_n : n < k\}$$

and hence $|S_\eta \cap S_\zeta| < \omega$.

It remains to verify (c). If X is a countable subset of ω^+ there is $\xi < \omega^+$ such that $X = A_\xi$ and condition (c) for X follows from (d). In what follows we assume $|X| = \omega^+$ and we show there is $Y \subset X$ with $|Y| = \omega$ such that Y satisfies (c) (ii).

Let $Y_0 \subset X$ with $|Y_0| = \omega$. If (c)(i) fails for Y_0 we set $Y = Y_0$; and if (c) (i) holds for Y_0 we choose $\eta_0 < \omega^+$ such that $|Y_0 \cap S_{\eta_0}| = \omega$ and we choose $Y_1 \subset X \setminus \bigcup_{\eta \le \eta_0} S_\eta$ such that $|Y_1| = \omega$. If (c)(i) fails for Y_1 set $Y = Y_1$; and if (c)(i) holds for Y_1 we choose $\eta_1 < \omega^+$ with $\eta_1 > \eta_0$ such that $|Y_1 \cap S_{\eta_1}| = \omega$.

We continue recursively. If there is (minimal) $n < \omega$ such that (c)(i) fails for Y_n we set $Y = Y_n$ and the proof is complete. Otherwise Y_n is defined for $n < \omega$ and we set

$$Y = \bigcup_{n < \omega} Y_n;$$

then $|Y| = \omega$, and since $|\{\eta < \omega^+ : |Y \cap S_\eta| = \omega\}| = \omega$ there is no finite $F \subset \omega^+$ such that $Y \subset \bigcup_{\eta \in F} S_\eta$. There is $\xi < \omega^+$ such that $Y = A_\xi$ and there is $\bar{\eta} < \omega^+$ such that $A_\xi \subset \bar{\eta}$. It follows from (d) that if $\bar{\eta} < \zeta < \omega^+$ then

$$Y \cap S_\zeta = A_\xi \cap S_\zeta \neq \emptyset,$$

so that Y satisfies (c)(ii), as required.

The proof is complete.

In passing we prove the following arrow relation; compare Theorem 1.7.

7.8 *Corollary*. Assume the continuum hypothesis. Then $\omega^+ \nrightarrow (\omega^+, \omega+2)$.

Proof. Let $\{S_\eta : \eta < \omega^+\}$ be a family of subsets of ω^+ satisfying conditions (a), (b) and (c) of Lemma 7.7, and set

$$P_1 = \{A \in [\omega^+]^2 : \min A \in S_{\max A}\} \text{ and}$$
$$P_0 = [\omega^+]^2 \backslash P_1.$$

Let $A \subset \omega^+$ with $|A| = \omega^+$. Since condition (c)(i) fails for A there is $\eta < \omega^+$ such that $A \cap S_\zeta \neq \varnothing$ for $\eta < \zeta < \omega^+$. We choose $\zeta \in A$ such that $\eta < \zeta < \omega^+$ and then $\xi \in A \cap S_\zeta$; we have $\xi < \zeta$ from condition (a) and hence $\{\xi, \zeta\} \in P_1$. It follows that $[A]^2 \not\subset P_0$.

Now let $B \subset \omega^+$ with type $B = \omega + 2$, let η and ζ be the last two elements of B, and set

$$C = B \backslash \{\eta, \zeta\}.$$

If $[B]^2 \subset P_1$ then for $\xi \in C$ we have ($\xi < \eta$ and) $\xi \in S_\eta$; similarly $\xi \in S_\zeta$. Thus if $[B]^2 \subset P_1$ then $S_\eta \cap S_\zeta$ contains the infinite set C, contrary to condition (b). If follows that $[B]^2 \not\subset P_1$.

The proof is complete.

7.9 *Theorem*. Assume the continuum hypothesis. There is an extremally disconnected, compact Hausdorff space X such that

(i) X does not have property (K), and

(ii) X has property P (hence, X is productively c.c.c.).

Proof. It follows from Corollary 2.22(b) and the remark preceding Theorem 7.4 that it is enough to find a space X with property P that does not have property (K); for then the Gleason space of X will have the required properties.

It follows from Lemma 7.7 that there is a family $\{S_\eta : \eta < \omega^+\}$ of subsets of ω^+ such that

(a) $S_\eta \subset \eta$ for $\eta < \omega^+$,

(b) $|S_\eta \cap S_{\eta'}| < \omega$ for $\eta < \eta' < \omega^+$, and

(c) if $X \subset \omega^+$ then either

(1) there is finite $F \subset \omega^+$ such that $X \subset \cup_{\eta \in F} S_\eta$, or

(2) there is $\bar{\eta} < \omega^+$ such that $X \cap S_\eta \neq \varnothing$ for $\bar{\eta} \leq \eta < \omega^+$.

Let X as a set be equal to $\{0, 1\}^{\omega^+}$, set $V_\eta = \{x \in X : x|S_\eta \equiv 0, x_\eta = 1\}$ for $\eta < \omega^+$, and give X the topology determined by the subbase $\{V_\eta : \eta < \omega^+\}$.

We verify that the space X satisfies (i) and (ii).

(i) We consider the family $\{V_\eta : \eta < \omega^+\}$ and we let $A \subset \omega^+$ with $|A| = \omega^+$. There is no finite $F \subset \omega^+$ with $A \subset \cup_{\eta \in F} S_\eta$; hence from condition (c) there is $\bar{\eta} < \omega^+$ such that

$$A \cap S_\eta \neq \varnothing \quad \text{for } \bar{\eta} \leq \eta < \omega^+.$$

There is $\eta \in A$ such that $\bar{n} \leq \eta$ and hence there is $\eta' \in A \cap S_\eta$. It is clear that

$$V_\eta \cap V_{\eta'} = \varnothing,$$

as required.

(ii) Let $\{U_\xi : \xi < \omega^+\}$ be a family of non-empty, basic open subsets of X; for $\xi < \omega^+$ there is a finite subset $F(\xi)$ of ω^+ such that

$$U_\xi = \bigcap_{\eta \in F(\xi)} V_\eta.$$

For $\xi < \alpha$ there is $X \in U_\xi$ and since $x|F(\xi) \equiv 1$ and $x|(\cup_{\eta \in F(\xi)} S_\eta) \equiv 0$ we have

$$(\bigcup_{\eta \in F(\xi)} S_\eta) \cap F(\xi) = \varnothing.$$

We assume without loss of generality, using the Erdős – Rado theorem for quasi-disjoint families (Theorem 1.4), that there is a set J such that

$$F(\xi) \cap F(\xi') = J \quad \text{for } \xi < \xi' < \omega^+;$$

we set

$$G(\xi) = F(\xi) \backslash J \quad \text{for } \xi < \omega^+.$$

We assume further without loss of generality that there is $n < \omega$ such that

$$|G(\xi)| = n \quad \text{for all } \xi < \omega^+.$$

If $n = 0$ we have

$$F(\xi) = F(\xi') \quad \text{for } \xi < \xi' < \omega^+$$

and the requirement of property P is satisfied with $A = \omega^+$, $\rho = 2$. We assume in what follows that $n > 0$ and for $\xi < \omega^+$ we enumerate $G(\xi)$ in its order inherited from ω^+:

$$G(\xi) = \{g_\xi(1) < g_\xi(2) < \ldots < g_\xi(n)\}.$$

We note that since $\{G(\xi) : \xi < \omega^+\}$ is a family of pairwise disjoint, finite subsets of ω^+ there is $A \subset \omega^+$ with $|A| = \omega^+$ such that if $\xi, \xi' \in A$ with $\xi < \xi'$ then $G(\xi) < G(\xi')$ (in the sense that if $\zeta \in G(\xi)$ and $\zeta' \in G(\xi')$ then $\zeta < \zeta'$). (Indeed set $\xi(0) = 0$. If $\zeta < \omega^+$ and $\xi(\eta)$ has been defined for all $\eta < \zeta$, note that

$$\sup\big(\bigcup_{\eta<\zeta} G(\xi(\eta))\big) < \omega^+$$

and choose for $\xi(\zeta)$ any ordinal $\xi < \omega^+$ such that

$$\sup\big(\bigcup_{\eta<\zeta} G(\xi(\eta))\big) < g_\xi(1).$$

This defines $\xi(\zeta)$ for all $\zeta < \omega^+$; set $A = \{\xi(\zeta) : \zeta < \omega^+\}$.)

We claim that the requirement of property P is satisfied with the set A defined above and with $\rho = \omega + n + 1$. Indeed we claim that if $B \subset A$ with type $B = \omega + n + 1$, say

$$B = \{\xi_0 < \xi_1 < \ldots < \xi_k < \ldots \xi_\omega < \xi_{\omega+1} < \ldots < \xi_{\omega+n}\},$$

then there are k, i with $k < \omega$ and $0 \le i \le n$ such that

$$F(\xi_k) \cup F(\xi_{\omega+i}) \cap (\cup\{S_\eta : \eta \in F(\xi_k) \cup F(\xi_{\omega+i})\}) = \varnothing$$

(and hence $V_{\xi_k} \cap V_{\xi_{\omega+i}} \neq \varnothing$). If the claim fails then for $k < \omega$, $0 \le i \le n$ there is $\eta \in F(\xi_k) \cup F(\xi_{\omega+i})$ such that

$$(F(\xi_k) \cup F(\xi_{\omega+i})) \cap S_\eta \neq \varnothing.$$

Since $F(\xi_k) \cap F(\xi_{\omega+i}) = J$ and

$$F(\xi_k) \cap \big(\bigcup_{\eta \in F(\xi_k)} S_\eta\big) = F(\xi_{\omega+i}) \cap \big(\bigcup_{\eta \in F(\xi_{\omega+i})} S_\eta\big) = \varnothing$$

it is clear that in fact $\eta \in G(\xi_k) \cup G(\xi_{\omega+i})$ and

$$(G(\xi_k) \cup G(\xi_{\omega+i})) \cap S_\eta \neq \varnothing.$$

Since for $\eta \in G(\xi_k)$ we have

$$G(\xi_{\omega+i}) \cap S_\eta \subset G(\xi_{\omega+i}) \cap \eta = \varnothing,$$

it follows that for $k < \omega, 0 \le i \le n$ there is $\eta \in G(\xi_{\omega+i})$ such that $G(\xi_k) \cap S_\eta \neq \varnothing$. We define functions

$$\psi, \varphi : \omega \times \{0, 1. \ldots, n\} \to \{1, 2, \ldots, n\}$$

by the rule

$$\eta = g_{\xi_{\omega+i}}(\psi(k, i)) \in G(\xi_{\omega+i}) \quad \text{and } g_{\xi_k}(\varphi(k, i)) \in G(\xi_k) \cap S_\eta.$$

We note that there is an infinite subset K of ω such that for each i

with $0 \leq i \leq n$ the two functions

$$\psi|K \times \{i\}, \ \ \varphi|K \times \{i\}$$

are constant functions. (Indeed there is infinite $K_0 \subset \omega$ such that $\psi|K_0 \times \{0\}$, $\varphi|K_0 \times \{0\}$ are constant. If $m < n$ and infinite K_m has been defined such that $\psi|K_m \times \{i\}$, $\varphi|K_m \times \{i\}$ are constant for $0 \leq i \leq m$, there is infinite $K_{m+1} \subset K_m$ such that $\psi|K_{m+1} \times \{m+1\}$, $\varphi|K_{m+1} \times \{m+1\}$ are constant. This defines K_m for $m \leq n$. The set $K = K_n$ is as required.)

Since $|\{0, 1, \ldots, n\}| = n + 1$ and $|\{1, 2, \ldots, n\}| = n$ there are $\bar{i}, \bar{j}$ such that $0 \leq \bar{i} < \bar{j} \leq n$ and the constant values of $\varphi|K \times \{\bar{i}\}$ and $\varphi|K \times \{\bar{j}\}$ are equal; that is, there is r such that $1 \leq r \leq n$ and

$$\varphi(k, \bar{i}) = \varphi(k, \bar{j}) = r \quad \text{for all } k \in K.$$

Let

$$\psi(k, \bar{i}) = m_0 \quad \text{and } \psi(k, \bar{j}) = m_1 \quad \text{for all } k \in k,$$

and set

$$\zeta(\bar{i}) = g_{\xi_{\omega+\bar{i}}}(m_0) \quad \text{and } \zeta(\bar{j}) = g_{\xi_{\omega+\bar{j}}}(m_1).$$

Since $\zeta(\bar{i}) \in G(\xi_{\omega+\bar{i}})$ and $\zeta(\bar{j}) \in G(\xi_{\omega+\bar{j}})$ and

$$G(\xi_{\omega+\bar{i}}) \cap G(\xi_{\omega+\bar{j}}) = \varnothing$$

we have $\zeta(\bar{i}) \neq \zeta(\bar{j})$.

Now for $k \in K$ we have

$$\zeta(\bar{i}) = g_{\xi_{\omega+\bar{i}}}(\psi(k, \bar{i}))$$

and hence

$$g_{\xi_k}(r) = g_{\xi_k}(\varphi(k, \bar{i})) \in G(\xi_k) \cap S_{\zeta(\bar{i})} \subset S_{\zeta(\bar{i})};$$

similarly for $k \in K$ we have

$$g_{\xi_k}(r) \in S_{\zeta(\bar{j})}.$$

Since $g_{\xi_k}(r) \in G(\xi_k)$ and

$$G(\xi_k) \cap G(\xi_{k'}) = \varnothing \quad \text{for } k, k' \in K, k \neq k'$$

we have

$$g_{\xi_k}(r) \neq g_{\xi_{k'}}(r) \quad \text{for } k, k' \in K, k \neq k'.$$

It follows that

$$|S_{\zeta(\bar{i})} \cap S_{\zeta(\bar{j})}| = \omega,$$

contrary to condition (b) for the family $\{S_\eta : \eta < \omega^+\}$. This contradiction completes the proof.

7.10 *Remark and Problem.* For $\beta \geq \omega$, we say that a space X has *property* P_β if for every set $\{U_\xi : \xi < \beta\}$ of non-empty, open subsets of X there are $A \dot{\subset} \beta$ with $|A| = \beta$ and an ordinal number ρ with $2 \leq \rho < \beta$ such that:

if $B \subset A$ and type $B = \rho$ then there are $\xi, \xi' \in B$ with $\xi \neq \xi'$ such that $U_\xi \cap U_{\xi'} \neq \varnothing$.

It is clear that the method of Theorem 7.9 proves the following statement: If α is a regular cardinal and $\alpha^+ = 2^\alpha$ then there is an extremally disconnected, compact Hausdorff space X such that

X has property P_{α^+} and

X does not have property $\mathrm{K}_{\alpha^+,2}$.

It is clear then that $S(X) \leq \alpha^+$, but we do not know if X has the stronger productive property: $S(X \times Y) \leq \alpha^+$ for every space Y with $S(Y) \leq \alpha^+$. This last must be left as a question.

The Laver–Galvin example

7.11 *Lemma.* Let $\alpha \geq \omega$ and A a set, and for $\xi < \alpha$ let $\{F^i_\xi : i \in I_\xi\}$ be a set of pairwise disjoint, finite subsets of A with $|I_\xi| = \alpha$. Then there are two subsets A_0, A_1 of A such that

$$A_0 \cap A_1 = \varnothing, \text{ and}$$
$$|\{i \in I_\xi : F^i_\xi \subset A_\varepsilon\}| = \alpha \text{ for } \xi < \alpha \text{ and } \varepsilon = 0, 1.$$

Proof. Without loss of generality we assume $I_\xi = \alpha$ for $\xi < \alpha$. We note that it follows easily from the assumptions that if B is a set with $|B| < \alpha$, if φ and ψ are functions from B to α and if $\xi < \alpha$, then there is $i < \alpha$ such that

$$(1) \quad F^i_\xi \cap (\cup_{\zeta \in B} F^{\psi(\zeta)}_{\varphi(\zeta)}) = \varnothing.$$

We define by recursion a function

$$\tau : \{\langle \xi, \eta \rangle : \xi \leq \eta < \alpha\} \to \alpha$$

such that

$$F^{\tau(\xi,\eta)}_\xi \cap F^{\tau(\xi',\eta)}_{\xi'} = \varnothing \text{ for } \langle \xi, \eta \rangle \neq \langle \xi', \eta' \rangle.$$

We set $\tau(0, 0) = 0$. If $\xi \leq \eta < \alpha$ and if τ has been defined on the set

$$\{\langle \xi', \eta' \rangle : \xi' \leq \eta' < \eta\} \cup \{\langle \xi', \eta \rangle : \xi' < \eta\},$$

we define $\tau(\xi, \eta)$, using (1), so that

$$F_\xi^{\tau(\xi,\eta)} \cap (\cup_{\xi' \leq \eta' < \eta} F_{\xi'}^{\tau(\xi',\eta')} \cup \cup_{\xi' < \xi} F_{\xi'}^{\tau(\xi',\eta)}) = \emptyset.$$

We set

$$J_\xi = \{\tau(\xi, \eta) : \xi \leq \eta < \alpha\} \text{ for } \xi < \alpha,$$

we choose sets $J_{\xi,0}$ and $J_{\xi,1}$ such that

$$J_{\xi,0} \cup J_{\xi,1} = J_\xi,$$
$$J_{\xi,0} \cap J_{\xi,1} = \emptyset, \text{ and}$$
$$|J_{\xi,0}| = |J_{\xi,1}| = \alpha,$$

and we set

$$A_\varepsilon = \cup \{F_\xi^\zeta : \xi < \alpha, \zeta \in J_{\xi,\varepsilon}\} \text{ for } \varepsilon = 0, 1.$$

The proof is complete.

Notation. For sets A, B we set

$$A \otimes B = \{\{a, b\} : a \in A, b \in B\}.$$

7.12 *Lemma.* Let $\alpha \geq \omega$ and $\alpha^+ = 2^\alpha$. There are two families $\{K_0(\eta) : \eta < \alpha^+\}, \{K_1(\eta) : \eta < \alpha^+\}$ such that (with $K_\varepsilon = \cup_{\eta < \alpha^+} (K_\varepsilon(\eta) \times \{\eta\})$ and $\varepsilon = 0, 1$):

(i) $K_0(\eta) \cup K_1(\eta) = \eta$ for $\eta < \alpha^+$, and

(ii) if $\mathscr{F} = \{F^i : i < \alpha\}$ is a set of finite subsets of α^+ such that

$$F^i \cap F^{i'} = \emptyset \text{ for } i < i' < \alpha,$$

then there is $\zeta < \alpha^+$ such that if $\zeta < \eta < \alpha^+$ and $\cup \mathscr{F} \subset \eta$ and $X \subset \eta$ with $|X| < \omega$ and $\varepsilon \in \{0, 1\}$ and

$$|\{i < \alpha : (F^i \otimes X) \cap K_\varepsilon = \emptyset\}| = \alpha,$$

then $|\{i < \alpha : F^i \otimes (X \cup \{\eta\}) \cap K_\varepsilon = \emptyset\}| = \alpha$.

Proof. For $\eta < \alpha^+$ we define sets $L_\varepsilon(\eta)$ for $\varepsilon = 0, 1$ such that

$L_\varepsilon(\eta) \subset \eta$,

$L_0(\eta) \cap L_1(\eta) = \emptyset$, and

if $\mathscr{F} = \{F^i : i < \alpha\}$ is a set of finite subsets of α such that

$F^i \cap F^{i'} = \emptyset$ for $i < i' < \alpha$

then there is $\zeta < \alpha^+$ such that if $\zeta < \eta < \alpha^+$ and $\cup \mathscr{F} \subset \eta$ and $X \subset \eta$ with $|X| < \eta$ and $\varepsilon \in \{0, 1\}$ and

$$|\{i < a : F^i \otimes X \subset \cup_{\eta' < \eta}(L_\varepsilon(\eta') \otimes \{\eta'\})\}| = \alpha,$$

then $|\{i < \alpha : F^i \otimes (X \cup \{\eta\}) \subset \cup_{\eta' < \eta}(L_\varepsilon(\eta') \otimes \{\eta'\})\}| = \alpha$.

We proceed by recursion.

We note that the set of all families $\mathscr{F} = \{F^i : i < \alpha\}$ of finite subsets of α such that

$$F^i \cap F^{i'} = \varnothing \text{ for } i < i' < \alpha$$

has cardinality $|\mathscr{P}_\omega(\alpha^+)|^\alpha = (\alpha^+)^\alpha = (2^\alpha)^\alpha = \alpha^+$. Let $\{\mathscr{F}_\zeta : \zeta < \alpha^+\}$ be a well-ordering of this set.

We set $L_\varepsilon(\eta) = \varnothing$ for $\varepsilon = 0, 1, \eta \leq \alpha$.

Let $\alpha < \eta < \alpha^+$, suppose that $L_\varepsilon(\eta')$ has been defined for $\varepsilon = 0, 1$ and $\eta' < \eta$, and let

$$\{(\varepsilon_\xi, \zeta_\xi, X_\xi) : \xi < \alpha\}$$

be a well-ordering of the set of ordered triples (ε, ζ, X) such that

$$\varepsilon \in \{0, 1\}, \zeta < \eta, \cup \mathscr{F}_\zeta \subset \eta, X \subset \eta, |X| < \omega \text{ and}$$
$$|\{i < \alpha : F^i_\zeta \otimes X \subset \cup_{\eta' < \eta}(L_\varepsilon(\eta') \otimes \{\eta'\})\}| = \alpha.$$

We set

$$I_\xi = \{i < \alpha : F^i_{\zeta_\xi} \otimes X_\xi \subset \cup_{\eta' < \eta}(L_{\varepsilon_\xi}(\eta') \otimes \{\eta'\})\} \text{ for } \xi < \alpha$$

and we apply Lemma 7.11 with A and F_ξ replaced by η and $F^i_{\zeta_\xi}$, respectively: there are two sets $L_0(\eta)$, $L_1(\eta)$ such that

$$L_0(\eta), L_1(\eta) \subset \eta,$$
$$L_0(\eta) \cap L_1(\eta) = \varnothing, \text{ and}$$
$$|\{i \in I_\xi : F^i_{\zeta_\xi} \subset L_{\varepsilon_\xi}(\eta)\}| = \alpha.$$

The definition of the family $\{L_\varepsilon(\eta) : \eta < \alpha^+, \varepsilon = 0, 1\}$ is complete. We finally set

$$K_\varepsilon(\eta) = \eta \backslash L_\varepsilon(\eta) \text{ for } \varepsilon = 0, 1 \text{ and } \eta < \alpha^+.$$

It is clear that the two families $\{K_\varepsilon(\eta) : \eta < \alpha^+\}$ $(\varepsilon \in \{0, 1\})$ are as required.

7.13 *Theorem.* Let $\alpha \geq \omega$ and $\alpha^+ = 2^\alpha$. There is an extremally disconnected, compact Hausdorff space X such that

$$S(X) \leq \alpha^+ \text{ and } S(X \times X) > \alpha^+.$$

Proof. We show first that there are two spaces X_0, X_1 such that $S(X_0) \leq \alpha^+, S(X_1) \leq \alpha^+$, and $S(X_0 \times X_1) > \alpha^+$.

For $\varepsilon \in \{0, 1\}$ let $\{K_\varepsilon(\eta) : \eta < \alpha^+\}$ be the family of Lemma 7.12, let X_ε as a set be equal to $\{0, 1\}^{(\alpha^+)}$, set

$$V_\varepsilon(\eta) = \{x \in X_\varepsilon : x | K_\varepsilon(\eta) \equiv 0, x(\eta) = 1\} \text{ for } \eta < \alpha^+,$$

and give X_ε the topology determined by the subbase $\{V_\varepsilon(\eta) : \eta < \alpha^+\}$.

(i) $S(X_0 \times X_1) > \alpha^+$. We claim the (indexed) family $\{V_0(\eta) \times V_1(\eta) : \eta < \alpha^+\}$ is a cellular family in $X_0 \times X_1$. Indeed if there are η, η' with $\eta < \eta' < \alpha^+$ and

$$\langle x_0, x_1 \rangle \in (V_0(\eta) \times V_1(\eta)) \cap (V_0(\eta') \times V_1(\eta'))$$

then since $\eta \in K_0(\eta') \cup K_1(\eta') = \eta'$ there is $\varepsilon \in \{0, 1\}$ such that $\eta \in K_\varepsilon(\eta')$; we have $x_\varepsilon(\eta) = 0$ and $x_\varepsilon(\eta) = 1$, a contradiction.

(ii) $S(X_0) \leq \alpha^+, S(X_1) \leq \alpha^+$. Let $\varepsilon \in \{0, 1\}$ and let $\{U_i : i < \alpha^+\}$ be a family of non-empty, basic open subsets of X_ε; for $i < \alpha^+$ there is a finite subset G^i of α^+ such that

$$U_i = \bigcap_{\eta \in G^i} V_\varepsilon(\eta).$$

For $i < \alpha^+$ there is $x \in U_i$ and since $x | G^i \equiv 1$ and $x | \bigcup_{\eta \in G^i} K_\varepsilon(\eta) \equiv 0$ we have

$$(\bigcup_{\eta \in G^i} K_\varepsilon(\eta)) \cap G^i = \emptyset.$$

We assume without loss of generality, using the Erdős–Rado theorem for quasi-disjoint families (Theorem 1.4), that there is a set J such that

$$G^i \cap G^{i'} = J \text{ for } i < i' < \alpha^+;$$

and we set $F^i = G^i \backslash J$ for $i < \alpha^+$. We note that if $i' < i < \alpha^+$ and $(F^i \otimes F^{i'}) \cap K_\varepsilon = \emptyset$ then $U_i \cap U_{i'} \neq \emptyset$. Thus to prove that $\{U_i : i < \alpha^+\}$ is not a cellular family in X_ε it is enough to show that there are i', i such that ($i' < \alpha < i < \alpha^+$ and) $F^i \otimes F^{i'} \cap K_\varepsilon = \emptyset$.

Set $\mathscr{F} = \{F^i : i < \alpha\}$ and let $\zeta < \alpha^+$ be an ordinal satisfying condition (ii) of Lemma 7.12 for $\mathscr{F}$; we assume without loss of generality that $\cup\mathscr{F}\subset\zeta$. There is $\bar{\imath}<\alpha^+$ such that

$$\alpha<\bar{\imath}\text{ and }F^{\bar{\imath}}\cap(\zeta+1)=\varnothing.$$

We note that if $\eta\in F^{\bar{\imath}}$ then $\eta>\zeta$ and $\cup_{i<\alpha}F^i\subset\eta$; further if $F^{\bar{\imath}}=\varnothing$ then the required relation $(F^{i'}\otimes F^{\bar{\imath}})\cap K_\varepsilon=\varnothing$ is satisfied for all $i'<\alpha$. We assume therefore that $F^{\bar{\imath}}\neq\varnothing$, we let

$$\{\eta_1<\eta_2<\ldots<\eta_n\}$$

be the elements of $F^{\bar{\imath}}$ in the order inherited from α^+, we note that

$$\{i<\alpha:(F^i\otimes\varnothing)\cap(\cup_{\eta'<\eta_1}(K_\varepsilon(\eta')\otimes\{\eta'\}))=\varnothing\}=\alpha,$$

and we apply condition (ii) of Lemma 7.12 successively n times, with

$$\eta=\eta_1,\eta_2,\ldots,\eta_n\text{ and}$$
$$X=\varnothing,\{\eta\},\{\eta_1,\eta_2\},\ldots,\{\eta_1,\eta_2,\ldots,\eta_{n-1}\}$$

correspondingly. We see that

$$|\{i<\alpha:(F^i\otimes F^{\bar{\imath}})\cap(\cup_{\eta'\leq\eta_n}(K_\varepsilon(\eta')\otimes\{\eta'\}))=\varnothing\}|=\alpha$$

and hence

$$|\{i<\alpha:(F^i\otimes F^{\bar{\imath}})\cap K_\varepsilon=\varnothing\}|=\alpha.$$

In particular there is $\bar{\imath}'<\alpha$ such that

$$(F^{\bar{\imath}'}\otimes F^{\bar{\imath}})\cap K_\varepsilon=\varnothing,$$

as required. The proof that $S(X_\varepsilon)\leq\alpha^+$ for $\varepsilon=0,1$ is complete.

Finally we set X equal to the discrete union (i.e., the disjoint union) of the Gleason spaces of X_0 and X_1. From Corollaries 2.18(a) and 2.21 and Theorem 2.20 it then follows, using the fact that $X\times X$ contains a homeomorph of $G(X_0)\times G(X_1)$ as an open-and-closed subspace, that

$$S(X)=S(G(X_0))+S(G(X_1))\leq\alpha^+\text{ and}$$
$$S(X\times X)\geq S(G(X_0)\times G(X_1))=S(G(X_0\times X_1))=S(X_0\times X_1)>\alpha^+.$$

The proof is complete.

We remark that if the extremally disconnected, compact Hausdorff space X of the statement of Theorem 7.13 is replaced by the space $\tilde{X}$ defined to be the disjoint union of X with $\beta(\alpha)$, then $\tilde{X}$ satisfies the properties required of X in Theorem 7.13 and in addition the equality

$$S(\tilde{X})=\alpha^+;$$

indeed $\tilde{X}$ is a compact, Hausdorff, extremally disconnected space with

$$S(\tilde{X}) = S(X) + S(\beta(\alpha)) = \alpha^+,$$

and since $X \times X$ is an open-and-closed subspace of $\tilde{X} \times \tilde{X}$ we have

$$S(\tilde{X} \times \tilde{X}) \geq S(X \times X) > \alpha^+.$$

We note further that from Corollary 5.16 it follows that for a space X as in Theorem 7.13 the space $X \times X$ (even with its natural P_{α^+}-space topology) has precalibre $(2^\alpha)^+$. Thus $S(X \times X) \leq (2^\alpha)^+$ and from the hypothesis $\alpha^+ = 2^\alpha$ we have for $X, \tilde{X}$ as above the relations

$$S(X) \leq S(\tilde{X}) = \alpha^+,$$
$$S(X \times X) = S(\tilde{X} \times \tilde{X}) = (2^\alpha)^+.$$

7.14 *Remarks.* (a) The space X constructed in the proof of Theorem 7.9 has in fact the following property, stronger than property P:

every (finite) power of X has property P.

(The proof is analogous to the proof in Theorem 7.9 that X has property P.)

Problem (Kalamidas). Assume the continuum hypothesis. Is there a space X with property P such that $X \times X$ does not have property P?

(b) The spaces X_ε ($\varepsilon \in \{0, 1\}$) constructed in the proof of Theorem 7.13 have in fact the following property, stronger than the property that $S(X_\varepsilon) \leq \alpha^+$:

$$S(X_\varepsilon^n) \leq \alpha^+ \text{ for } 1 \leq n < \omega.$$

(The proof is analogous to the proof in Theorem 7.13 that $S(X_\varepsilon) \leq \alpha^+$.) It then follows from Theorem 3.27 that $S(X_\varepsilon^I) \leq \alpha^+$ for every non-empty index set I, and we have the following consequence of Theorem 7.13.

Corollary. Let $\alpha \geq \omega$ and $\alpha^+ = 2^\alpha$. There are extremally disconnected, compact Hausdorff spaces X_0, X_1 such that

$S(X_\varepsilon^I) \leq \alpha^+$ for $\varepsilon \in \{0, 1)$ and I a set, and

$S(X_0 \times X_1) > \alpha^+$.

We note in particular for $\alpha = \omega$ that the spaces X_ε are not productively c.c.c. spaces; see in this connection Problem 7.6(a) above.

7.15 *Problem* (Argyros and Negrepontis). Assume the generalized continuum hypothesis and let α, β and κ be cardinals with α and κ regular and with $\mathrm{cf}(\beta) < \kappa < \beta < \beta^+ = \alpha$.

(a) Are there spaces X and Y such that $S(X) = \alpha$, $S(Y) = \kappa$, and $S(X \times Y) > \alpha$? The answer is unknown even for $\mathrm{cf}(\beta) = \omega$, $\kappa = \omega^+$. We conjecture that there are such spaces. Such a space X will (from Theorem 2.2(a), with α, β, γ and δ replaced by α, κ, 2 and 2, respectively) satisfy $S(X) = \alpha$ and will not have calibre α (nor even calibre $(\alpha, \kappa, 2)$), thus in a sense improving the space of Theorem 5.28.

(b) Is there a space X such that $S(X \times Y) \leq \alpha$ for every space Y with $S(Y) \leq \kappa$, but X does not have calibre $(\alpha, \kappa, 2)$? We do not even know, for $\alpha = \beta^+$ with $\mathrm{cf}(\beta) = \omega$, if there is a space X such that $S(X \times Y) \leq \alpha$ for every c.c.c. space Y, but X does not have calibre $(\alpha, \alpha, 2)$.

7.16 *Problems* (Negrepontis). Assume the generalized continuum hypothesis, and let α be a regular limit cardinal that is not weakly compact.

(a) Is there a space X such that $S(X) = \alpha$ and $S(X \times X) > \alpha$? (Jensen (1972) has defined such a space using Gődel's axiom V = L.)

(b) Is there a space X such that $S(X \times Y) = \alpha$ for every space Y such that $S(Y) \leq \alpha$, but X does not have calibre $(\alpha, \alpha, 2)$?

(c) Is there a space X such that $S(X) = \alpha$ but X does not have calibre α?

Of course a positive answer to either (a) or (b) implies a positive answer to (c).

7.17 *Problem.* If in the examples of Kunen, and of Laver and Galvin, we strengthen the topologies so that they include the usual product topology, do the resulting spaces retain the essential topological features of the original spaces?

7.18 *Remark.* We have given five examples of compact spaces with no strictly positive measure.

(1) The Galvin–Hajnal space (6.32); this has calibre ω^+.

(2) Gaifman's space (6.23); this has property K_n for $2 \le n < \omega$ and, assuming the continuum hypothesis, it does not have calibre ω^+.

(3) The examples of Argyros (6.26); for $2 \le m < \omega$ there is an Argyros space with property K_m which, assuming the continuum hypothesis, does not have property K_{m+1}.

(4) Kunen's space (7.9); assuming the continuum hypothesis, this is productively c.c.c. and does not have property (K).

(5) The Laver–Galvin space (7.13); assuming the continuum hypothesis, this is a c.c.c. space that is not productively c.c.c.

On the other hand a compact space with a strictly positive measure has, by the result of Argyros and Kalamidas (6.15), property K_n for every natural number $n \ge 2$; the Erdős space of 6.18 has a strictly positive measure and, assuming the continuum hypothesis, it does not have calibre ω^+.

7.19 *Remarks.* (a) The diagram of implications that emerges (for a space X) from the foregoing positive and negative results is as follows (p. 198). The numbers in parenthesis refer to the appropriate section of this book, the symbol (o) indicates that the implication is obvious, and square brackets indicate special set-theoretic assumptions.

(b) The diagram on page 199 summarizes those results of the preceding chapters which respond to the following question.

Assume the generalized continuum hypothesis, let α and κ be regular cardinal numbers with $\alpha \ge \omega$ and $\kappa \ge \omega^+$, and let X be a space such that $S(X) = \kappa$; does X have precalibre α?

Here the symbol N means 'no, never'; Y means 'yes, always'; and S means 'sometimes'. When S appears it is accompanied by the symbols + and − ; these indicate respectively instances in which X does and does not have precalibre α.

7.20 *Problem.* We indicate below in the Notes to this chapter that the conclusion of theorem 7.13 cannot be established in ZFC if the assumption $\alpha^+ = 2^\alpha$ is omitted: there are models of ZFC in which the product of (arbitrarily many) c.c.c. spaces is a c.c.c. space. It is fascinating nevertheless to wonder if a partial analogue of

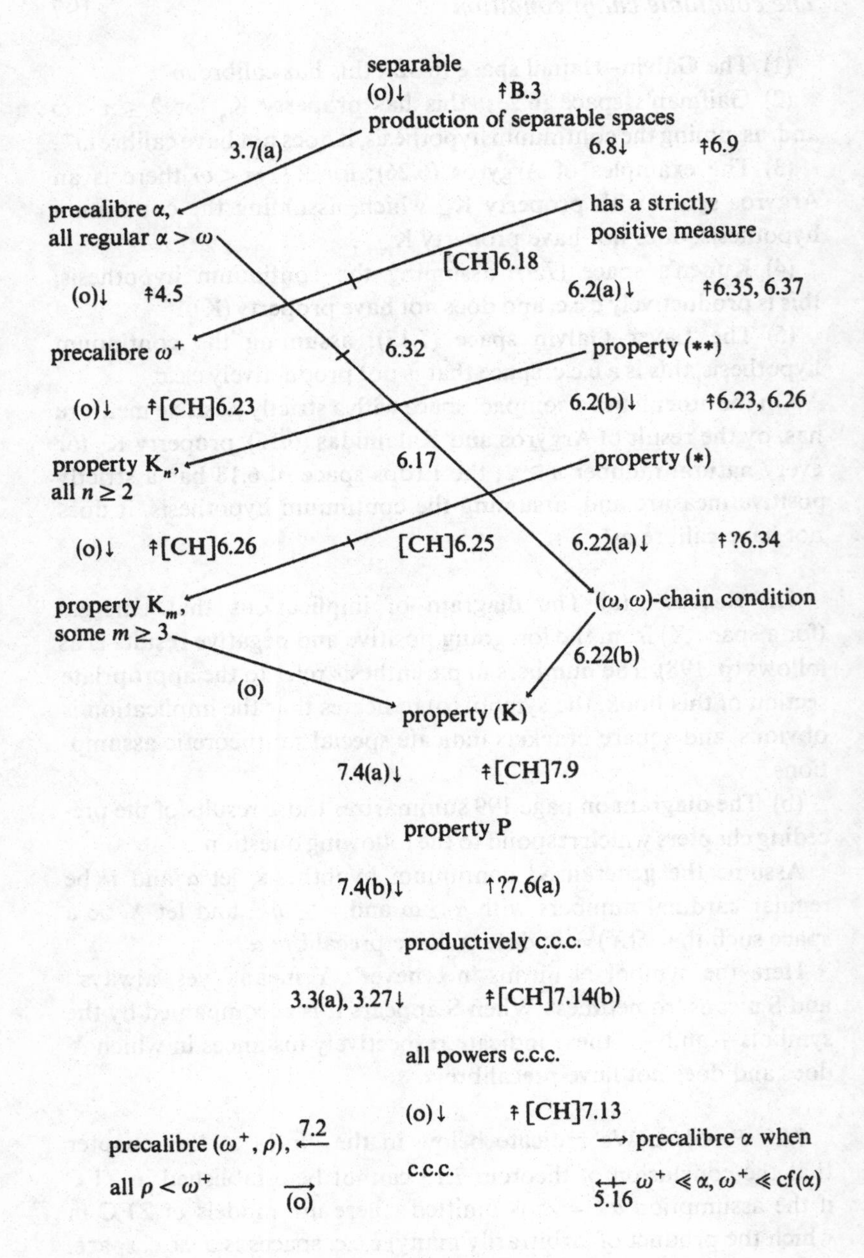

separable
(o)↓
†B.3
production of separable spaces
3.7(a)
6.8↓
†6.9
precalibre α,
all regular α > ω
has a strictly
positive measure
[CH]6.18
(o)↓
†4.5
6.2(a)↓
†6.35, 6.37
precalibre ω⁺
6.32
property (**)
(o)↓
†[CH]6.23
6.2(b)↓
†6.23, 6.26
property Kₙ,
all n ≥ 2
6.17
property (*)
(o)↓
†[CH]6.26
[CH]6.25
6.22(a)↓
†?6.34
property Kₘ,
some m ≥ 3
(ω, ω)-chain condition
6.22(b)
(o)
property (K)
7.4(a)↓
†[CH]7.9
property P
7.4(b)↓
↑?7.6(a)
productively c.c.c.
3.3(a), 3.27↓
†[CH]7.14(b)
all powers c.c.c.
(o)↓
†[CH]7.13
precalibre (ω⁺, ρ),
all ρ < ω⁺
7.2
(o)
c.c.c.
precalibre α when
ω⁺ ≪ α, ω⁺ ≪ cf(α)
5.16

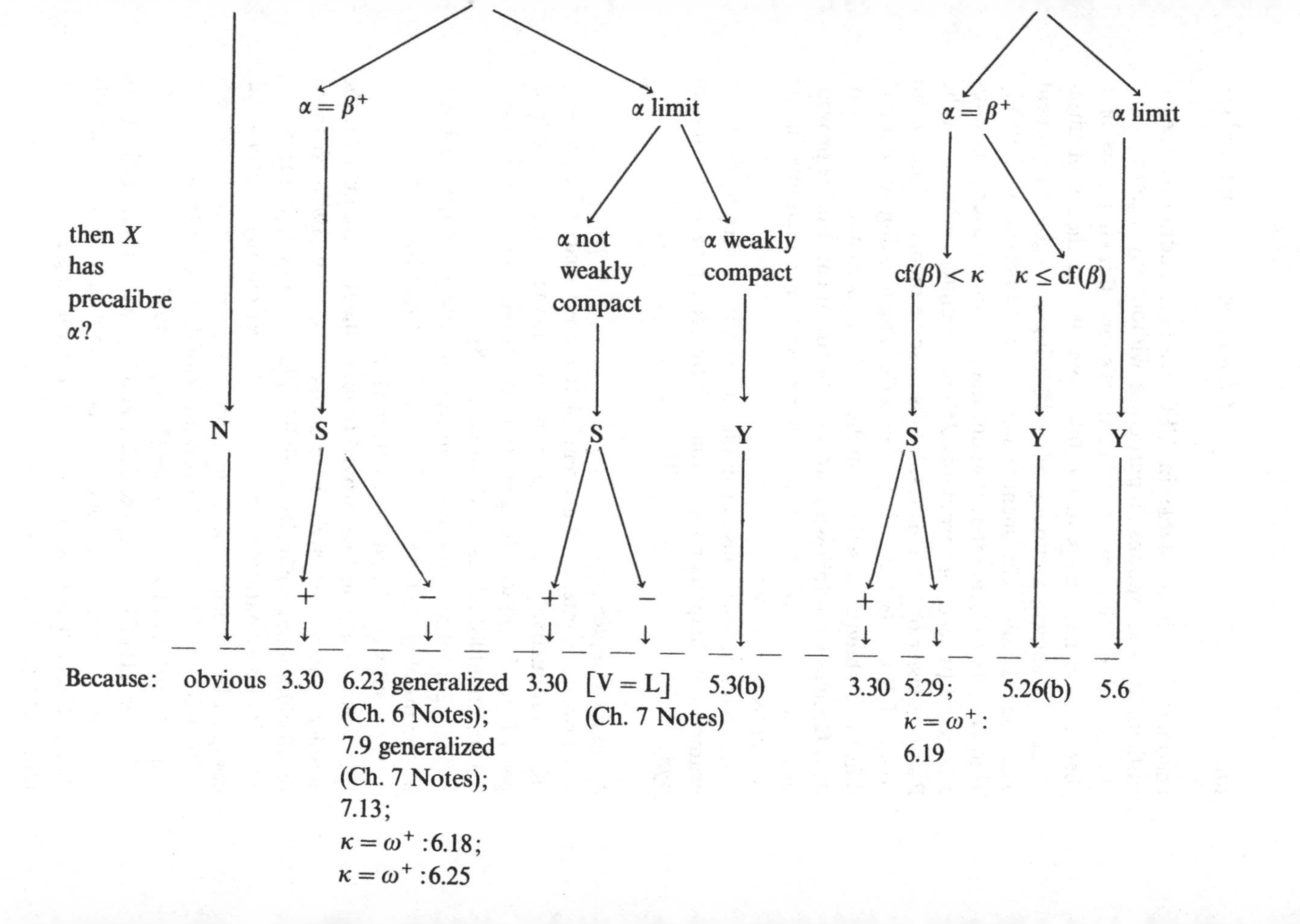

then X has precalibre α?
α = β⁺
α limit
α not weakly compact
α weakly compact
α = β⁺
α limit
cf(β) < κ
κ ≤ cf(β)
N
S
S
Y
S
Y
Y
+
−
+
−
+
−
Because:
obvious
3.30
6.23 generalized (Ch. 6 Notes); 7.9 generalized (Ch. 7 Notes); 7.13; κ = ω⁺ :6.18; κ = ω⁺ :6.25
3.30
[V = L] (Ch. 7 Notes)
5.3(b)
3.30
5.29; κ = ω⁺: 6.19
5.26(b)
5.6

Theorem 7.13 is available in ZFC without additional special set-theoretic assumptions (in particular, without any segment of the generalized continuum hypothesis). We are thinking here of a phenomenon analogous to the following. In the theory of ultrafilters it is shown, assuming $\alpha^+ = 2^\alpha$, using a diagonal argument based on a disjoint refinement lemma not very different qualitatively from Lemma 7.11, that the set of ultrafilters uniform over α contains a large set (indeed, of cardinality 2^{2^α}) of ultrafilters minimal in the Rudin–Keisler order. This result fails in some models of the axiom system $[\mathrm{ZFC} + (\alpha^+ < 2^\alpha)]$, but in ZFC alone, using an entirely different technique based on families of large oscillation ('independent families'), the following weaker statement can still be proved: there do exist uniform ultrafilters pairwise incomparable in the Rudin–Keisler order.

Of course we can consider similar questions concerning the other examples of Chapters 6 and 7 that use the (generalized) continuum hypothesis.

Notes for Chapter 7

The principal content of Theorem 7.4, that every space with property (K) is productively c.c.c., is due to Galvin; the intermediate 'property P' used here was designed to facilitate efficient use of Theorem 7.3, an unpublished result of Kunen (1978).

Theorem 7.9, also due to Kunen, appears in Wage (1979) (Theorem 6); the proof given here, based on Lemma 7.7, is Galvin's.

The results of 7.7 and 7.8 are from Hajnal (1960).

It was Laver who first proved (unpublished) Theorem 7.13; the simpler construction given here is due to Galvin (1980). Without much additional difficulty Galvin (1980) shows that for $\alpha \geq \omega$, $\alpha^+ = 2^\alpha$ and n a natural number, there is an extremally disconnected, compact Hausdorff space X such that $S(X^n) = \alpha^+$ and $S(X^{n+1}) > \alpha^+$.

We thank Fred Galvin for several helpful, detailed letters and for unpublished, handwritten notes (34 pages, 1976).

The proofs of Theorems 6.32, 7.9 and 7.13 were given originally in the context of graph theory and the generic topology of partially ordered sets. Perhaps our (equivalent) topological presentation makes the proofs more transparent.

Extending Theorems 7.9 and 7.13, Kalamidas (1981) has proved the following result.

Theorem. Assume the generalized continuum hypothesis, and let λ be an (infinite) regular cardinal.

(a) There is a space X such that X does not have calibre $(\lambda^+, \lambda^+, 2)$, and $S_\lambda(X \times Y) = \lambda^+$ for every space Y such that $S_\lambda(Y) = \lambda^+$; and

(b) there are spaces X and Y such that $S_\lambda(X) = S_\lambda(Y) = \lambda^+$, and $S_\lambda(X \times Y) > \lambda^+$.

We indicate how Martin's axiom (together with the denial of the continuum hypothesis) affects the chain conditions investigated in Chapters 6 and 7. Briefly, it trivializes or collapses the spectrum of properties between the countable chain condition and (pre-) calibre ω^+, and it has no effect on questions concerning strictly positive measures.

We give the (standard) topological equivalent of Martin's axiom. The original formulation, given by Solovay & Tennenbaum (1971), Martin & Solovay (1970) and Kunen (1968), was motivated by the forcing conditions of Cohen (1966).

Martin's axiom. If X is a compact Hausdorff space with c.c.c., $\alpha < 2^\omega$ and $\{D_\xi : \xi < \alpha\}$ is a set of dense, open subsets of X, then $\cap_{\xi < \alpha} D_\xi$ is a dense subset of X.

The basic consistency result concerning Martin's axiom is the following statement (cf. Solovay & Tennenbaum 1971).

Theorem. The axiom system [ZFC] is equi-consistent with the system [ZFC + Martin's axiom + $\omega^+ < 2^\omega$].

The effect of Martin's axiom on the countable chain condition is described by the following result.

Theorem. Assume Martin's axiom and assume $\omega^+ < 2^\omega$.

(a) If α is an infinite cardinal such that $\omega^+ \leq \mathrm{cf}(\alpha) \leq \alpha < 2^\omega$ and X is a c.c.c. space, then X has precalibre $(\alpha, \mathrm{cf}(\alpha))$.

(b) Every c.c.c. space has precalibre ω^+.

(c) The class of c.c.c. spaces is closed under the formation of products.

Proof. (a) It is by Corollary 2.22 enough to consider the case that X is the (compact, Hausdorff) space $G(X)$. If $\{U_\eta : \eta < \alpha\}$ is a set of non-empty, open subsets of X and if

$$V_\xi = \cup\{U_\eta : \xi \leq \eta < \alpha\} \text{ for } \xi < \alpha$$

then from Lemma 7.1 there is $\bar{\xi} < \alpha$ such that

$$V_\xi \text{ is dense in cl } V_{\bar{\xi}} \text{ for } \bar{\xi} \leq \xi < \alpha.$$

Since $\text{cl}\, V_{\bar{\xi}}$ is a compact, Hausdorff c.c.c. space and $\alpha < 2^\omega$, we have from Martin's axiom that there is

$$p \in \cap\{V_\xi : \bar{\xi} \leq \xi < \alpha\};$$

it is then clear that there is $A \subset \alpha$ with $|A| = \text{cf}(\alpha)$ such that $\cap_{\eta \in A} U_\eta \neq \emptyset$, as required.

(b) is the case $\alpha = \omega^+$ of (a).

(c) From (b) and Theorem 7.4 it follows that (under the assumptions of the theorem) every c.c.c. space is productively c.c.c. Thus the product of finitely many c.c.c. spaces is a c.c.c. space, and the required statement (for arbitrary products) follows from (the case $\lambda = 2$, $\alpha = \omega^+$ of) Theorem 3.25 or, alternatively, from (the case $\alpha = \omega^+$, $\beta = \gamma = 2$, $\kappa = \omega$, $Y = X_I$ of) Theorem 3.5(a).

All essential parts of the preceding argument appear in Juhász (1970); they are attributed there to Kunen, and also to Solovay & Rowbottom.

For expository surveys of some uses of Martin's axiom, the reader may consult Martin & Solovay (1970), M. E. Rudin (1975, section IV, 1977, and Shoenfield (1975). Some specific applications and related axioms are given by Malyhin & Šapirovskiĭ (1973), Tall (1974, 1977, 1979), Juhász (1977), Herink (1977), Juhász & Weiss (1978), Kunen & Tall (1979), Wage (1979), Argyros (1981b), Zachariades (1981), and Bell (1982a). David Fremlin is currently completing a comprehensive monograph (provisionally) titled Consequences of Martin's Axiom.

In classical work K. Gődel (1938, 1940) showed that his axiom V = L (the statement that every set is constructible, that is, roughly speaking, the statement that there is in the universe no set not required by the axioms) is consistent with ZFC; further, V = L implies the generalized continuum hypothesis. Jensen (1968, 1972) derived from V = L (in fact from his 'diamond principle', a combinatorial consequence of V = L) the existence of a Souslin line (that is, a non-separable *linearly ordered* space with c.c.c.), and earlier Kurepa (1950, 1952) had shown that if X is a Souslin line then $X \times X$ does not have c.c.c. Thus we have the following diagram of implications.

	V = L	$\Longrightarrow$ Gődel (1938, 1940)	$\omega^+ = 2^\omega$	
Jensen (1968, 1972)	$\Downarrow$		$\Downarrow$ Laver, Galvin (7.13)
	there is a Souslin line	$\Longrightarrow$ Kurepa (1950, 1952)	there is X such that $S(X) = \omega^+ < S(X \times X)$.

It is clear from this and from the theorem proved above that the question 'Is the product of c.c.c. spaces a c.c.c. space?' is not settled by the axioms of ZFC; like Souslin's problem, this question is answered in one way by Martin's axiom (together with the denial of the continuum hypothesis) and quite differently by Gődel's axiom V = L.

We note that Tennenbaum (1968) has proved the consistency with ZFC of the existence of a Souslin line together with the denial of the continuum hypothesis; see also Jech (1967). In a major result whose proof occupies the latter half of Devlin & Johnsbråten (1974), Jensen shows the consistency with ZFC and $\omega^+ = 2^\omega$ of the statement that there is no Souslin line.

Jensen (1972, Theorem 6.2) has shown further, still assuming V = L, that if α is a strongly inaccessible cardinal that is not weakly compact, then there is a linearly ordered space X such that $S(X) = \alpha$ and $S(X \times X) > \alpha$. It is unknown whether this implication can be proved in ZFC, or even in ZFC + GCH.

We are indebted to K. Prikry for the following remarks. It has been proved independently by Silver and Lévy (cf. Jech (1978, Exercise 35.5)) that if α is a strongly inaccessible, non-measurable cardinal with an α-complete, uniform filter p for which the Souslin number of the Stone space of the quotient Boolean algebra $2^\alpha/p$ is equal to α, then α is not weakly compact and in fact there is a space

X such that $S(X) = \alpha$ and $S(X \times X) > \alpha$. It has been shown by Kunen (1978) that the hypothesis of this theorem is consistent with ZFC.

For properties of the Rudin–Keisler order cited in Problem 7.20, see Comfort & Negrepontis (1974), especially Theorems 9.13 and 10.4. It was Kunen (1972) who showed that there is a set S (with $|S| = 2^{\alpha}$) of uniform ultrafilters on α, pairwise incomparable in the Rudin–Keisler order. That one may choose $|S| = 2^{2^{\alpha}}$ was shown subsequently by Shelah (cf. Shelah & Rudin 1978).

Other recent applications of the technique of large independent families can be found in Kunen (1980).

Corson-compact spaces

We consider the following classes of spaces, inspired from Banach space theory. Let X be a compact Hausdorff space.

(a) X is called an *Eberlein-compact* space if there are a Banach space B and a homeomorphic embedding of X into B, where B has the weak topology (i.e., the topology induced on B by its dual B^*).

(b) X is called a *Talagrand-compact* space if the Banach space $C(X)$ is $\mathscr{K}$-analytic in its weak topology. (A topological space is *$\mathscr{K}$-analytic* if it is the continuous image of a $K_{\sigma\delta}$-space Y, i.e., a space Y of the form

$$Y = \bigcap_{n<\omega} \bigcup_{m<\omega} K_{m,n} \quad \text{with } K_{m,n} \text{ compact.)}$$

(The introduction of Talagrand-compact spaces is due to Talagrand (1977, 1979a), and was inspired in part by Rosenthal's (1974a) beautiful example of a non-weakly compactly generated subspace of $L^1\{0,1\}^{2^{\omega}}$.)

(c) X is called a *Gul'ko-compact* space if the Banach space $C(X)$ is $\mathscr{K}$-countably determined in its weak topology. (A topological space Y is *$\mathscr{K}$-countably determined* if there is a sequence $\{A_n : n < \omega\}$ of (closed) subsets of Y such that for every $x \in Y$ there is a non-empty subset $N(x)$ of ω such that $x \in N_{n \in N(x)} A_n$, and $\bigcap_{n \in N(x)} A_n$ is compact.)

(d) X is a *Corson-compact* space if there are a set Γ and a homeomorphic embedding of X into the ω^+-Σ-product of $\mathbb{R}^{\Gamma}$ based at

0, i.e., into

$$\Sigma(\mathbb{R}^\Gamma) = \{x \in \mathbb{R}^\Gamma : |\{\gamma \in \Gamma : x_\gamma \neq 0\}| \leq \omega\}.$$

We set

$$c_0(\Gamma) = \{x \in \mathbb{R}^\Gamma : \varepsilon > 0 \text{ implies } |\{\gamma \in \Gamma : |x_\gamma| \geq \varepsilon\}| < \omega\} \subset \Sigma(\mathbb{R}^\Gamma).$$

The fundamental result of Amir & Lindenstrauss (1968) states that every Eberlein-compact has a homeomorphic embedding into $c_0(\Gamma)$ for some set Γ (and hence is Corson-compact).

Talagrand (1975) proved that every Eberlein-compact is Talagrand-compact, thus establishing a conjecture of Corson (1961). It is clear that every Talagrand-compact is Gul'ko-compact; Gul'ko (1979) has proved the deep result that every Gul'ko-compact is Corson-compact (see also Vašak (1980), who independently in 1976 proved (in an implicit form) the same result). That the Gul'ko-compact spaces (as defined here) coincide with those considered by Gul'ko (1979) follows from general topological results (cf. Nagata (1972, Theorems 3, 5)).

Furthermore, there is a Talagrand-compact space that is not Eberlein-compact (Talagrand (1977, 1979a)), and there is a Corson-compact space that is not Gul'ko-compact (Alster & Pol (1980), Argyros, Mercourakis & Negrepontis (1983)).

It is an open and interesting question whether there is a Gul'ko-compact space that is not Talagrand-compact.

The survey papers of Arhangel'skiĭ (1976, 1978) and Wage (1980) provide information on Eberlein-compact and Corson-compact spaces.

The behavior of these classes of spaces with respect to the chain conditions is as follows:

(i) A c.c.c. Eberlein-compact space is separable (and hence metrizable).

(ii) A Corson-compact space with calibre $\aleph_1$ is separable (and hence metrizable). Hence, assuming MA + CH, every c.c.c. Corson-compact space is separable (and hence metrizable).

(iii) It has been proved by Argyros, Mercourakis & Negrepontis (1983), answering a problem posed by Talagrand (1977, 1979a), that every Gul'ko-compact c.c.c. space is separable (and hence metrizable). In fact, they proved that if X is a Gul'ko-compact space, then $S(X) = (w(X))^+$.

It is not known if a Gul'ko-compact (or even a Talagrand-compact) space has a dense G_δ metrizable subset consisting of G_δ-points. (This is a property enjoyed by Eberlein-compact spaces; cf. Namioka (1974) and Benyamini, Rudin & Wage (1977).) These questions are due to Talagrand (1977, 1979a, 1979b), who proved a weaker result; if answered in the positive they would establish properties stronger than the chain condition proved by Argyros and Negrepontis for Talagrand-compact spaces.

(iv) It has been observed that, assuming the continuum hypothesis, there is a non-separable Corson-compact space with a strictly positive measure. In fact, one may take the example of Haydon (1978) and use Šapirovskiĭ's (1977, 1979) result: every compact space of countable tightness can be mapped by an irreducible function into a Corson-compact space (cf. Alster & Pol (1980)).

(v) Assuming the continuum hypothesis, Argyros, Mercourakis & Negrepontis (1983) have proved:

1. There is a Corson-compact space that does not have a strictly positive measure and does not have calibre $\aleph_1$ (in fact, does not have measure-calibre $\aleph_1$), but does have property (*) and property K_n for every natural number $n \geq 2$. (This is a variation of Gaifman's construction using, as in 6.23 above, the existence of Lusin sets.)

2. There is a Corson-compact c.c.c. space X such that $X \times X$ does not have c.c.c. (cf. 7.13).

3. For every $n \geq 2$, there is a Corson-compact space K such that K has property K_n but not property K_{n+1} (cf. 7.9).

4. There is a Corson-compact space K such that K does not have Knaster's property (K) but K has property P (and hence is productively c.c.c.) (cf. 6.26).

5. The result of (iv) can be proved by similar elementary methods, i.e., without recourse to Haydon's 'sophisticated' example or Šapirovskiĭ's result, but using only e.g. the Erdős example given in 6.19 (the case $\beta = \omega$).

6. It follows from the result of Šapirovskiĭ (1979, Corollary 15), on the existence of a dense set with a point-countable base in every Corson-compact space, that there are (not necessarily compact)

c.c.c. spaces with a point-countable base with chain properties as in 1 through 5 above. (This improves, for example, van Douwen's result given in Galvin (1980) of a first-countable c.c.c. space whose square is not c.c.c.

It is an open question, due to van Douwen and Negrepontis, whether there are compact, first-countable spaces with the chain properties of 1 through 5 above.

It has been proved further by Argyros, Mercourakis & Negrepontis (1983) that if X is a Corson-compact space, then there is a one-to-one bounded linear operator $T: C(X) \rightarrow c_0(\Gamma)$ (for some set Γ). Hence for every Corson-compact space X the Banach space $C(X)$ has an equivalent strictly convex norm (and in fact, using the method of Troyanski (1971), it even has an equivalent locally uniformly convex norm). From this and the results of (v) above on chain conditions there follows a negative answer, assuming the continuum hypothesis, to the question posed earlier by Argyros and Negrepontis (see the Notes for Chapter 6): does every compact c.c.c. space X such that $C(X)$ has an equivalent strictly convex norm have a strictly positive measure?

8

Classes of Compact-calibres, Using Spaces of Ultrafilters

In parallel with the investigation of calibres in Chapter 4, we here examine the extent of classes of compact-calibres of spaces. We determine those classes **T** of infinite cardinals for which there is a Hausdorff space $X(\mathbf{T})$ such that $X(\mathbf{T})$ does not have compact-calibre α if and only if $\alpha \in \mathbf{T}$; and we show that such a space $X(\mathbf{T})$, when it exists, may be chosen completely regular.

In contrast with its behavior concerning calibre, the cardinal number ω plays no special role with respect to compact-calibre: there are infinite, non-compact, Hausdorff spaces which have compact-calibre ω, and others which do not. Accordingly, the format of Corollary 8.4 is simpler than that of Corollary 4.5 (to which it is similar), and 8.5 and 8.6 are simpler than 4.6 and 4.7, respectively.

The examples we give of spaces with compact-calibre properties prescribed in advance are defined by beginning with the Stone–Čech compactification of a discrete space (which has, like every compact space, compact-calibre α for all $\alpha \geq \omega$) and discarding from it the (non-principal) ultrafilters with various degrees of non-uniformity. The spaces defined in this way certainly do not have Souslin number equal to ω^+, and in fact we do not know whether the spaces $X(\mathbf{T})$ described above can be chosen to satisfy in addition $S(X(\mathbf{T})) = \omega^+$.

In Corollary 8.12 we note a curiosity: there are completely regular, Hausdorff, non-compact spaces with compact-calibre α for all $\alpha \geq \omega$.

8.1 *Theorem.* Let α, κ and λ be infinite cardinals, and $X = N_\kappa(\lambda)$. The following statements are equivalent.

(a) $\kappa \leq \alpha \leq \lambda$ or $\kappa \leq \mathrm{cf}(\alpha) \leq \lambda$;

(b) X is not pseudo-(α, α)-compact;

(c) X does not have compact-calibre α.

Proof. (a) $\Rightarrow$ (b). We show first that X is not pseudo-(λ, κ)-

compact. Set $U_\eta = \{\eta\}$ for $\eta < \lambda$ and let $x \in X$. If $x \in \lambda$, then $\{x\}$ is a neighborhood V of x such that

$$|\{\eta < \lambda : V \cap U_\eta \neq \varnothing\}| = 1 < \kappa;$$

and if $x \in X \backslash \lambda$, there is $A \in \mathscr{P}_\kappa^*(\lambda)$ such that $x \in \mathrm{cl}_{\beta(\lambda)} A \subset X$, and then $\mathrm{cl}_{\beta(\lambda)} A$ is an (open-and-closed) neighborhood V of x such that

$$|\{\eta < \lambda : U \cap U_\eta \neq \varnothing\}| = |A| < \kappa.$$

Now if either $\kappa \leq \alpha \leq \lambda$ or $\kappa \leq \mathrm{cf}(\alpha) \leq \lambda$, then (from Theorem 2.1 (c)) either X is not pseudo-(α, α)-compact or X is not pseudo-$(\mathrm{cf}(\alpha), \mathrm{cf}(\alpha))$-compact. It follows from Theorem 2.3 (c) that X is not pseudo-(α, α)-compact.

(b) $\Rightarrow$ (c). This follows from Theorem 2.1(f).

(c) $\Rightarrow$ (a). We consider four cases.

Case 1. $\kappa > \lambda$. The space X is then the compact space $\beta(\lambda)$; hence X has compact-calibre α.

Case 2. $\kappa \leq \lambda$ and $\alpha < \kappa$. Let $\{U_\xi : \xi < \alpha\} \subset \mathscr{T}^*(X)$, for $\xi < \alpha$ let $\eta(\xi) \in U_\xi \cap \lambda$, set $A = \{\eta(\xi) : \xi < \alpha\}$ and set

$$K = \mathrm{cl}_{\beta(\alpha)} A.$$

Then K is a compact subspace of X, and

$$\eta(\xi) \in U_\xi \cap K \text{ for } \xi < \alpha.$$

Case 3. $\kappa \leq \lambda$ and $\lambda < \mathrm{cf}(\alpha)$. Let $\{U_\xi : \xi < \alpha\} \subset \mathscr{T}^*(X)$ and set

$$A(\eta) = \{\xi < \alpha : \eta \in U_\xi\} \quad \text{for } \eta < \lambda.$$

Since $\lambda < \mathrm{cf}(\alpha)$, there is $\bar{\eta} < \lambda$ such that $|A(\bar{\eta})| = \alpha$.

Case 4. $\kappa \leq \lambda$ and $\mathrm{cf}(\alpha) < \kappa$ and $\lambda < \alpha$. It follows from Case 2 (applied to $\mathrm{cf}(\alpha)$ in place of α) that X has compact-calibre $\mathrm{cf}(\alpha)$. Since $\pi w(X) = \lambda < \alpha$, it follows from Theorem 2.3(b) that X has compact-calibre α.

In each of the four cases the space X has compact-calibre α.

The proof is complete.

We see next that in Theorem 2.3(b′) the condition $\pi w(X) < \alpha$ cannot be omitted.

8.2 *Theorem.* Let α be an (infinite) singular cardinal. There is a completely regular, Hausdorff space X such that X has compact-calibre $\mathrm{cf}(\alpha)$, X does not have compact-calibre α, and $\pi w(X) = \alpha$.

Proof. We set $X = N_{(\mathrm{cf}(\alpha))^+}(\alpha)$. It is immediate from Theorem 8.1 that X has the required properties.

We saw in Corollary 3.7(b) that if α is an (infinite) regular cardinal and $\{X_i : i \in I\}$ a set of spaces such that X_i has compact-calibre α for each $i \in I$, then X_I has compact-calibre α. We see next that the analogous statement fails for singular cardinals α. Theorem 8.3 supplements and complements Corollary 3.11(ii), which treats products of spaces with compact-calibre (α, β) with α singular and $\alpha > \beta$.

8.3 *Theorem.* Let α be an (infinite) singular cardinal. There is a set $\{X_i : i \in I\}$ of completely regular, Hausdorff spaces such that X_i has compact-calibre α for all $i \in I$, and X_I does not have compact-calibre α.

Proof. There is a set $\{\alpha_\sigma : \sigma < \mathrm{cf}(\alpha)\}$ of cardinal numbers such that

$$\alpha_0 = \mathrm{cf}(\alpha),$$
$$\mathrm{cf}(\alpha) < \alpha_{\sigma'} < \alpha_\sigma < \alpha \text{ for } 0 < \sigma' < \sigma < \mathrm{cf}(\alpha), \text{ and}$$
$$\sum_{\sigma < \mathrm{cf}(\alpha)} \alpha_\sigma = \alpha.$$

We set

$$X_\sigma = N_{(\mathrm{cf}(\alpha))^+}(\alpha_\sigma) \text{ for } \sigma < \mathrm{cf}(\alpha), \text{ and}$$
$$X = \prod_{\sigma < \mathrm{cf}(\alpha)} X_\sigma.$$

That X_σ has compact-calibre α for all $\sigma < \mathrm{cf}(\alpha)$ follows from Theorem 8.1 (with κ and λ replaced by $(\mathrm{cf}(\alpha))^+$ and α_σ, respectively).

We are to show that X does not have compact-calibre α. We show that in fact X does not have compact-calibre $(\alpha, (\mathrm{cf}(\alpha))^+)$.

We set

$$U(\sigma, \eta) = \{x \in X : x_0 = \sigma, x_\sigma = \eta\} \text{ for } 0 < \sigma < \mathrm{cf}(\alpha), \eta < \alpha_\sigma,$$

and we note that

$$\pi_\sigma[U(\sigma, \eta)] = \{\eta\} \quad \text{for } 0 < \sigma < \mathrm{cf}(\alpha), \eta < \alpha_\sigma.$$

Let K be a compact subspace of X, and set

$$A(\sigma) = \{\eta < \alpha_\sigma : U(\sigma, \eta) \cap K \neq \emptyset\} \quad \text{for } 0 < \sigma < \mathrm{cf}(\alpha).$$

For $\eta \in A(\sigma)$ we have $\pi_\sigma[U(\sigma,\eta)] \cap \pi_\sigma[K] \neq \varnothing$, and hence $\eta \in \pi_\sigma[K]$. It follows from Lemma B.10 of the Appendix (with λ, κ and K replaced by α_σ, $(\mathrm{cf}(\alpha))^+$ and $\pi_\sigma[K]$, respectively) that

$$|A(\sigma)| \leq \mathrm{cf}(\alpha) \quad \text{for } 0 < \sigma < \mathrm{cf}(\alpha);$$

hence

$$|\{\langle\sigma,\eta\rangle : U(\sigma,\eta) \cap K \neq \varnothing\}| = \sum\{|A(\sigma)| : 0 < \sigma < \mathrm{cf}(\alpha)\} \leq \mathrm{cf}(\alpha),$$

as required.

The proof is complete.

We give now a consequence of Theorem 8.1, and a theorem which allows us to determine those sets of infinite cardinal numbers which can arise as the class of non-compact-calibres of a space.

8.4 *Corollary.* Let λ and α be infinite cardinals and $X = N_\lambda(\lambda)$. Then X has compact-calibre α if and only if $\alpha \neq \lambda$ and $\mathrm{cf}(\alpha) \neq \lambda$.

8.5 *Theorem.* Let $\mathbf{S}$ be a non-empty set of infinite cardinal numbers, $X(\lambda) = N_\lambda(\lambda)$ for $\lambda \in \mathbf{S}$ and $X(\mathbf{S}) = \prod_{\lambda \in \mathbf{S}} X(\lambda)$.

For $\alpha \geq \omega$, the following statements are equivalent.

(a) $X(\mathbf{S})$ does not have compact-calibre α;

(b) $\alpha \in \mathbf{S}$ or $\mathrm{cf}(\alpha) \in \mathbf{S}$.

Proof. (a) $\Rightarrow$ (b). We suppose that $\alpha \notin \mathbf{S}$ and $\mathrm{cf}(\alpha) \notin \mathbf{S}$. It follows from Theorem 6.1 that $X(\lambda)$ has compact-calibre α for all $\lambda \in \mathbf{S}$.

We show that $X(\mathbf{S})$ has compact-calibre α. If α is regular this follows from Corollary 3.7(b). We assume in what follows that α is singular.

Let $\{\alpha_\sigma : \sigma < \mathrm{cf}(\alpha)\}$ be a set of cardinal numbers such that

$$\alpha_\sigma \text{ is regular for } \sigma < \mathrm{cf}(\alpha),$$
$$\mathrm{cf}(\alpha) < \alpha_{\sigma'} < \alpha_\sigma < \alpha \text{ for } \sigma' < \sigma < \mathrm{cf}(\alpha), \text{ and}$$
$$\sum_{\sigma < \mathrm{cf}(\alpha)} \alpha_\sigma = \alpha.$$

We claim that if $J \in \mathscr{P}^*_\omega(\mathbf{S})$, then X_J has compact-calibre α and compact-calibre $(\alpha_{\sigma+1}, \alpha_\sigma)$. That X_J has compact-calibre α follows from Theorem 2.2(b); to prove that X_J has compact-calibre $(\alpha_{\sigma+1}, \alpha_\sigma)$ we consider two cases.

Case 1. $\alpha_\sigma \notin J$. Then $X(\lambda)$ has compact-calibre α_σ for all $\lambda \in J$.

It follows that X_J has compact-calibre α_σ and hence compact-calibre $(\alpha_{\sigma+1}, \alpha_\sigma)$.

Case 2. $\alpha_\sigma \in J$. If $J \backslash \{\alpha_\sigma\} = \emptyset$ then X_J, which is $X(\alpha_\sigma)$, has compact-calibre $\alpha_{\sigma+1}$ and hence compact-calibre $(\alpha_{\sigma+1}, \alpha_\sigma)$. If $J \backslash \{\alpha_\sigma\} \neq \emptyset$ then, since $X_{J \backslash \{\alpha_\sigma\}}$ has compact-calibre $(\alpha_{\sigma+1}, \alpha_\sigma)$ (by Case 1) and $X(\alpha_\sigma)$ has compact-calibre $\alpha_{\sigma+1}$, the space X_J, which is $X_{J \backslash \{\alpha_\sigma\}} \times X(\alpha_\sigma)$, has compact-calibre $(\alpha_{\sigma+1}, \alpha_\sigma)$.

The proof of the claim is complete. It follows from Corollary 3.11 (b) that $X(\mathbf{S})$ has compact-calibre α, as required.

(b) $\Rightarrow$ (a). From Corollary 8.4 it follows that there is $\lambda \in \mathbf{S}$ such that $X(\lambda)$ does not have compact-calibre α. It is then clear that $X(\mathbf{S})$ does not have compact-calibre α.

8.6 *Corollary*. Let $\mathbf{T}$ be a class of infinite cardinal numbers. Then statements (a) and (b) are equivalent.

(a) There is an infinite space X such that X does not have compact-calibre α if and only if $\alpha \in \mathbf{T}$.

(b) (i) If $\mathrm{cf}(\alpha) \in \mathbf{T}$ then $\alpha \in \mathbf{T}$,

(ii) $\{\mathrm{cf}(\alpha) : \alpha \in \mathbf{T}\}$ is a set, and

(iii) if κ is a regular cardinal and $\kappa \notin \mathbf{T}$, then $\{\alpha \in \mathbf{T} : \mathrm{cf}(\alpha) = \kappa\}$ is a set.

Proof. (a) $\Rightarrow$ (b). (i) follows from Theorem 2.3 (b).

(ii) If α is an infinite cardinal and $\mathrm{cf}(\alpha) > d(X)$, then it follows from Theorems 2.6 and 2.1 (d) that X has compact-calibre α. Hence

$$\{\mathrm{cf}(\alpha) : \alpha \in \mathbf{T}\} \subset (d(X))^+.$$

(iii) From Theorem 2.3 (b′) we have

$$\{\alpha \in \mathbf{T} : \mathrm{cf}(\alpha) = \kappa\} \subset (\pi w(X))^+ \quad \text{for } \kappa \notin \mathbf{T}.$$

(b) $\Rightarrow$ (a). For $\kappa \in \{\mathrm{cf}(\alpha) : \alpha \in \mathbf{T}\}$ we set

$$\begin{aligned} \mathbf{S}(\kappa) &= \{\kappa\} && \text{if } \kappa \in \mathbf{T}, \\ &= \{\alpha \in \mathbf{T} : \mathrm{cf}(\alpha) = \kappa\} && \text{if } \kappa \notin \mathbf{T}, \end{aligned}$$

we set

$$\mathbf{S} = \cup \{\mathbf{S}(\kappa) : \kappa \in \{\mathrm{cf}(\alpha) : \alpha \in \mathbf{T}\}\},$$

we note that $\mathbf{S}$ is a set and $\mathbf{S} \subset \mathbf{T}$, and we set

$$X(\lambda) = N_\lambda(\lambda) \text{ for } \lambda \in \mathbf{S} \text{ and}$$
$$X = \prod_{\lambda \in \mathbf{S}} X(\lambda).$$

Suppose next that $\alpha \geq \omega$ and that X does not have compact-calibre α. From Theorem 8.5 it follows that $\alpha \in \mathbf{S}$ or $\mathrm{cf}(\alpha) \in \mathbf{S}$, and since $\mathbf{S} \subset \mathbf{T}$ we have $\alpha \in \mathbf{T}$ from (i).

Suppose next that $\alpha \in \mathbf{T}$. If $\mathrm{cf}(\alpha) \in \mathbf{T}$ then

$$\mathrm{cf}(\alpha) \in \mathbf{S}(\mathrm{cf}(\alpha)) \subset \mathbf{S},$$

and if $\mathrm{cf}(\alpha) \notin \mathbf{T}$ then $\alpha \in \mathbf{S}(\mathrm{cf}(\alpha)) \subset \mathbf{S}$; it follows from Theorem 8.5 that X does not have calibre α.

8.7 *Remark.* The space X defined in the proof of the implication (b) $\Rightarrow$ (a) in Corollary 8.6 is a completely regular, Hausdorff space. This shows that if for a class $\mathbf{T}$ of infinite cardinal numbers there is a space X such that X does not have compact-calibre α if and only if $\alpha \in \mathbf{T}$, then X may be chosen to be a completely regular, Hausdorff space. The spaces X defined above do not, however, satisfy the condition $S(X) = \omega^+$ (cf. Remark 4.8). We do not know whether, given a class $\mathbf{T}$ of cardinals satisfying the conditions of Corollary 8.6(b), it is possible to define a completely regular Hausdorff space X satisfying the conditions of Corollary 8.6(a) and in addition the condition $S(X) = \omega^+$.

The remainder of this chapter is preparatory for Theorem 8.11, where we determine the compact-calibres and pseudo-compactness numbers of generalized Σ-products.

8.8 *Lemma.* Let $\{X_i : i \in I\}$ be a set of Hausdorff spaces, $p \in X_I$, and $\{x(\xi) : \xi < \alpha\}$ a subset of X_I such that

$$E_p(x(\xi)) \cap E_p(x(\xi')) = \varnothing \quad \text{for } \xi' < \xi < \alpha, \text{ and}$$
$$Y = \{x(\xi) : \xi < \alpha\} \cup \{p\}.$$

Then Y is closed in X_I.

Proof. Let $x \in X_I \backslash Y$. Since $x \neq p$ there is $i \in E(x)$ and there is a neighborhood U_i of x_i in X_i such that $p_i \notin U_i$. If $i \notin \cup_{\xi<\alpha} E(x(\xi))$, we set $W = \pi_i^{-1}(U_i)$; and if there is $\xi < \alpha$ such that $i \in E(x(\xi))$, then (ξ is unique and) we choose a neighborhood V of x such that $x(\xi) \notin V$ and we set $W = V \cap \pi_i^{-1}(U_i)$. Then W is a neighborhood of x and $W \cap Y = \varnothing$.

8.9 *Lemma.* Let $\{X_i : i \in I\}$ be a set of σ-compact, Hausdorff spaces, $p_i \in X_i$ for $i \in I$, and $X = \Sigma_\omega(p, X_I)$. Then X is σ-compact.

Proof. For $i \in I$ there is a set $\{X(n,i): n < \omega\}$ of compact subspaces of X_i such that $X_i = \cup_{n<\omega} X(n,i)$. We assume without loss of generality that

$$p_i \in X(n,i) \text{ for } i \in I, n < \omega, \text{ and}$$
$$X(n',i) \subset X(n,i) \text{ for } n' < n < \omega.$$

For $n < \omega$ we set

$$X(n) = \{x \in X_I : |E(x)| \leq n \text{ and } x \in \prod_{i \in I} X(n,i)\}.$$

Since $\{x \in X_I : |E(x)| \leq n\}$ is closed in X_I and $\prod_{i \in I} X(n,i)$ is compact, the space $X(n)$ is compact for $n < \omega$.

We show $\Sigma_\omega(p, X_I) = \cup_{n<\omega} X(n)$. The inclusion $\supset$ is clear. Let $x \in \Sigma_\omega(p, X_I)$ with $|E(x)| = k < \omega$, let $x_i \in X(n(i), i)$ for $i \in E(x)$, and set

$$\bar{n} = \max\{k, \max\{n(i): i \in E(x)\}\};$$

then $x \in X(\bar{n})$.

8.10 *Theorem.* Let α be an (infinite) singular cardinal with $\mathrm{cf}(\alpha) = \omega$ and let $\kappa > \omega, \lambda \geq \omega$, and $X = \Sigma_\kappa\{0,1\}^\lambda$. Then X has compact-calibre α.

Proof. The space $\Sigma_{\omega^+}\{0,1\}^\lambda$ is dense in X; it follows from Theorem 2.5(a)(ii) that it is sufficient to prove the theorem in the case $\kappa = \omega^+$. In what follows we assume that $\kappa = \omega^+$ and we consider three cases.

Case 1. $\lambda < \alpha$. It follows from Theorem 3.8 that X has compact-calibre ω. Since

$$\pi w(X) \leq w(X) \leq w(\{0,1\}^\lambda) = \lambda < \alpha,$$

the space X has compact-calibre α by Theorem 2.3(b').

Case 2. $\lambda = \alpha$. Let $\{U^\xi : \xi < \alpha\}$ be a set of non-empty basic subsets of $\{0,1\}^\alpha$, and let $\{\alpha_n : n < \omega\}$ be a set such that

$$\alpha_n \text{ is a regular cardinal for } n < \omega,$$
$$\omega < \alpha_{n'} < \alpha_n < \alpha \quad \text{for } n' < n < \omega, \text{ and}$$
$$\sum_{n<\omega} \alpha_n = \alpha.$$

It follows from Lemma 1.9 and Theorem A.7 of the Appendix

(with $\kappa = \omega$) that there are $\{A_n : n < \omega\}$ and $\{J_n : n < \omega\}$ such that

$$\begin{aligned} & A_n \in [\alpha]^{\alpha_n} \quad \text{for } n < \omega, \\ & A_{n'} \cap A_n = \varnothing \quad \text{for } n' < n < \omega, \text{ and} \\ & R(U^{\xi'}) \cap R(U^{\xi}) = J_n \quad \text{for } \xi, \xi' \in A_n, \xi' \neq \xi \\ & \qquad\qquad \subset J_n \quad \text{for } \xi' \in A_{n'}, \xi \in A_n, n' < n < \omega. \end{aligned}$$

We set

$$A = \bigcup_{n<\omega} A_n, \quad J = \bigcup_{n<\omega} J_n, \quad I = \alpha \backslash J, \text{ and}$$

$$V^{\xi} = \pi_I[U^{\xi}] \quad \text{for } \xi \in A,$$

and we note that

$$\begin{aligned} & |A| = \alpha, |J| \leq \omega, \text{ and} \\ & R(V^{\xi'}) \cap R(V^{\xi}) = \varnothing \quad \text{for } \xi', \xi \in A, \xi' \neq \xi. \end{aligned}$$

For $\xi < \alpha$ we define $x(\xi)$ by

$$\begin{aligned} x(\xi)_\eta & \in V^{\xi}_\eta \quad \text{if } \eta \in R(V^{\xi}) \\ & = 0 \quad \text{if } \eta \in I \backslash R(V^{\xi}). \end{aligned}$$

We note that if $\xi', \xi \in A$ and $\xi' \neq \xi$, then

$$E(x(\xi')) \cap E(x(\xi)) \subset R(V^{\xi'}) \cap R(V^{\xi}) = \varnothing;$$

we note further that if $\xi \in A$, then

$$|E(x(\xi))| \leq |R(V^{\xi})| < \omega$$

and hence $x(\xi) \in \Sigma_\omega \{0, 1\}^I$. We define $p \in \{0, 1\}^I$ by

$$p_\eta = 0 \quad \text{for } \eta \in I,$$

we set $K_I = \{x(\xi) : \xi \in A\} \cup \{p\}$, and we note from Lemma 8.8 that K_I is closed in $\{0, 1\}^I$ and hence compact.

We set $K = \{0, 1\}^J \times K_I$. Then K is compact, and since $|J| \leq \omega$ and $K_I \subset \Sigma_\omega \{0, 1\}^I$, we have $K \subset \Sigma_{\omega^+} \{0, 1\}^\alpha$. Finally, for $\xi \in A$ there is $y(\xi)$ such that

$$\begin{aligned} & y(\xi)_J \in \pi_J[U^{\xi}], \text{ and} \\ & y(\xi)_I = x(\xi); \end{aligned}$$

it is then clear that $y(\xi) \in U^{\xi} \cap K$, as required.

Case 3. $\lambda > \alpha$. Let $\{U^{\xi} : \xi < \alpha\}$ be a set of non-empty basic subsets of $\{0, 1\}^\lambda$, and set $F = \bigcup_{\xi < \alpha} R(U^{\xi})$. Since $|F| \leq \alpha$ it follows from

Cases 1 and 2 that there are $A \in [\alpha]^{\alpha}$ and compact $K_F \subset \Sigma_{\omega^+}\{0,1\}^F$ such that $\pi_F[U^{\xi}] \cap K_F \neq \emptyset$ whenever $\xi \in A$. We define

$$p = \langle p_\eta : \eta \in \alpha \backslash F \rangle \in \{0,1\}^{\alpha \backslash F}$$

by

$$p_\eta = 0 \quad \text{for } \eta \in \alpha \backslash F$$

and we set $K = K_F \times \{p\}$. Then K is a compact subspace of X, and $U^{\xi} \cap K \neq \emptyset$ whenever $\xi \in A$, as required.

The proof is complete.

8.11 *Theorem.* Let α, κ and λ be infinite cardinals and $X = \Sigma_\kappa \{0,1\}^\lambda$. The following statements are equivalent.

(a) $\mathrm{cf}(\alpha) = \omega$ and $\kappa = \omega$;

(b) X is not pseudo-(α, α)-compact;

(c) X does not have compact-calibre α.

Proof. (a) $\Rightarrow$ (b). We have $\omega \subset \lambda$. We set

$$U_k = \{x \in X : x_n = 1 \quad \text{for } n \leq k\} \text{ for } k < \omega,$$

we note that $\{U_k : k < \omega\} \subset \mathscr{T}^*(X)$, and we show that the (indexed) family $\{U_k : k < \omega\}$ is locally finite in X. For $x \in X$ we choose $\bar{n}$ such that $x_{\bar{n}} = 0$, we set $U = \pi_{\bar{n}}^{-1}(\{0\})$, and we note that

$$\{k < \omega : U \cap U_k \neq \emptyset\} = \bar{n}.$$

It follows that X is not pseudo-(ω, ω)-compact. Since $\mathrm{cf}(\alpha) = \omega$, it follows from Theorem 2.3(c) that X is not pseudo-(α, α)-compact.

(b) $\Rightarrow$ (c). This follows from Theorem 2.1(f).

(c) $\Rightarrow$ (a). We show that (c) fails if (a) fails.

We consider three cases.

Case 1. $\mathrm{cf}(\alpha) > \omega$. From Lemma 8.9 there is a set $\{K_n : n < \omega\}$ of compact spaces such that

$$\Sigma_\omega \{0,1\}^\lambda = \bigcup_{n<\omega} K_n.$$

Let $\{U^{\xi} : \xi < \alpha\} \subset \mathscr{T}^*(X)$, and set

$$A_n = \{\xi < \alpha : U^{\xi} \cap K_n \neq \emptyset\} \quad \text{for } n < \omega.$$

Since $\Sigma_\omega \{0,1\}^\lambda$ is dense in X, we have $\cup_{n<\omega} A_n = \alpha$, and since $\mathrm{cf}(\alpha) > \omega$, there is $\bar{n} < \omega$ such that $|A_{\bar{n}}| = \alpha$, as required.

Case 2. $\alpha = \omega$ and $\kappa > \omega$. From Theorems 3.8 and 4.1 (a) it follows that $\Sigma_{\omega^+}\{0,1\}^\lambda$ has compact-calibre α. Since $\kappa > \omega$ the space $\Sigma_{\omega^+}\{0,1\}^\lambda$ is dense in X, and hence X has compact-calibre ω by Theorem 2.5(a) (ii).

Case 3. $\alpha > \omega$, $\mathrm{cf}(\alpha) = \omega$, and $\kappa > \omega$. It follows from Theorem 8.10 that X has compact-calibre α.

The proof is complete.

We note the following surprising consequence of Theorem 8.11.

8.12. *Corollary.* There are completely regular, Hausdorff, non-compact spaces X with compact-calibre α for all $\alpha \geq \omega$.

Proof. We set $X = \Sigma_\kappa\{0,1\}^\lambda$, with $\kappa > \omega$ (to ensure that X has compact-calibre α) and with $\lambda \geq \kappa$ (to ensure that X is not compact).

For a specific example, we set $X = \Sigma_{\omega^+}\{0,1\}^{(\omega^+)}$.

We note that from Theorem 2.1(f) it follows that the (non-compact) spaces X of Corollary 8.12 are pseudo-(α, α)-compact for all $\alpha \geq \omega$.

Notes for Chapter 8

We announced most of the results of this chapter in the abstract of Comfort & Negrepontis (1978b); see also Comfort (1980).

9

Pseudo-compactness Numbers: Examples

Among the chain conditions considered in this monograph are these three: calibre, compact-calibre, and pseudo-compactness. For the first two of these we have presented, in Corollaries 4.7 and 8.6, respectively, quite satisfactory characterizations of those classes of infinite cardinals which can arise, for some (completely regular, Hausdorff) space X, as the class of infinite cardinals for which the chain condition in question fails for X.

We believe that an equally definitive result concerning classes of pseudo-compactness may be available, but we abandoned the search when we noticed that the general methods used for calibre and compact-calibre do not work here. The difficulty is not in finding, say for a given regular cardinal λ, a space which is pseudo-(α, α)-compact if and only if $\mathrm{cf}(\alpha) \neq \lambda$ (cf. Theorem 8.1). The difficulty is in finding a space which avoids (as pseudo-compactness numbers) the elements of a given set of cardinals. In brief: pseudo-compactness properties are not preserved by finite products. In the present chapter we document this failure in various ways. There is, for example, Corollary 9.8: for $\alpha \geq \beta$ with $\alpha \geq \omega$ and $1 < m < \omega$ there is a completely regular, Hausdorff space X such that X^{m-1} is pseudo-(α, β)-compact and X^m is not pseudo-(α, β)-compact.

The examples depend heavily on elementary properties (discussed and derived in the Appendix) of the Stone–Čech compactification $\beta(\alpha)$ of the (discrete) space α and, more particularly, of the type $T(p)$ of p for $p \in \beta(\alpha)$.

We cite a natural question left unsettled (cf. Remarks 9.11). If $\alpha \geq \beta \geq \omega$ with β singular and $1 < m < \omega$, is there a set $\{X_k : k < m\}$ of completely regular, Hausdorff spaces such that $\prod_{k \in A} X_k$ is pseudo-(β, β)-compact for every non-empty $A \subset m$ with $A \neq m$, but $\prod_{k < m} X_k$ is not pseudo-(α, ω)-compact?

We see first that in Theorem 2.3(c′) the condition $\pi w(X) < \alpha$ cannot be omitted.

9.1 *Theorem.* Let α be an (infinite) singular cardinal. There is a completely regular, Hausdorff space X such that X is pseudo-$(\mathrm{cf}(\alpha), \mathrm{cf}(\alpha))$-compact, X is not pseudo-(α, α)-compact, and $\pi w(X) = \alpha$.

Proof. We set $X = N_{(\mathrm{cf}(\alpha))^+}(\alpha)$. It is immediate from Theorem 8.1 that X has the required properties.

We begin our investigation into the product of pseudo-(α, β)-compact spaces with a simple result used in the proof of Lemma 9.5 below.

9.2 *Lemma.* Let $\alpha \geq \beta \geq \omega$, let X be a pseudo-(α, β)-compact space and K a compact space. Then $X \times K$ is pseudo-(α, β)-compact.

Proof. Let $\{U_\xi \times V_\xi : \xi < \alpha\} \subset \mathscr{T}^*(X \times K)$ and let $\bar{x}$ be an element of X such that

$$|\{\xi < \alpha : U \cap U_\xi \neq \varnothing\}| \geq \beta$$

for every neighborhood U of $\bar{x}$. If for every $y \in K$ there is a neighborhood $U(y) \times V(y)$ of $\langle \bar{x}, y \rangle$ such that

$$|\{\xi < \alpha : (U(y) \times V(y)) \cap (U_\xi \times V_\xi) \neq \varnothing\}| < \beta,$$

we choose $F \in \mathscr{P}_\omega(K)$ such that $K = \cup_{y \in F} V(y)$, we set $U = \cap_{y \in F} U(y)$, and we note that

$$|\{\xi < \alpha : U \cap U_\xi \neq \varnothing\}| < \beta.$$

Hence there is $\bar{y} \in K$ such that

$$|\{\xi < \alpha : (U \times V) \cap (U_\xi \times V_\xi) \neq \varnothing\}| \geq \beta$$

for every neighborhood $U \times V$ of $\langle \bar{x}, \bar{y} \rangle$.

The following lemma will be used in 9.4 (for β regular) and in 9.5 (for β singular).

9.3. *Lemma.* Let $\alpha \geq \beta \geq \omega$, $p \in \beta(\alpha)$ with $\|p\| = \beta$, $0 < m < \omega$,

$$\alpha \cup T(p) \subset X_k \subset \beta(\alpha) \quad \text{for } k < m,$$

and $X = \prod_{k<m} X_k$.

If $s : \beta \to X$ satisfies

$$|(\pi_k \circ s)[B]| = \beta \quad \text{for all } B \in [\beta]^\beta, k < m,$$

then there is $q \in X$ such that $|\{\eta < \beta : s(\eta) \in V\}| = \beta$ for every neighborhood V of q.

Proof. There is $A \in p \cap [\alpha]^\beta$; we write

$$A = \{\xi(\zeta) : \zeta < \beta\}$$

with $\xi(\zeta') \neq \xi(\zeta)$ for $\zeta' < \zeta < \beta$.

By induction over $\{k : 0 \leq k \leq m-1\}$ we define a set $\{B_k : k < m\}$ of sets such that

(i) $B_k \in [\beta]^\beta$ for $k < m$,

(ii) $B_{k+1} \subset B_k$ for $k < m-1$, and

(iii) $\pi_k \circ s | B_k$ is one-to-one for $k < m$.

We set $B = B_{m-1}$ and we write

$$B = \{\eta(\zeta) : \zeta < \beta\}$$

with $\eta(\zeta') \neq \eta(\zeta)$ for $\zeta' < \zeta < \beta$.

For $k < m$ we define $f_k : \alpha \to \alpha \subset X_k$ by

$$f_k(\xi(\zeta)) = \pi_k \circ s(\eta(\zeta)) \quad \text{for } \xi(\zeta) \in A,$$

$$f_k(\xi) = 0 \quad \text{for } \xi \in \alpha \backslash A,$$

we set $q_k = \bar{f}_k(p)$, we note (from Lemma B.7 of the Appendix) that $q_k \in T(p) \subset X_k$, and we set

$$q = \langle q_k : k < m \rangle \in X.$$

Let $V = \prod_{k<m} V_k$ be a basic neighborhood of q and set $U = \cap_{k<m} \bar{f}_k^{-1}(V_k)$ and

$$C = \{\zeta < \beta : \xi(\zeta) \in U\}.$$

For $\zeta \in C$ and $k < m$ we have

$$(\pi_k \circ s)(\eta(\zeta)) = f_k(\xi(\zeta)) = \bar{f}_k(\xi(\zeta)) \in V_k$$

and hence $s(\eta(\zeta)) \in V$. Since $\| p \| = \beta$ we have $|C| = \beta$, and for $\zeta', \zeta \in C$ with $\zeta' \neq \zeta$ we have $\eta(\zeta') \neq \eta(\zeta)$ and hence

$$|\{\eta < \beta : s(\eta) \in V\}| = \beta,$$

as required.

9.4 *Lemma*. Let $\alpha \geq \beta \geq \omega$ with β regular, $p \in \beta(\alpha)$ with $\| p \| = \beta$,

$0 < m < \omega$,

$$\alpha \cup T(p) \subset X_k \subset \beta(\alpha) \text{ for } k < m,$$

and $X = \prod_{k<m} X_k$. Then X is pseudo-(β, β)-compact.

Proof. We proceed by induction. In each case we let $\{U^\eta : \eta < \beta\}$ be a set of non-empty basic subsets of X, we let $s(\eta) \in U^\eta \cap \alpha^m$ for $\eta < \beta$, and we show there is $q \in X$ such that

$$|\{\eta < \beta : s(\eta) \in V\}| = \beta$$

for every neighborhood V of q.

We suppose either that $m = 1$, or that $m = n + 1$ and the statement holds for $m = n$. We consider two cases.

Case 1. There are $\bar{k} < m, \bar{B} \in [\beta]^\beta$ such that $|(\pi_{\bar{k}} \circ s)[\bar{B}]| < \beta$. Since β is regular there is $\bar{\xi} < \alpha$ such that

$$|\{\eta \in \bar{B} : (\pi_{\bar{k}} \circ s)(\eta) = \bar{\xi}\}| = \beta.$$

If $m = 1$, then ($\bar{k} = 0$ and) the statement holds with $q = \bar{\xi}$. If $m = n + 1$, the statement follows from the inductive hypothesis and the fact that $\{q \in X : q_{\bar{k}} = \bar{\xi}\}$ is (a copy of) $\prod\{X_k : k < m, k \neq \bar{k}\}$.

Case 2. $|(\pi_k \circ s)[B]| = \beta$ for all $B \in [\beta]^\beta, k < m$. The statement then follows from Lemma 9.3.

9.5 *Lemma.* Let $\alpha \geq \beta \geq \omega$ with β singular, $p \in \beta(\alpha)$ with $\|p\| = \beta$, $0 < m < \omega$,

$$N_{(\mathrm{cf}(\beta))^+}(\alpha) \cup T(p) \subset X_k \subset \beta(\alpha) \quad \text{for } k < m,$$

and $X = \prod_{k<m} X_k$. Then X is pseudo-(β, β)-compact.

Proof. We proceed by induction. In each case we let $\{U^\eta : \eta < \beta\}$ be a set of non-empty basic subsets of X, we let $s(\eta) \in U^\eta \cap \alpha^m$ for $\eta < \beta$, and we show there is $q \in X$ such that

$$|\{\eta < \beta : s(\eta) \in V\}| = \beta$$

for every neighborhood V of q.

We suppose either that $m = 1$, or that $m = n + 1$ and the statement holds for $m = n$. We consider two cases.

Case 1. There are $\bar{k} < m, \bar{B} \in [\beta]^\beta$ such that $|(\pi_{\bar{k}} \circ s)[\bar{B}]| \leq \mathrm{cf}(\beta)$. We set

$$A = (\pi_{\bar{k}} \circ s)[\bar{B}] \text{ and } K = \mathrm{cl}_{\beta(\alpha)} A$$

and we note that K is a compact subspace of $X_{\bar{k}}$. If $m = 1$ then ($\bar{k} = 0$ and) $\{s(\eta) : \eta \in \beta\} \subset K$; it is then clear that there is $q \in K$ such that

$$|\{\eta \in \bar{B} : s(\eta) \in V\}| = \beta$$

for every neighborhood V of q. If $m = n + 1$, then

$$\{s(\eta) : \eta \in \bar{B}\} \subset (\prod\{X_k : k < m, k \neq \bar{k}\}) \times K$$

and the statement follows from the inductive hypothesis and Lemma 9.2.

Case 2. $|(\pi_k \circ s)[B]| > \mathrm{cf}(\beta)$ for all $k < m$, $B \in [\beta]^\beta$. We claim that $|(\pi_k \circ s)[B]| = \beta$ for all $k < m$, $B \in [\beta]^\beta$. For such k and B set $C = (\pi_k \circ s)[B]$, suppose $|C| < \beta$, and set

$$B(\xi) = \{\eta \in B : (\pi_k \circ s)(\eta) = \xi\} \text{ for } \xi \in C.$$

Since $\cup_{\xi \in C} B(\xi) = B$ we have $\sup_{\xi \in C} |B(\xi)| = \beta$. Let $\{\beta_\sigma : \sigma < \mathrm{cf}(\beta)\}$ be a set of cardinal numbers such that

$$\beta_\sigma < \beta \quad \text{for } \sigma < \mathrm{cf}(\beta), \text{ and}$$

$$\sum_{\sigma < \mathrm{cf}(\beta)} \beta_\sigma = \beta,$$

for $\sigma < \mathrm{cf}(\beta)$ choose $\xi(\sigma) \in C$ such that $|B(\xi(\sigma))| \geq \beta_\sigma$, and set

$$\bar{B} = \cup_{\sigma < \mathrm{cf}(\beta)} B(\xi(\sigma)).$$

Then $|\bar{B}| = \beta$ and $(\pi_k \circ s)[\bar{B}] = \{\xi(\sigma) : \sigma < \mathrm{cf}(\beta)\}$, a contradiction. The claim is established. The required statement follows from Lemma 9.3.

The proof is complete.

9.6 *Theorem.* Let $\alpha \geq \beta \geq \omega$ and $1 < m < \omega$.

(a) If β is regular there is a set $\{X_k : k < m\}$ of completely regular, Hausdorff spaces such that

$\prod_{k \in A} X_k$ is pseudo-(β, β)-compact for $A \subset m$, $A \neq m$, and
$\prod_{k < m} X_k$ is not pseudo-(α, ω)-compact.

(b) If β is singular, there is a set $\{X_k : k < m\}$ of completely regular, Hausdorff spaces such that

$\prod_{k \in A} X_k$ is pseudo-(β, β)-compact for $A \subset m$, $A \neq m$, and
$\prod_{k < m} X_k$ is not pseudo-$(\alpha, (\mathrm{cf}(\beta))^+)$-compact.

Proof. From Corollary B.11 there is $\{p_n : n < m\} \subset \beta(\alpha)$ such that

$$\|p_n\| = \beta \quad \text{for } n < m, \text{ and}$$
$$T(p_{n'}) \cap T(p_n) = \varnothing \quad \text{for } n' < n < m.$$

In (a) we set

$$Y_n = \alpha \cup T(p_n) \quad \text{for } n < m,$$

and in (b) we set

$$Y_n = N_{(\mathrm{cf}(\beta))^+}(\alpha) \cup T(p_n) \quad \text{for } n < m,$$

and (using these definitions) we set

$$X_k = \cup \{Y_n : n < m, n \neq k\} \quad \text{for } k < m.$$

Now let $A \in \mathscr{P}^*_m(m)$. There is $n \in m \backslash A$. In (a) we have

$$\alpha \cup T(p_n) \subset X_k \subset \beta(\alpha) \quad \text{for } k \in A,$$

and hence $\prod_{k \in A} X_k$ is pseudo-(β, β)-compact (by Lemma 9.4); and in (b) we have

$$N_{(\mathrm{cf}(\beta))^+}(\alpha) \cup T(p_k) \subset X_k \subset \beta(\alpha) \quad \text{for } k \in A,$$

and hence $\prod_{k \in A} X_k$ is pseudo-(β, β)-compact (by Lemma 9.5).

Now for $\xi < \alpha$ we define $x(\xi) \in \prod_{k<m} X_k$ by

$$x(\xi)_k = \xi \quad \text{for } k < m,$$

we set $U^\xi = \{x(\xi)\}$ for $\xi < \alpha$, and we note that $\{U^\xi : \xi < \alpha\}$ is an (indexed) family of non-empty open subsets of $\prod_{k<m} X_k$.

In (a) we have $\cap_{k<m} X_k = \alpha$ and hence for $x \in \prod_{k<m} X_k$ there is a neighborhood V of x such that

$$|\{\xi < \alpha : V \cap U^\xi \neq \varnothing\}| < \omega;$$

it follows that $\prod_{k<m} X_k$ is not pseudo-(α, ω)-compact.

In (b) we have $\cap_{k<m} X_k = N_{(\mathrm{cf}(\beta))^+}(\alpha)$, and hence for $x \in \prod_{k<m} X_k$ there is a neighborhood V of x such that

$$|\{\xi < \alpha : V \cap U^\xi \neq \varnothing\}| \leq \mathrm{cf}(\beta);$$

it follows that $\prod_{k<m} X_k$ is not pseudo-$(\alpha, (\mathrm{cf}(\beta))^+)$-compact.

The proof is complete.

9.7 *Theorem.* Let $\alpha \geq \beta \geq \omega$ and $1 < m < \omega$.

(a) If β is regular there is a completely regular, Hausdorff space X such that

X^{m-1} is pseudo-(β, β)-compact, and

X^m is not pseudo-(α, ω)-compact.

(b) If β is singular there is a completely regular, Hausdorff space X such that

X^{m-1} is pseudo-(β, β)-compact, and

X^m is not pseudo-$(\alpha, (\text{cf}(\beta))^+)$-compact.

Proof. To prove (a) we define Y_n and X_k as in (the proof of) Theorem 9.6(a), and to prove (b) we define Y_n and X_k as in (the proof of) Theorem 9.6(b); and using this definition we let X be the set

$$X = \bigcup_{k<m} (X_k \times \{k\})$$

with the disjoint union topology. (The singleton spaces $\{k\}$ are introduced to ensure that

$$(X_{k'} \times \{k'\}) \cap (X_k \times \{k\}) = \emptyset \text{ for } k' < k < m.)$$

The space X^{m-1} is the union of $m^{m-1} < \omega$ pairwise disjoint, open-and-closed subspaces S of the form

$$S = \prod_{k<m} (X_k \times \{k\})^{d_k},$$

with $d_k \geq 0$ for $k < m$, and $\sum_{k<m} d_k = m - 1$, and for each such space S there is $n(S) < m$ such that $d_{n(S)} = 0$. In (a) we have

$$\alpha \cup T(p_{n(S)}) \subset X_k \subset \beta(\alpha) \text{ for } d_k \neq 0,$$

and hence S is pseudo-(β, β)-compact (by Lemma 9.4); and in (b) we have

$$N_{(\text{cf}(\beta))^+}(\alpha) \cup T(p_n) \subset X_k \subset \beta(\alpha) \text{ for } d_k \neq 0,$$

and hence S is pseudo-(β, β)-compact (by Lemma 9.5). Thus in each case X^{m-1} is the union of finitely many pairwise disjoint open-and-closed pseudo-(β, β)-compact subspaces. It is then clear that X^{m-1} is pseudo-(β, β)-compact.

The space X^m contains an open-and-closed subspace homeomorphic to $\prod_{k<m} X_k$. In (a) the space $\prod_{k<m} X_k$ is not pseudo-(α, ω)-compact (by Theorem 9.6(a)) and hence X^m is not pseudo-(α, ω)-compact (by Theorem 2.5(b)(iv)); and in (b) the space $\prod_{k<m} X_k$ is not pseudo-$(\alpha, (\text{cf}(\beta))^+)$-compact (by Theorem 9.6(b)) and hence X^m is not pseudo-$(\alpha, (\text{cf}(\beta))^+)$-compact (by Theorem 2.5(b)(iv)).

The proof is complete.

9.8 *Corollary*. Let $\alpha \geq \beta \geq 2$ with $\alpha \geq \omega$ and $1 < m < \omega$. There is a completely regular, Hausdorff space X such that

X^{m-1} is pseudo-(α, β)-compact, and

X^m is not pseudo-(α, β)-compact.

Proof. From Theorem 2.1 it follows that a regular, Hausdorff space is pseudo-(α, n)-compact for $n < \omega$ if and only if it is pseudo-(α, ω)-compact. Thus it is enough to consider the case $\beta \geq \omega$.

For regular $\beta \geq \omega$ the statement follows from Theorem 9.7(a), and for singular $\beta \geq \omega$ the statement follows from Theorem 9.7(b).

We saw in Theorem 3.16 that if α is a cardinal such that $\mathrm{cf}(\alpha) > \omega$ and $\{X_i : i \in I\}$ is a set of spaces such that X_J is pseudo-(α, α)-compact for all $J \in \mathscr{P}^*_\omega(I)$, then X_I is pseudo-(α, α)-compact. We see now that the analogous statement fails for cardinals α such that $\mathrm{cf}(\alpha) = \omega$. Theorem 9.9 supplements and complements Corollary 3.11(iii), which treats the product of pseudo-(α, β)-compact spaces with α singular and $\alpha > \beta$.

9.9 *Theorem*. There is a set $\{X_k : k < \omega\}$ of completely regular, Hausdorff spaces such that

$\prod_{k<m} X_k$ is pseudo-(α, α)-compact for $0 < m < \omega$ if $\mathrm{cf}(\alpha) = \omega$, and

$\prod_{k<\omega} X_k$ is not pseudo-(α, α)-compact if $\mathrm{cf}(\alpha) = \omega$.

Proof. From Corollary B.11 of the Appendix there is $\{p_k : k < \omega\} \subset U(\omega)$ such that

$$T(P_{k'}) \cap T(p_k) = \varnothing \text{ for } k' < k < \omega.$$

We set

$$Y_n = \omega \cup T(p_n) \text{ for } n < \omega, \text{ and}$$
$$X_k = \cup \{Y_n : n < \omega, n \neq k\} \text{ for } k < \omega.$$

For $k < m < \omega$ we have

$$Y_m = \omega \cup T(p_m) \subset X_k \subset \beta(\omega),$$

and hence $\prod_{k<m} X_k$ is pseudo-(ω, ω)-compact by Lemma 9.4.

We note further that $\pi w(\prod_{k<m} X_k) = \omega$; hence from Theorem 2.3(c′) it follows that if $\alpha > \omega$ and $\mathrm{cf}(\alpha) = \omega$ then $\prod_{k<m} X_k$ is pseudo-(α, α)-compact.

Now set $X = \prod_{k<\omega} X_k$; we show that X is not pseudo-(ω, ω)-compact. We set

$$U_k = \{x \in X : x_n = k \text{ for } n \leq k\} \text{ for } k < \omega,$$

we note that $\{U_k : k < \omega\} \subset \mathscr{T}^*(X)$, and we show that the (indexed) family $\{U_k : k < \omega\}$ is locally finite in X.

If $x \in U_n$, then U_n is a neighborhood of x and

$$|\{k < \omega : U_n \cap U_k \neq \emptyset\}| = 1 < \omega.$$

If $x \notin \cup_{n<\omega} U_n$, there is $\bar{n} < \omega$ such that $x_{\bar{n}} \neq x_0$; indeed, otherwise there is $q \in \beta(\omega)$ such that $x_n = q$ for all $n < \omega$, and from $\cap_{n<\omega} X_n = \omega$ we have $q \in \omega$ and hence $x \in U_q$. Let V_0 and $V_{\bar{n}}$ be disjoint neighborhoods in $\beta(\omega)$ of x_0 and $x_{\bar{n}}$, respectively, and set

$$U = (\pi_0^{-1}(V_0) \cap \pi_{\bar{n}}^{-1}(V_{\bar{n}})) \backslash (\cup_{k<\bar{n}} U_k).$$

Then U is a neighborhood of x, and

$$U \cap (\cup_{k<\omega} U_k) = \emptyset.$$

The proof that X is not pseudo-(ω, ω)-compact is complete. It follows from Theorem 2.3(c) that if $\alpha > \omega$ and $\mathrm{cf}(\alpha) = \omega$, then X is not pseudo-(α, α)-compact.

We note that the space $X = \prod_{k<\omega} X_k$ defined in (the proof of) Theorem 9.9 is pseudo-(α, ω)-compact for all $\alpha > \omega$. Indeed, for $J \in \mathscr{P}_\omega^*(\omega)$ the space X_J is pseudo-(ω, ω)-compact and hence pseudo-(ω^+, ω)-compact; thus X is pseudo-(ω^+, ω)-compact (by Corollary 3.6(c)) and hence pseudo-(α, ω)-compact for all $\alpha > \omega$.

We combine the essential features of the spaces $X_k (k < \omega)$ into a single space.

9.10 *Theorem.* There is a completely regular, Hausdorff space X such that

X^m is pseudo-(α, α)-compact for $0 < m < \omega$ if $\mathrm{cf}(\alpha) = \omega$, and

X^ω is not pseudo-(α, α)-compact if $\mathrm{cf}(\alpha) = \omega$.

Proof. From Corollary B.11 of the Appendix there is $\{p_k : k < \omega\} \subset \beta(\omega)$ such that

$$T(p_{k'}) \cap T(p_k) = \emptyset \text{ for } k' < k < \omega.$$

We choose $\{A_k : k < \omega\}$ such that

$$\bigcup_{k<\omega} A_k = \omega,$$

$$|A_k| = \omega \text{ for } k < \omega, \text{ and}$$

$$A_{k'} \cap A_k = \emptyset \text{ for } k' < k < \omega,$$

and we set

$$B_k = A_k \cup \bigcup_{n \neq k} (\hat{A}_k \cap T(p_n)),$$

$$C = \beta(\omega) \setminus \bigcup_{k<\omega} \hat{A}_k, \text{ and}$$

$$X = C \cup \left(\bigcup_{k<\omega} B_k \right).$$

We show that X^m is pseudo-(ω, ω)-compact for $0 < m < \omega$. For $I \in \mathscr{P}^*(m)$ and $f \in \omega^I$ we set

$$V(f) = \prod_{i \in I} B_{f(i)}$$

and we claim that $V(f)$ is pseudo-(ω, ω)-compact. Indeed, there is $\bar{n} \in \omega \setminus f[I]$ and for $i \in I$ there is

$$p_{\bar{n},i} \in \hat{A}_{f(i)} \cap T(p_{\bar{n}}),$$

and since

$$T_{A_{f(i)}}(p_{\bar{n},i}) = \hat{A}_{F(i)} \cap T(p_{\bar{n}}, i) \text{for } i \in I$$

by Corollary B.9 of the Appendix and $A_{f(i)} \cup T_{A_{f(i)}}(p_{\bar{n},i})$ is homeomorphic to $\omega \cup T(p_{\bar{n}})$, the space $V(f)$ is pseudo-(ω, ω)-compact for $f \in \omega^I$ by Lemma 9.4.

Now let $\{U(n) : n < \omega\}$ be a set of basic subsets of $(\beta(\omega))^m$, let I be a subset of m maximal with respect to the property that there are $f \in \omega^I$ and $A \in [\omega]^\omega$ such that

$$U(n)_i \cap B_{f(i)} \neq \emptyset \text{ for } i \in I, n \in A,$$

and set $J = m \setminus I$. (Our notation in what follows indicates that I, $J \neq \emptyset$. We omit the (routine) modifications required in case $I = \emptyset$ or $J = \emptyset$.) Since $V(f)$ is pseudo-(ω, ω)-compact and $\pi^I[U(n)] \cap V(f) \neq \emptyset$ for $n \in A$, there is $x \in X^I$ such that

$$|\{n \in A : V_I \cap \pi_I[U(n)] \neq \emptyset\}| = \omega$$

for every neighborhood V_I of x. We note that there is $y \in (\beta(\omega))^J$ such that

$$|\{n \in A : V \cap U(n) \neq \emptyset\}| = \omega$$

for every neighborhood $V = V_I \times V_J$ of $\langle x, y \rangle$. Indeed, otherwise, since $(\beta(\omega))^J$ is compact, there are finite $F \subset (\beta(\omega))^J$ and for $y \in F$ a neighborhood $V(y) = V(y)_I \times V(y)_J$ of $\langle x, y\rangle$ such that

$$(\beta(\omega))^J = \bigcup_{y\in F} U(y)_J, \text{ and}$$

$$|\{n \in A : V(y) \cap U(n) \neq \emptyset\}| < \omega;$$

we set $V = \bigcap_{y\in F} V(y)_I$, we have that V is a neighborhood of x, and we have

$$|\{n \in A : V \cap \pi_I[U(n)] \neq \emptyset\}| < \omega,$$

a contradiction.

We claim that $y_i \in X$ for $i \in J$. If $y_i \notin X$, then there is $k < \omega$ such that $y_i \in \hat{A}_k$, and since

$$|\{n \in A : V_i \cap U(n)_i \neq \emptyset\}| = \omega$$

for every neighborhood V_i of y_i and $B_k \cap \hat{A}_k$ is dense in $\hat{A}_k$, we have

$$|\{n \in A : U(n)_i \cap B_k \neq \emptyset\}| = \omega.$$

This contradicts the maximality of I.

We have shown that there is $\langle x, y\rangle \in X^I \times X^J = X^m$ such that

$$|\{n < \omega : (V_I \times V_J) \cap U(n) \neq \emptyset\}| = \omega$$

for every neighborhood $V = V_I \times V_J$ of $\langle x, y\rangle$ in $(\beta(\omega))^m$. The proof that X^m is pseudo-(ω, ω)-compact is complete. Since $\pi w(X^m) = \omega$, it follows from Theorem 2.3(c′) that if $\alpha > \omega$ and $\mathrm{cf}(\alpha) = \omega$, then X^m is pseudo-(α, α)-compact.

We show that X^ω is not pseudo-(α, α)-compact if $\mathrm{cf}(\alpha) = \omega$. For $k < \omega$ the space B_k is an open-and-closed subspace of X; hence there is a continuous function g_k from X onto B_k. The function g, defined on X^ω by the rule

$$(g(x))_k = g_k(x_k),$$

is a continuous function from X^ω onto $\prod_{k<\omega} B_k$. Since B_k is homeomorphic to the space X_k of (the proof of) Theorem 9.9 and $\prod_{k<\omega} X_k$ is not pseudo-(α, α)-compact, the space $\prod_{k<\omega} B_k$ is not pseudo-(α, α)-compact. It follows from Theorem 2.5(b)(i) that X^ω is not pseudo-(α, α)-compact.

The proof is complete.

9.11 *Remarks.* The theorems proved above leave certain natural questions unanswered.

(a) Let $\alpha \geq \beta \geq \omega$ with β singular and let $1 < m < \omega$. Is there a set $\{X_k : k < m\}$ of completely regular, Hausdorff spaces such that

$\prod_{k \in A} X_k$ is pseudo-(β, β)-compact for $A \in \mathscr{P}^*_m(m)$, and

$\prod_{k < m} X_k$ is not pseudo-(α, ω)-compact?

We note that for regular β the corresponding question is answered (in the affirmative) by Theorem 9.6(a).

Now let $\alpha \geq \beta \geq \omega$ with β singular, let φ be a function from β onto $\mathrm{cf}(\beta)$ such that $|\varphi^{-1}(\{\sigma\})| < \beta$ for $\sigma < \mathrm{cf}(\beta)$, let $p \in \beta(\alpha)$ with $\| p \| = \beta$ and set $q = \bar{\varphi}(p)$ and $X = \alpha \cup T(p) \cup T(q)$. It can then be shown, by an argument similar to that of Lemma 9.5 but more delicate, that X^m is pseudo-(β, β)-compact for $0 < m < \omega$. It then follows, choosing $\{p_k : k < m\} \subset \beta(\alpha)$ such that

$$\| p \| = \beta \text{ for } k < m, \text{ and}$$
$$T(p_{k'}) \cap T(p_k) = \emptyset \text{ for } k' < k < m,$$

and setting

$$q_k = \bar{\varphi}(p_k) \text{ for } k < m,$$
$$Y_n = \alpha \cup T(p_n) \cup T(q_n) \text{ for } n < m, \text{ and}$$
$$X_k = \cup\{Y_n : n < m, n \neq k\} \text{ for } k < n,$$

that

$\prod_{k \in A} X_k$ is pseudo-(β, β)-compact for $A \in \mathscr{P}^*_m(m)$, and

$\prod_{k < m} X_k$ is not pseudo-$(\alpha, (\mathrm{cf}(\beta))^+)$-compact.

The desired stronger conclusion, that $\prod_{k < m} X_k$ is not pseudo-(α, ω)-compact, seems unavailable because we are uncertain whether $\cap_{k < \omega} X_k = \alpha$; in particular we are uncertain whether (φ may be chosen so that)

$$T(q_{k'}) \cap T(q_k) = \emptyset \text{ for } k' < k < m.$$

Thus the following question, of interest in itself, is relevant to (a) above.

(b) Let β be a singular cardinal, $p \in \beta(\beta)$ with $\| p \| = \beta$ and φ a function from β onto $\mathrm{cf}(\beta)$ such that $|\varphi^{-1}(\{\sigma\})| < \beta$ for $\sigma < \mathrm{cf}(\beta)$. Does it follow that $\bar{\varphi}[T(\mathrm{p})] = U(\mathrm{cf}(\beta))$?

Notes for Chapter 9

The spaces we have denoted $T(p)$ (with $p \in \beta(\alpha)$) were introduced by Frolík (1967a) in connection with his proofs (see Frolík 1967a, 1968) of the statement, given without appealing to the continuum hypothesis or to any other special set-theoretic assumption, that the space $\beta(\omega)\backslash\omega$ is not homogeneous.

The constructions of the present chapter, several of which appear in Comfort (1979, 1980), are for the most part simple extensions of the arguments used by Comfort (1967) and Frolík (1967b) to prove (independently) the case $\alpha = \omega$ of Theorems 9.6 and 9.9.

10

Continuous Functions on Product Spaces

We begin this chapter with a proof that under suitable hypotheses a function f continuous on a product space X_I (or even on some of its dense subspaces) is effectively defined on a 'small' subproduct (in the sense that there are 'small' $J \subset I$ and continuous g on X_J such that $f = g \circ \pi_J$). We apply this result, together with some of the product-space theorems of Chapter 3, to obtain an identification in concrete form of the Stone–Čech compactification, the Hewitt realcompactification, and the Dieudonné topological completion of certain completely regular Hausdorff spaces (these terms are defined in Appendix B). Such an opportunity occurs infrequently since the Stone–Čech compactification of a space X, whose elements are normally described as maximal filters in the class of zero-sets of X, is defined by an appeal to (a variant of) Zorn's lemma; it does not arise in practice as a readily identifiable, familiar space.

Among the specific results proved are these (cf. Corollary 10.7 and Theorem 10.17): in a product of metrizable spaces, and in a product of compact spaces, every ω^+-Σ-space is C-embedded. We conclude with an example, one of several due to Ulmer, showing that there do exist ω^+-Σ-products $\Sigma_{\omega^+}(p, X_I)$ not C-embedded in X_I.

Definition. Let $Y \subset X_I$, $J \subset I$, let Z be a set and $f: Y \to Z$. Then f *depends on* J (in Y) if $f(y) = f(y')$ whenever $y, y' \in Y$ and $y_J = y'_J$.

If α is a cardinal and there is $J \in \mathscr{P}_\alpha(I)$ such that f depends on J, then we say that f *depends on fewer than* α *coordinates*, or that f *depends on* $< \alpha$ *coordinates*; the expression f *depends on* $\leq \alpha$ *coordinates* is defined analogously.

We note that if f depends on J and $J \subset J' \subset I$ then f depends on J'.

We will show that under certain circumstances a (continuous) function will depend on a small set of coordinates. We note in passing that for sufficiently ill-behaved sets Y it will occur that every function $f: Y \to Z$, regardless of the nature of Z and regardless of whether or not f is continuous, depends on any set J. For an example, let $X_\xi = \mathbb{R}$ for $\xi < 2^\omega$ and let $D = \{x(n) : n < \omega\}$ be a (countable) dense subset of $\prod_{\xi < 2^\omega} X_\xi$ such that

$$x(n)_\xi \neq x(n')_\xi \text{ whenever } \xi < 2^\omega, n, n' < \omega, n \neq n';$$

then for every set $Z, f: D \to Z$ and $J \subset 2^\omega$, the function f depends on J.

The following lemma, crucial to the results of this chapter, does not depend on the Erdős–Rado theorem.

10.1 *Lemma.* Let $\omega \leq \kappa \leq \alpha$ with either $\kappa < \alpha$ or α regular, let $\{X_i : i \in I\}$ be a set of non-empty spaces, Y a dense, pseudo-(α, κ)-compact subspace of $(X_I)_\kappa$, (M, ρ) a metric space and f a continuous function from Y to M. Then for every $\varepsilon > 0$ there is $J \in \mathscr{P}_\alpha(I)$ such that

$$\rho(f(x), f(y)) \leq \varepsilon \quad \text{if } x, y \in Y, x_J = y_J.$$

Proof. We suppose the result fails.

We claim that for $J \in \mathscr{P}_\alpha(I)$ there are $x, y \in Y$ such that $x_J = y_J$, $\rho(f(x), f(y)) > \varepsilon$, and

$$|\{i \in I : x_i \neq y_i\}| < \kappa.$$

Indeed, there are $\tilde{x}, \tilde{y} \in Y$ such that $\tilde{x}_J = \tilde{y}_J$ and $\rho(f(\tilde{x}), f(\tilde{y})) > \varepsilon$, and since f is continuous there are basic neighborhoods U and V of $\tilde{x}$ and $\tilde{y}$, respectively, such that

$$\rho(f(x), f(y)) > \varepsilon \quad \text{for } x \in U \cap Y, y \in V \cap Y.$$

We set $A = R(U) \cup R(V)$, and we define x, y by

$$\begin{aligned} x_i &= \tilde{x}_i && \text{if } i \in J \cup A, \\ y_i &= \tilde{y}_i && \text{if } i \in J \cup A, \\ x_i = y_i &= \tilde{x}_i && \text{if } i \in I \backslash (J \cup A). \end{aligned}$$

Then x, y are as required. The claim is established.

For $\xi < \alpha$ we define $x(\xi), y(\xi), U^\xi, V^\xi, J(\xi)$ and $A(\xi)$ so that

(i) $x(\xi), y(\xi) \in Y$ for $\xi < \alpha$;

(ii) $A(\xi) = \{i \in I : x(\xi)_i \neq y(\xi)_i\}$ for $\xi < \alpha$;

(iii) $A(\xi) \in \mathscr{P}_\kappa(I)$ for $\xi < \alpha$;

(iv) U^ξ and V^ξ are basic neighborhoods in $(X_I)_\kappa$ of $x(\xi)$ and $y(\xi)$, respectively, for $\xi < \alpha$;

(v) $\rho(f(x), f(y)) > \varepsilon$ if $x \in U^\xi \cap Y$, $y \in V^\xi \cap Y$, for $\xi < \alpha$;

(vi) $R(U^\xi) = R(V^\xi)$ for $\xi < \alpha$;

(vii) $U_i^\xi \cap V_i^\xi = \emptyset$ if $i \in A(\xi)$, for $\xi < \alpha$;

(viii) $J(0) = \emptyset$;

(ix) $J(\xi) = (\cup_{\eta<\xi} A(\eta)) \cup (\cup_{\eta<\xi} R(U^\eta))$ for $0 < \xi < \alpha$; and

(x) $x(\xi)_{J(\xi)} = y(\xi)_{J(\xi)}$ for $\xi < \alpha$.

We proceed by recursion. If $\xi < \alpha$ and $x(\eta)$, $y(\eta)$, U^η, V^η, $J(\eta)$ and $A(\eta)$ are defined for $\eta < \xi$, we define $J(\xi)$ by (viii) or (ix) and we use the claim just established to define $x(\xi)$, $y(\xi)$ and $A(\xi)$ satisfying (i), (x), (ii) and (iii). There are neighborhoods $\tilde{U}^\xi$ and $\tilde{V}^\xi$ of $x(\xi)$ and $y(\xi)$, respectively, such that

$$\rho(f(x), f(y)) > \varepsilon \quad \text{for } x \in \tilde{U}^\xi \cap Y, y \in \tilde{V} \cap Y.$$

We define neighborhoods U_i^ξ and V_i^ξ in X_i of $x(\xi)_i$ and $y(\xi)_i$, respectively so that

$$\begin{aligned} &U_i^\xi \cap V_i^\xi = \emptyset \quad \text{for } i \in A(\xi), \\ &U_i^\xi \subset \tilde{U}_i^\xi \quad \text{and } V_i^\xi \subset \tilde{V}_i^\xi \quad \text{for } i \in A(\xi), \text{ and} \\ &U_i^\xi = V_i^\xi = \tilde{U}_i^\xi \cap \tilde{V}_i^\xi \quad \text{for } i \in I \backslash A(\xi); \end{aligned}$$

it is then clear that (iv), (v), (vi) and (vii) are satisfied.

Now for $i \in I$ set $B(i) = \{\xi < \alpha : U_i^\xi \neq V_i^\xi\}$. Since $\xi \in B(i)$ if and only if $i \in A(\xi)$, and since

$$A(\eta) \cap A(\xi) = \emptyset \quad \text{for } \eta < \xi < \alpha$$

by (ix), we have $|B(i)| \leq 1$ for $i \in I$.

For $p \in Y$, let $\mathscr{A}(p)$ denote the set of basic neighborhoods in $(X_I)_\kappa$ of p, and for $W \in \mathscr{A}(p)$ set

$$A(p, W) = \{\xi < \alpha : W \cap U^\xi \neq \emptyset\}.$$

Since Y is pseudo-(α, κ)-compact, there is $\bar{p} \in Y$ such that $|A(\bar{p}, W)| \geq \kappa$ for all $W \in \mathscr{A}(\bar{p})$. We claim that for $W \in \mathscr{A}(\bar{p})$ there is $\bar{\xi} \in A(\bar{p}, W)$ such that $W \cap V^{\bar{\xi}} \neq \emptyset$. Indeed, since $|A(\bar{p}, W)| \geq \kappa$ and $|R(W)| < \kappa$ and $|B(i)| \leq 1$ for all $i \in R(W)$, there is $\bar{\xi} \in A(p, W)$ such that $\bar{\xi} \notin B(i)$ for $i \in R(W)$; from $W \cap U^{\bar{\xi}} \neq \emptyset$ and $U_i^{\bar{\xi}} = V_i^{\bar{\xi}}$ for all $i \in R(W)$ it follows that $W \cap V^{\bar{\xi}} \neq \emptyset$. The claim is proved.

Since Y is dense in $(X_I)_\kappa$, we have

$$W \cap U^\xi \cap Y \neq \varnothing \text{ and } W \cap V^\xi \cap Y \neq \varnothing$$

for all $W \in \mathscr{A}(\bar{p})$. From (v) it follows that f is not continuous at $\bar{p}$. This contradiction completes the proof.

10.2 *Corollary.* Let $\omega \leq \kappa \leq \alpha$ with either $\kappa < \alpha$ or α regular, let $\{X_i : i \in I\}$ be a set of non-empty spaces, Y a dense, pseudo-(α, κ)-compact subspace of $(X_I)_\kappa$, (M, ρ) a metric space and f a continuous function from Y to M. Then

(a) if $\mathrm{cf}(\alpha) = \omega$ then f depends on $\leq \alpha$ coordinates;

(b) if $\mathrm{cf}(\alpha) > \omega$ then f depends on $< \alpha$ coordinates.

Proof. For $1 \leq n < \omega$ there is (by Lemma 5.1) $J_n \in \mathscr{P}_\alpha(I)$ such that

$$\rho(f(x), f(y)) \leq 1/n \quad \text{if } x, y \in Y, x_{J_n} = y_{J_n}.$$

We set $J = \cup \{J_n : 1 \leq n < \omega\}$. Then f depends on J, and

$$|J| \leq \alpha \quad \text{in (a)}$$
$$< \alpha \quad \text{in (b)}.$$

10.3 *Lemma.* Let $\alpha \geq \kappa \geq \omega$, let $\{X_i : \in I\}$ be a set of non-empty spaces, Y a subspace of $(X_I)_\kappa$ such that $\pi_J[Y] = X_J$ for every non-empty $J \subset I$ with $|J| < \alpha$, Z a space and f a continuous function from Y to Z such that f depends on $< \alpha$ coordinates. Then there is continuous $\bar{f} : (X_I)_\kappa \to Z$ such that $f \subset \bar{f}$.

Proof. There is $J \in \mathscr{P}^*(I)$ such that f depends on J. For $x \in X_I$ we choose $\bar{x} \in Y$ such that $x_J = \bar{x}_J$ (such $\bar{x}$ exists because $x_J \in X_J = \pi_J[Y]$), and we set

$$\bar{f}(x) = f(\bar{x}).$$

For $y \in Y$ we have $y_J = \bar{y}_J$ and hence

$$\bar{f}(y) = f(\bar{y}) = f(y)$$

(since f depends on J); thus $f \subset \bar{f}$. It remains to show that $\bar{f}$ is continuous.

Let $x \in X$ and W a neighborhood in Z of $\bar{f}(x)$ and let U and V be neighborhoods of $\bar{x}_J$ and $\bar{x}_{I \setminus J}$ in $(X_J)_\kappa$ and $(X_{I \setminus J})_\kappa$, respectively, such that

$$f[(U \times V) \cap Y] \subset W.$$

Since $x_J = \bar{x}_J \in U$, the set $\pi_J^{-1}(U)$ is a neighborhood in $(X_I)_\kappa$ of x. We claim that if $x' \in \pi_J^{-1}(U)$ then $\bar{f}(x') \in W$. Indeed, since $|J \cup R(V)| < \alpha$ there is $y' \in Y$ such that

$$\begin{aligned} y'_i &= x'_i \quad \text{if } i \in J \\ &\in V_i \quad \text{if } i \in R(V) \backslash J \\ &\in X_i \quad \text{if } i \in I \backslash (J \cup R(V)), \end{aligned}$$

and from $y' \in (U \times V) \cap Y$ we have

$$\bar{f}(x') = f(\overline{x'}) = f(y') \in W,$$

as required.

The proof is complete.

Even in the case $\kappa = \omega$ and $Z = [0, 1]$ of Lemma 10.3 it is not known whether the hypothesis that $\pi_J[Y] = X_J$ for all $J \in \mathscr{P}^*_\alpha(I)$ can be replaced by the weaker hypothesis that Y is dense in $(X_J)_\kappa$ and (for all $J \in \mathscr{P}^*_\alpha(I)$) every continuous $f : \pi_J[Y] \to Z$ extends to continuous $\bar{f} : (X_J)_\kappa \to Z$.

10.4 *Theorem.* Let α and κ be infinite cardinals, $\{X_i : i \in I\}$ a set of spaces, and Y a subspace of $(X_I)_\kappa$ such that $\pi_J[Y] = (X_J)_\kappa$ for every non-empty $J \subset I$ with $|J| < \alpha$. Assume that either

(i) $\omega \leq \kappa \ll \alpha$ with α regular and $(X_J)_\kappa$ is pseudo-(α, κ)-compact for every non-empty $J \subset I$ with $|J| < \kappa$, or

(ii) $\omega \leq \kappa \ll \alpha$, $\kappa \ll \text{cf}(\alpha)$ and $(X_J)_\kappa$ is pseudo-(α, α)-compact for every non-empty $J \subset I$ with $|J| < \kappa$.

Then Y is M-embedded in $(X_I)_\kappa$.

Proof. It follows in (i) from Theorem 3.5(c) that Y is pseudo-(α, κ)-compact, and it follows in (ii) from Theorem 3.13 that Y is pseudo-(α, α)-compact. Thus Y is a dense, pseudo-(α, κ)-compact subspace of $(X_I)_\kappa$, and from Corollary 10.2 we have that every continuous function from Y to a metric space depends on $< \alpha$ coordinates. That Y is M-embedded in $(X_I)_\kappa$ then follows from Lemma 10.3.

10.5 *Corollary.* Let α be an infinite cardinal, $\{X_i : i \in I\}$ a set of spaces, and Y a subspace of X_I such that $\pi_J[Y] = X_J$ for all non-empty $J \subset I$ with $|J| < \alpha$. Assume that either

(i) $\omega < \alpha$ with α regular, X_i is a regular Hausdorff space for $i \in I$ and $S(X_J) \leq \alpha$ for all non-empty $J \subset I$ with $|J| < \omega$, or

(ii) $\omega < \text{cf}(\alpha)$ and $d(X_i) < \text{cf}(\alpha)$ for $i \in I$.

Then Y is M-embedded in X_I.

Proof. In (i) for $J \in \mathscr{P}^*_\omega(I)$ the space X_J has calibre $(\alpha, 2)$, hence by Theorem 2.1(e) is pseudo-$(\alpha, 2)$-compact, hence by Theorem 2.1(i) is pseudo-(α, ω)-compact.

In (ii) for $J \in \mathscr{P}^*_\omega(I)$ we have $d(X_J) < \text{cf}(\alpha)$ and therefore by Theorem 2.7 the space X_J has calibre (α, α); thus by Theorem 2.1(e) for $J \in \mathscr{P}^*_\omega(I)$ the space X_J is pseudo-(α, α)-compact.

The required conclusion now follows from (the case $\kappa = \omega$ of) Theorem 10.4.

It follows from the uniqueness of the Dieudonné topological completion of a completely regular, Hausdorff space (cf. Theorem B.2 of the Appendix) that if Y and $(X_I)_\kappa$ are related as in Theorem 10.4 then $\gamma(Y) = \gamma((X_I)_\kappa)$. This makes it possible to identify specifically and in concrete form the Hewitt realcompactification and the Stone–Čech compactification of certain spaces (cf. Corollary 10.7 below).

For $\alpha \geq \omega$ we say that a space X is *weakly α-compact* if for every cover $\{V_i : i \in I\}$ of X by open sets there is $J \subset I$ with $|J| < \alpha$ such that $\cup_{i \in I} V_i$ is dense in X.

10.6 *Lemma.* Let $\alpha \geq \omega$ and let X be a space.

(a) If $S(X) \leq \alpha$, then X is weakly α-compact; and

(b) if X is weakly α-compact, then X is pseudo-α-compact.

Proof. (a) Let $\{V_i : i \in I\}$ be a cover of X by open sets and let $\{U_j : j \in J\}$ be a family of non-empty open subsets of X such that

(i) $U_j \cap U_{j'} = \varnothing$ for $j, j' \in J, j \neq j'$ (i.e., the indexed family $\{U_j : j \in J\}$ is a cellular family),

(ii) for $j \in J$ there is $i \in I$ such that $U_j \subset V_i$, and

(iii) the family $\{U_j \in J\}$ is maximal with respect to (i) and (ii).

It follows from (iii) that $U_{j \in J} U_j$ is dense in X, and from (i) that $|J| < \alpha$. For $j \in J$ we choose $i(j) \in I$ such that $U_j \subset V_{i(j)}$, and from

$$\bigcup_{j \in J} U_j \subset \bigcup_{j \in J} V_{i(j)} \subset X$$

we note that $\cup_{j \in J} V_{i(j)}$ is dense in X, as required.

(b) Let $\{U_\xi : \xi < \beta\}$ be a locally finite (indexed) family of non-empty open subsets of X; we show $\beta < \alpha$. There is an open cover $\{V_i : i \in I\}$ of X such that $|\{\xi < \beta : V_i \cap U_\xi \neq \varnothing\}| < \omega$ for $i \in I$. Since X is weakly α-compact, there is $J \subset I$ with $|J| < \alpha$ such that $\cup_{i \in J} V_i$ is dense in X. We set

$$B(i) = \{\xi < \beta : V_i \cap U_\xi \neq \varnothing\} \text{ for } i \in J,$$

and we note that since $\cup_{i \in J} V_i$ is dense in X we have $\cup_{i \in J} B(i) = \beta$. From $|B(i)| < \omega$ and $|J| < \alpha$ it now follows that $\beta < \alpha$, as required.

10.7 *Corollary*. Let $\{X_i : i \in I\}$ be a set of realcompact spaces and Y a subspace of X_I such that $\pi_J[Y] = X_J$ for every non-empty $J \subset I$ with $|J| \leq \omega$.

(a) If either

(i) $d(X_i) \leq \omega$ for all $i \in I$, or

(ii) $S(X_J) \leq \omega^+$ for every non-empty $J \subset I$ with $|J| < \omega$, or

(iii) X_J is a Lindelőf space for every non-empty $J \subset I$ with $|J| < \omega$, or

(iv) X_J is weakly ω^+-compact for every non-empty $J \subset I$ with $|J| < \omega$, or

(v) X_J is pseudo-ω^+-compact for every non-empty $J \subset I$ with $|J| < \omega$,

then $X_I = \upsilon(Y)$.

(b) If each space X_i is compact, then $X_I = \beta(Y)$.

Proof. (a) Properties (i) through (v) are related by the implications (i)$\Rightarrow$(ii)$\Rightarrow$(iv)$\Rightarrow$(v) and (iii)$\Rightarrow$(iv), so it is sufficient to show from (v) that $X_I = \upsilon(Y)$. From (the case $\kappa = \omega, \alpha = \omega^+$ of) Theorem 10.4(i) it follows that Y is M-embedded (and *a fortiori* C-embedded) in X_I. Since X_I is realcompact, we have

$$\upsilon(Y) = \upsilon(X_I) = X_I$$

from Theorem B.4 of the Appendix.

(b) Again from Theorem 10.4(i) it follows that Y is M-embedded (and *a fortiori* C^*-embedded) in X_I. Since X_I is compact we have

$$\beta(Y) = \beta(X_I) = X_I$$

from Theorem B.4 of the Appendix.

We continue with another consequence of Theorem 10.4: under

appropriate (and quite stringent) conditions, homeomorphic subspaces of a product space have homeomorphic complements.

10.8 *Theorem.* Let $\omega < \mathrm{cf}(\alpha)$, let $\{X_i : i \in I\}$ be a set of non-compact metric spaces with $|I| \geq \alpha$ and with $d(X_i) < \mathrm{cf}(\alpha)$ for $i \in I$, and let A, A' be subsets of X_I such that either

$|A \cup A'| < 2^{|I|}$, or

there is a set $\{K_\xi : \xi < \beta\}$ of compact subspaces of X_I with $\beta \leq \alpha$ such that $A \cup A' \subset \cup_{\xi < \beta} K_\xi$.

Then if $X_I \backslash A$ and $X_I \backslash A'$ are homeomorphic, then A and A' are homeomorphic.

Proof. We set $Y = X_I \backslash (A \cup A')$ and we claim that if $J \in \mathscr{P}^*_\alpha(I)$ then $\pi_J[Y] = X_J$. If the claim fails there is $p \in X_J \backslash \pi_J[Y]$ and we have $\{p\} \times X_{I \backslash J} \subset A \cup A'$. We consider two cases.

Case 1. $|A \cup A'| < 2^{|I|}$. Since X_i is non-compact we have $|X_i| > 1$ for $i \in I \backslash J$, and from $|I \backslash J| = |I|$ we have

$$|\{p\} \times X_{I \backslash J}| \geq 2^{|I|},$$

a contradiction.

Case 2. There is a set $\{K_\xi : \xi < \beta\}$ of compact subspaces of X_I with $\beta \leq \alpha$ such that $A \cup A' \subset \cup_{\xi < \beta} K_\xi$. We assume without loss of generality that $\beta = \alpha$ and we choose $\{i(\xi) : \xi < \alpha\} \subset I \backslash J$ such that

$$i(\xi') \neq i(\xi) \text{ for } \xi' < \xi < \alpha.$$

There is $x \in X_{I \backslash J}$ such that

$$x_{i(\xi)} \in X_{i(\xi)} \backslash \pi_{i(\xi)}[K_\xi] \quad \text{for } \xi < \alpha,$$

and we have $\langle p, x \rangle \in X_J \times X_{I \backslash J} = X_I$ and $\langle p, x \rangle \notin \cup_{\xi < \alpha} K_\xi$; this contradicts the relations

$$\{p\} \times X_{I \backslash J} \subset A \cup A' \subset \bigcup_{\xi < \alpha} K_\xi.$$

The proof that $\pi_J[Y] = X_J$ for $J \in \mathscr{P}^*_\alpha(I)$ is complete.

It follows from Corollary 10.5(ii) that Y is M-embedded in X_I, and since Y is dense in X_I and X_i is metric for $i \in I$, we have

$$\gamma(Y) = \gamma(X_I) = X_I$$

(cf. Theorem B.2). Since $Y \subset X_I \backslash A \subset X_I$, we have $\gamma(X_I \backslash A) = X_I$; that $\gamma(X_I \backslash A') = X_I$ follows similarly.

Now let f be a homeomorphism of $X_I \backslash A$ onto $X_I \backslash A'$. Since

$\gamma(X_I\backslash A) = X_I$ and $\gamma(X_I\backslash A') = X_I$ both f and f^{-1} extend to continuous functions $\bar{f}$ and $\overline{f^{-1}}$ from X_I into X_I. Since

$$(\bar{f}\circ\overline{f^{-1}})(x) = x = (\overline{f^{-1}}\circ f)(x) \quad \text{for } x \in Y$$

and Y is dense in X_I, we have

$$(\bar{f}\circ\overline{f^{-1}})(x) = x = (\overline{f^{-1}}\circ\bar{f})(x) \quad \text{for } x \in X_I.$$

It follows that $\bar{f}$ and $\overline{f^{-1}}$ are (mutually inverse) homeomorphisms of X_I onto X_I. Since

$$\bar{f}[X_I\backslash A] = f[X_I\backslash A] = X_I\backslash A',$$

we have $\bar{f}[A] = A'$, so that A is homeomorphic to A'.

10.9 *Remarks.* (a) The assumption in the statement of Theorem 10.8 that the sets K_ξ (with $\xi < \beta$) are compact may be replaced by the weaker assumption that these sets are pseudocompact. To verify this it is sufficient, according to the considerations of Case 2 in the proof of Theorem 10.8, to show that $\pi_{i(\xi)}[K_\xi] \neq X_{i(\xi)}$ for $\xi < \alpha$; for this it is enough to show that every pseudocompact metric space M is compact. Let the indexed family $\{U_n : n < \omega\}$ be a locally finite family of non-empty open subsets of M, and let $x_n \in U_n$ for $n < \omega$. Then the infinite set $\{x_n : n < \omega\}$ has no accumulation point in M. Thus M is not countably compact; it follows that the metric space M is not compact.

(b) We note the following special case of Theorem 10.8 with $\alpha = \omega^+$. Let β be an uncountable cardinal and let p and q be distinct elements of the space $\mathbb{R}^\beta$. Then no two of the three spaces $\mathbb{R}^\beta$, $\mathbb{R}^\beta\backslash\{p\}$, $\mathbb{R}^\beta\backslash\{p, q\}$ are homeomorphic.

(c) The results the Theorem 10.8 are in contrast to the situation in the space $\mathbb{R}^\omega$. Indeed, Anderson (1966) has shown, en route to the important result that $\mathbb{R}^\omega$ is homeomorphic to sequential Hilbert space, that if Y is a compact subset of R^ω, then $R^\omega\backslash Y$ is homeomorphic to $\mathbb{R}^\omega$; earlier Klee (1956) had shown that if Y is a compact subspace of an infinite-dimensional normed linear space E, then $E\backslash Y$ and E are homeomorphic.

We continue with two consequences of Corollary 10.7.

10.10 *Corollary.* Let $\{X_i : i \in I\}$ be a set of spaces, $p \in X_I$ and $\Sigma_{\omega^+}(p, X_I) \subset Y \subset X_I$.

(a) If $X_i = \mathbb{R}$ for $i \in I$ then $X_I = \upsilon(Y)$;

(b) if X_i is a compact Hausdorff space for $i \in I$, then $X_I = \beta(Y)$.

Definition. Let X be a space, $Y \subset X$ and $x \in X$.

(a) Y is a *G_δ-set of* X if there is $\{U_n : n < \omega\} \subset \mathscr{T}^*(X)$ such that $Y = \cap_{n<\omega} U_n$;

(b) x is a *G_δ-point* if $\{x\}$ is a G_δ-set;

(c) Y is *G_δ-dense* in X if $Y \cap U \neq \emptyset$ for every non-empty G_δ-set U of X.

We note that if every element of a space X is a G_δ-point (in particular, if X is metrizable), then for $Y \subset X$ we have: Y is G_δ-dense in X if and only if $Y = X$.

10.11 *Corollary.* Let $\{X_i : i \in I\}$ be a set of non-empty metric spaces and $Y \subset X_I$.

(a) If each space X_i is separable, then $X_I = \upsilon(Y)$ if and only if Y is G_δ-dense in X_I;

(b) if Y is pseudocompact and dense in X_I, then each X_i is compact and $X_I = \beta(Y)$.

Proof. (a) Like every completely regular, Hausdorff space, the space Y is G_δ-dense in $\upsilon(Y)$; indeed, otherwise there is $f \in C(\upsilon(Y))$ such that $f > 0$ on Y and $f^{-1}(\{0\}) \neq \emptyset$, and then the (continuous) function g defined on Y by the rule $g = 1/(f|Y)$ does not extend continuously over $\upsilon(Y)$. If, conversely, Y is G_δ-dense in X_I, then for $J \in \mathscr{P}^*_\omega(I)$ we have $\pi_J[Y]$ G_δ-dense in X_J and hence $\pi_J[Y] = X_J$ (since X_J is metrizable). The (separable, metric) spaces X_i are Lindelőf spaces and hence realcompact, so the required conclusion $X_I = \upsilon(Y)$ now follows from Corollary 10.7(a).

(b) For $J \in \mathscr{P}^*_\omega(I)$, the space $\pi_J[Y]$ is pseudocompact and dense in X_J; since a pseudocompact metrizable space is compact (cf. Remark 10.9(a)), we have $\pi_J[Y] = X_J$, each space X_i is compact, and Corollary 10.7(b) applies.

In 10.12 – 10.14 below we offer an alternative approach, not dependent on the Erdős–Rado theorem on quasi-disjoint families, to some of the results presented earlier in this chapter.

10.12 *Lemma.* Let β be a cardinal, $\{X_i : i \in I\}$ a set of spaces, and $A = \cup_{\eta<\beta} A_\eta \subset X_I$.

(a) If A depends on $J \subset I$ then cl A depends on J; and

(b) if A_η depends on $J_\eta \subset I$ for $\eta < \beta$ then A depends on $\cup_{\eta<\beta} J_\eta$.

Proof. (a) Let $x \in \text{cl } A, y \in X_I$ and $x_J = y_J$. There is a net $\langle x(\lambda): \lambda \in \Lambda \rangle$ of elements of A such that $\lim_\lambda x(\lambda) = x$. For $\lambda \in \Lambda$ we define $y(\lambda) \in X_I$ by

$$\begin{aligned} y(\lambda)_i &= x(\lambda)_i \quad \text{if } i \in J \\ &= y_i \quad\quad \text{if } i \in I \backslash J. \end{aligned}$$

Since

$$\lim_\lambda y(\lambda)_i = \lim_\lambda x(\lambda)_i = x_i = y_i \quad \text{for } i \in J, \text{ and}$$
$$y(\lambda)_i = y_i \quad \text{for } i \in I \backslash J,$$

we have $\lim_\lambda y(\lambda) = y$; and from $y(\lambda)_J = x(\lambda)_J$ we have $y(\lambda) \in A$. It follows that $y \in \text{cl } A$, as required.

(b) Set $\tilde{J} = \cup_{\eta<\beta} J_\eta$ and let $x \in A$ and $y \in X_I$ with $x_{\tilde{J}} = y_{\tilde{J}}$. There is $\bar{\eta} < \beta$ such that $x \in A_{\bar{\eta}}$ and from $x_{J_{\bar{\eta}}} = y_{J_{\bar{\eta}}}$ we have $y \in A_{\bar{\eta}} \subset A$, as required.

10.13 *Lemma.* Let $\alpha \geq \omega$, let $\{X_i : i \in I\}$ be a set of spaces with $d(X_i) \leq \alpha$ for $i \in I$, and U an open subspace of X_I. Then cl U depends on $\leq \alpha$ coordinates.

Proof. We assume $U \neq \emptyset$. Let $\{U_\eta : \eta < \beta\}$ be a maximal cellular family of basic (in X_I) subsets of U. Since $d(X_i) \leq \alpha$, for $i \in I$ we have

$$S(U) \leq S(X_I) \leq \alpha^+$$

from Theorem 3.28 and hence $\beta \leq \alpha$. From the maximality of the family $\{U_\eta : \eta < \beta\}$ it follows that $\cup_{\eta<\beta} U_\eta$ is dense in U and hence cl $U = \text{cl}(\cup_{\eta<\beta} U_\eta)$. Since the (basic) set U_η depends on $< \omega$ coordinates, it follows from Lemma 10.12 that $\cup_{\eta<\beta} U_\eta$, and hence its closure, depends on $\leq \alpha$ coordinates.

10.14 *Theorem.* Let $\alpha \geq \omega$, let $\{X_i : i \in I\}$ be a set of spaces with $d(X_i) \leq \alpha$ for $i \in I$, Y a Hausdorff space with $w(Y) \leq \alpha$ and f a continuous function from X_I into Y. Then f depends on $\leq \alpha$ coordinates.

Proof. Let $\{B_\xi : \xi < \alpha\}$ be a basis for Y, and set $U_\xi = f^{-1}(B_\xi)$ for $\xi < \alpha$. From Lemma 10.13 there is $J_\xi \subset I$ with $|J_\xi| \leq \alpha$ such that cl U_ξ depends on J_ξ. We set $J = \cup_{\xi<\alpha} J_\xi$, we note that $|J| \leq \alpha$, and we show that f depends on J.

If $x, y \in X_I$ with $x_J = y_J$ and $f(x) \neq f(y)$, then since Y is a Hausdorff space there is $\xi < \alpha$ such that $f(x) \in B_\xi$ and $f(y) \notin \text{cl } B_\xi$, and we have

$$x \in f^{-1}(B_\xi) = U_\xi \subset \text{cl } U_\xi, \text{ and}$$
$$f(y) \notin \text{cl } B_\xi \supset \text{cl} f[U_\xi] \supset f[\text{cl } U_\xi],$$

and hence $y \notin \text{cl } U_\xi$. This contradiction completes the proof.

The following result is derived from Theorem 10.14 or, alternatively, from Corollary 10.2.

As usual, we say that space X is *normal* if for every pair $\langle A, B \rangle$ of disjoint, closed subsets of X there are disjoint open subsets U and V of X such that $A \subset U$ and $B \subset V$. In the following proof we use Urysohn's lemma: if A and B are are disjoint, closed subsets of a normal space X, then there is a continuous function f from X to $[0, 1]$ such that

$$\begin{aligned} f(x) &= 0 \quad \text{for } x \in A, \\ &= 1 \quad \text{for } x \in B. \end{aligned}$$

10.15 *Theorem.* (a) The space $\omega^{(\omega^+)}$ is not normal.

(b) If $\{X_i : i \in I\}$ is a set of non-empty spaces with X_I normal, then

$$|\{i \in I : X_i \text{ is not countably compact }\}| \leq \omega.$$

Proof. (a) For $k = 0, 1$ we set

$$A_k = \{x \in \omega^{(\omega^+)} : \text{if } n \neq k \text{ then } |\{\xi < \omega^+ : x_\xi = n\}| \leq 1\}.$$

It is clear that A_0, A_1 are disjoint, closed subsets of $\omega^{(\omega^+)}$; hence if $\omega^{(\omega^+)}$ is normal there is $f \in C(\omega^{(\omega^+)}, [0, 1])$ such that

$$\begin{aligned} f(x) &= 0 \quad \text{for } x \in A_0, \\ &= 1 \quad \text{for } x \in A_1. \end{aligned}$$

From Theorem 10.14 (with $\alpha = \omega$) or Corollary 10.2 (with $\kappa = \omega$ and $\alpha = \omega^+$) there is $J \subset \omega^+$ with $|J| = \omega$ such that f depends on J; we write $J = \langle \xi_n : n < \omega \rangle$ with

$$\xi_{n'} \neq \xi_n \quad \text{for } n' < n < \omega.$$

We define $p, q \in \omega^{(\omega^+)}$ by

$$\begin{aligned} p_\xi = q_\xi &= n \quad \text{for } \xi = \xi_n \in J \\ p_\xi &= 0 \quad \text{for } \xi \in \omega^+ \backslash J, \\ q_\xi &= 1 \quad \text{for } \xi \in \omega^+ \backslash J. \end{aligned}$$

Since $p \in A_0$ and $q \in A_1$, we have $f(p) = 0$ and $f(q) = 1$, and since $p_J = q_J$, we have $f(p) = f(q)$. This contradiction completes the proof.

(b) If the result fails there is $J \subset I$ with $|J| = \omega^+$ such that X_i is not countably compact for $i \in J$. The normal space X_I then contains a closed subspace homeomorphic to the non-normal space ω^J, a contradiction.

We conclude this chapter with three miscellaneous results which answer questions raised naturally by what has appeared above. The first is a partial converse to Corollary 10.2; the second shows that an ω^+-Σ-product in a product of first-countable spaces is C-embedded, thus substantially generalizing an important special case of Theorem 10.4; and the last shows that not every ω^+-Σ-product is C-embedded.

10.16 *Theorem.* Let $\alpha > \omega$ with α regular and let $\{X_i : i \in I\}$ be a set of completely regular, Hausdorff spaces with $|I| \geq \alpha$ and $|X_i| \geq 2$ for $i \in I$. If every continuous function from X_I to $[0, 1]$ depends on $< \alpha$ coordinates, then X_I is pseudo-α-compact.

Proof. If the statement fails, then by Theorem 3.6(c) there is $J \in \mathscr{P}^*_\omega(I)$ such that X_J is not pseudo-α-compact. Let $\{U_\xi : \xi < \alpha\}$ be a locally finite, cellular family in X_J, let $\{i_\xi : \xi < \alpha\}$ be a faithfully indexed subset of $I \backslash J$, let V_ξ be an open subset of X_{i_ξ} for $\xi < \alpha$ with $V_\xi \neq X_{i_\xi}$, and set

$$W_\xi = U_\xi \times V_\xi \times X_{I \backslash (J \cup \{i_\xi\})} \quad \text{for } \xi < \alpha.$$

The (indexed) family $\{W_\xi : \xi < \alpha\}$ is a locally finite, cellular family in X_I. For $\xi < \alpha$ we choose $x(\xi) \in W_\xi$ and $f_\xi \in C(X_I, [0, 1])$ such that

$$f_\xi(x(\xi)) = 1, \quad \text{and}$$
$$f_\xi(x) = 0 \quad \text{for } x \in X_I \backslash W_\xi,$$

and we set $f = \sum_{\xi < \alpha} f_\xi$. Since

$$W_{\xi'} \cap W_\xi = \emptyset \quad \text{for } \xi' < \xi < \alpha,$$

we have $f[X_I] \subset [0, 1]$, and since $\{W_\xi : \xi < \alpha\}$ is locally finite and $f_\xi(x) = 0$ for $x \notin W_\xi$, the function f is continuous.

To complete the proof it is enough to show that if $J' \subset I$ and f depends on J', then $\{i_\xi : \xi < \alpha\} \subset J'$. We fix $\xi < \alpha$. Since $V_\xi \subsetneqq X_{i_\xi}$, there is $y(\xi)\in X_I$ such that

$$y(\xi)_i = x(\xi)_i \quad \text{for } i \neq i_\xi$$
$$y(\xi)_{i_\xi} \in X_{i_\xi}\backslash V_\xi.$$

Now for $\eta < \alpha$ with $\eta \neq \xi$ we have $x(\xi)\notin W_\eta$ and also

$$y(\xi)_J = x(\xi)_J \in U_\xi \backslash U_\eta,$$

and hence $y(\xi)\notin W_\eta$; it follows that $f_\eta(x(\xi)) = 0$ and $f_\eta(y(\xi)) = 0$ for $\eta \neq \xi$, and from $y(\xi)_{i_\xi} \notin V_\xi$ (and hence $y(\xi)\notin W_\xi$) we have

$$f(x(\xi)) = f_\xi(x(\xi)) = 1, \quad \text{and}$$
$$f(y(\xi)) = f_\xi(y(\xi)) = 0.$$

It is clear that if f depends on J' then $i_\xi \in J'$, as required.

A space X is *first-countable* if for $x\in X$ there is a countable neighborhood base at x.

10.17 *Theorem.* Let $\{X_i : i\in I\}$ be a set of first-countable spaces and $p\in X_I$. Then $\Sigma_{\omega^+}(p)$ is C-embedded in X_I.

Proof. If the statement fails there are $f\in C(\Sigma_{\omega^+}(p))$ and $q\in X_I\backslash\Sigma_{\omega^+}(p)$ such that f does not extend continuously to q; since $\mathbb{R}$ is complete there is $\varepsilon > 0$ such that the oscillation of f is at least ε on every neighborhood in X_I of q. For U a neighborhood of q in X_I we choose $x_U, y_U \in U\cap\Sigma_{\omega^+}(p)$ such that

$$|f(x_U) - f(y_U)| > \varepsilon;$$

we assume without loss of generality, since f is continuous and $\Sigma_\omega(p)$ is a dense subspace of $\Sigma_{\omega^+}(p)$, that $x_U, y_U \in \Sigma_\omega(p)$.

Now for $i\in I$ and $z\in X_i$ let $\{U_n(z) : n < \omega\}$ be a neighborhood base at z with $U_{n+1}(z) \subset U_n(z)$ for $n < \omega$. (To avoid ambiguity in this choice we assume without loss of generality that $X_{i'}\cap X_i = \varnothing$ for $i, i'\in I$ with $i \neq i'$.) For $n < \omega$ we define $J(n)$, $U(n)$, $x(n)$, $y(n)$ as follows:

(i) $J(0) = \varnothing$;
(ii) $U(0) = X_I$;
(iii) $x(0) = x_{U(0)}$, $y(0) = y_{U(0)}$,
(iv) $J(n+1) = J(n)\cup\{i\in I : x(n)_i \neq p_i \text{ or } y(n)_i \neq p_i\}$ for $n < \omega$;
(v) $U(n+1) = \{z\in X_I : z_i \in U_{n+1}(q_i) \text{ for } i\in J(n)\}$ for $n < \omega$; and
(vi) $x(n+1) = x_{U(n+1)}$, $y(n+1) = y_{U(n+1)}$ for $n < \omega$.

We define $\bar{z} \in X_I$ by

$$\begin{aligned} \bar{z}_i &= q_i \quad \text{for } i \in \bigcup_{n<\omega} J(n) \\ &= p_i \quad \text{for } i \in I \backslash \bigcup_{n<\omega} J(n) \end{aligned}$$

Since $|J(n)| < \omega$ for $n < \omega$ we have $\bar{z} \in \Sigma_{\omega^+}(p)$. We claim further that

$$\lim_n x(n) = \lim_n y(n) = \bar{z}.$$

Indeed, if $i \in J(k)$ with $k < \omega$, then for $k < m < n < \omega$ we have

$$\begin{aligned} &x(n)_i \in U_n(q_i) \subset U_m(q_i) = U_m(\bar{z}_i), \quad \text{and} \\ &y(n)_i \in U_n(q_i) \subset U_m(q_i) = U_m(\bar{z}_i) \end{aligned}$$

and hence

$$\lim_n x(n)_i = \lim_n y(n)_i = \bar{z},$$

and if $i \notin \cup_{n<\omega} J(n)$, then

$$x(n)_i = y(n)_i = p_i = \bar{z}_i$$

and hence

$$\lim_n x(n)_i = \lim_n y(n)_i = \bar{z}_i.$$

The claim is established. Since

$$|f(x(n)) - f(y(n))| > \varepsilon \quad \text{for } n < \omega,$$

the function f is not continuous at the point $\bar{z}$. This contradiction completes the proof.

It follows from Theorem 10.17 that in a product of metrizable spaces, and in particular in a product of discrete spaces, every ω^+-Σ-product is C-embedded. Together with the results of Corollary 10.5, this raises the question whether every ω^+-Σ-product is C-embedded. In this connection we note a curiosity. If $f \in C(\Sigma_{\omega^+}(p))$ with $p \in X_I$ and if $q \in X_I \backslash \Sigma_{\omega^+}(p)$, there is a natural way to attempt to extend f continuously to q: give the spaces X_i the discrete topology, note that f remains continuous on $\Sigma_{\omega^+}(p)$ in the stronger topology now inherited from the (new) product space X_I, and extend f to q using Theorem 10.17. It is clear that if relative to the original product topology the function f extends continuously to q, it is with the value just obtained; and the question of determining whether a function continuous on an ω^+-Σ-product extends continuously over X_I becomes the question whether a particular function, known to be continuous relative to a certain strong topology, is continuous relative to the usual product topology.

We show now that not every ω^+-Σ-product is C-embedded.

10.18 *Theorem.* Let $\alpha \geq \omega$. There are a set $\{X_i : i \in I\}$ of completely regular, Hausdorff spaces, $p, q \in X_I$ and $f \in C(\Sigma_\alpha(p), \{0, 1\})$ such that no continuous function from $\Sigma_\alpha(p) \cup \{q\}$ to $[0, 1]$ extends f.

Proof. We note first that there is a family $\mathscr{F}$ of subsets of α such that

(i) $|\mathscr{F}| = \alpha$;
(ii) $|A| = \alpha$ for A in $\mathscr{F}$;
(iii) if $\mathscr{G} \subset \mathscr{F}$ and $|\mathscr{G}| < \omega$, then $\cap \mathscr{G} \neq \emptyset$; and
(iv) if $\mathscr{G} \subset \mathscr{F}$ and $|\mathscr{G}| \geq \omega$, then $\cap \mathscr{G} = \emptyset$.

(To define such a family $\mathscr{F}$, let $\{B_\xi : \xi < \alpha\}$ be a faithful indexing of $\mathscr{P}^*_\omega(\alpha)$, and set $A_\zeta = \{\xi < \alpha : \zeta \in B_\xi\}$ for $\zeta < \alpha$ and $\mathscr{F} = \{A_\zeta : \zeta < \alpha\}$.) Let $q_0 \in \beta(\alpha)$ with $\mathscr{F} \subset q_0$ and set $X_0 = \alpha \cup \{q_0\}$ with the topology inherited from $\beta(\alpha)$.

For $A \in \mathscr{F}$ let X_A denote the one-point compactification $A \cup \{q_A\}$ of the (discrete) space A, set

$$X = X_0 \times \prod_{A \in \mathscr{F}} X_A,$$

and let $p \in X$ satisfy

$$p_0 \neq q_0,$$
$$p_A \neq q_A \text{ for } A \in \mathscr{F}$$

Now we set

$$K = \{x \in \Sigma_\alpha(p) : x_0 \neq q_0, \text{ and } x_A = x_0 \text{ whenever } x_0 \in A \in \mathscr{F}\}.$$

Because no element of α lies in infinitely many elements of $\mathscr{F}$, the set K is open in $\Sigma_\alpha(p)$. (In detail: if $x \in K$ and $x_0 = \xi \in \alpha$, then

$$\pi_0^{-1}(\{\xi\}) \cap \cap \{\pi_A^{-1}(\{\xi\}) : \xi \in A \in \mathscr{F}\}$$

is a neighborhood in X of x whose intersection with $\Sigma_\alpha(p)$ is a subset of K.)

We claim also that K is closed in $\Sigma_\alpha(p)$. Let $x \in \Sigma_\alpha(p) \backslash K$. If $x_0 \neq q_0$, there is $A \in \mathscr{F}$ such that $x_A \neq x_0$ and then

$$\pi_0^{-1}(\{x_0\}) \cap \pi_A^{-1}(\{x_A\})$$

is a neighborhood in X of x disjoint from K. If $x_0 = q_0$, then since $|\mathscr{F}| = \alpha$ and $x \in \Sigma_\alpha(p)$ there is $A \in \mathscr{F}$ such that $x_A = p_A$ and there is $B \in \mathscr{F}$ such that $x_A \in \alpha \backslash B$; we set

$$U = \{q_0\} \cup (A \cap B).$$

Then U is a neighborhood in X_0 of q_0 and hence

$$\pi_0^{-1}(U) \cap \pi_A^{-1}(\{x_A\})$$

is a neighborhood in X of x; this misses K, for if $y_0 \in U$ then either $y_0 \neq q_0$ (hence $y \notin K$) or $y_0 \in A \cap B$ (hence $y_A \neq x_A$). The proof that K is closed in $\Sigma_\alpha(p)$ is complete.

To complete the proof it is enough to show that every neighborhood in X of q meets both K and $\Sigma_\alpha(p) \backslash K$, for then the function which is 1 on K and 0 on $\Sigma_\alpha(p) \backslash K$ will admit no continuous extension to q. Let $V = V_0 \times \prod_{A \in \mathscr{F}} V_A$ be a neighborhood in X of q with (finite) restriction set $R(V)$, and set

$$\begin{aligned} y_0 &= q_0 \\ y_A &= q_A \quad \text{for } A \in R(V) \\ &= p_A \quad \text{for } A \in \mathscr{F} \backslash R(V); \end{aligned}$$

then $y \in V \cap (\Sigma_\alpha(p) \backslash K)$.

We define $z \in V \cap K$. Since $|A \backslash V_A| < \omega$ for $A \in \mathscr{F}$, we have $V_A \in q_0$ for $A \in \mathscr{F} \subset q_0$; hence there is

$$z_0 \in V_0 \cap \cap \{V_A : A \in R(V) \cap \mathscr{F}\}.$$

We set $\mathscr{F}' = \{A \in \mathscr{F} : z_0 \in A\}$, we note that $\mathscr{F}'$ is a finite subset of $\mathscr{F}$ such that $R(V) \cap \mathscr{F} \subset \mathscr{F}'$, and we set

$$\begin{aligned} z_A &= z_0 \quad \text{for } A \in \mathscr{F}' \\ &= p_A \quad \text{for } A \in \mathscr{F} \backslash \mathscr{F}'. \end{aligned}$$

It is then clear that $z \in V \cap K$.

The proof is complete.

Notes for Chapter 10

From Corollary 10.2 it follows that under certain conditions every continuous function from a product space X_I to a metric space depends on countably many coordinates. When each X_i is compact, this is due to Mibu (1944) and Bishop (1959); when each is separable and metrizable, to Mazur (1952) (with an additional cardinality hypothesis, but for sequentially continuous functions) and Corson & Isbell (1960); when each has property (K), to Ross & Stone (1964); when each is separable, to Gleason (see Ross & Stone (1964) or Isbell (1964, pp. 130–2)); when each finite product of the X_i is a Lindelőf space, to Engelking (1966); and when X (or each X_F with $F \in \mathscr{P}^*_\omega(I)$) is pseudo-$\omega^+$-compact, to Noble & Ulmer (1972). Using his own generalization of the concept of calibre as introduced by Shanin (1946b), Miščenko (1966a, b) has given analogous statements in the context of uniform spaces (for uniformly continuous functions). Some of these authors consider functions defined on subspaces of X_I, and some impose hypotheses weaker than metrizability on the range space.

That every continuous function from a product of compact, Hausdorff spaces to the real line or to the complex plane depends on countably many coordinates is a direct consequence of the Stone–Weierstrass theorem noted by many authors: the algebra of continuous functions depending on finitely many coordinates separates points and distinguishes points from closed sets (and is conjugate-closed), and clearly every function in its uniform closure depends on $\leq \omega$ coordinates. For a related argument using partitions of unity, for functions defined on the product of two compact spaces, see Dieudonné (1937).

In addition to the new contributions they contain, the works of Engelking (1966) and Hušek (1976) are valuable surveys of the literature of continuous functions defined on product spaces and their subspaces; Hušek (1976) has classified into six groupings the patterns of proof used by the authors he cites.

The fact that continuous functions from a product to spaces subjected to a countability condition may fail to depend on countably many coordinates has been noted by Miščenko (1966a), Engelking (1968, p. 20), and Marty (1969, Example 20). The argument we use in Theorem 10.16 is suggested by Noble & Ulmer (1972, Theorem 2.3).

The argument of 10.12–10.14 was introduced by Bockstein (1948) to show that disjoint, open subsets in a product of separable metric spaces remain disjoint when projected into a suitable countable subproduct; see also Corson & Isbell (1960). In our treatment of 10.12–10.16 we have followed Ross & Stone (1964); Theorem 10.15, first proved by Stone (1948), has been generalized by Noble (1967).

It should be noted that statements weaker than Theorem 10.15 can be proved easily and directly. To see for example that $\mathbb{R}^{2^\omega}$ is not normal, let X be a separable, realcompact space with a closed discrete subspace D such that $|D| = 2^\omega$ (e.g., let $X = S \times S$, with S the 'Sorgenfrey line' as described by Sorgenfrey (1947) or Kelley (1955, Problem 4I)), and embed X into $\mathbb{R}^{C(X)}$ using the function $e: X \to \mathbb{R}^{C(X)}$ defined by $e(x)_f = f(x)$. Since $e[X]$, which is homeomorphic to X, is realcompact and C-embedded in $\mathbb{R}^{C(X)}$, the space $e[X]$ is closed in $\mathbb{R}^{C(X)}$ (cf. Theorem B.4 of the Appendix). Since $|C(D)| = 2^{2^\omega}$ and $|C(X)| = 2^\omega$, it follows from Tietze's extension theorem that X is not normal; hence $\mathbb{R}^{C(X)}$, which is $\mathbb{R}^{2^\omega}$, is not normal.

The recursive argument of Lemma 10.1 was introduced by Glicksberg (1959) to prove the case $\kappa = \alpha = \omega$; see also Kister (1962). Noble & Ulmer (1972) proved Lemma 10.1 and Corollary 10.2(b) in the case that $\kappa = \omega$ and Y contains an ω^+-Σ-product. The present treatment of 10.1–10.4 follows closely Comfort & Negrepontis (1972b).

ω^+-Σ-products were introduced and systematically studied by Corson (1959), who established the relation $\upsilon(\Sigma_{\omega^+}(p, X_I)) = X_I$ for X_i separable metric; Corollary 10.7(b) (for ω^+-Σ-products) was given independently by Glicksberg (1959) and Kister (1962). Parts (a) and (b) of Corollary 10.11 are from Efimov & Engelking (1965), and Noble (1972), respectively.

Van der Slot (1972) showed that two compact subspaces of $\mathbb{R}^\alpha$ (with $\alpha > \omega$) are homeomorphic if and only if their complements are homeomorphic. Our proof of Theorem 10.8 follows arguments suggested by Hušek (1972, Corollary 3) and Hušek (1976, (5), p. 39).

Theorems 10.17 and 10.18 are from the thesis of Ulmer (1970), which contains several results more general than those given here. For example: every ω^+-Σ-product in a product of P_{ω^+}-spaces is C-embedded, but there are spaces $\{X_i : i \in I\}$, all but one a P_{ω^+}-space and each containing exactly one non-isolated point, such that some ω^+-Σ-product in X_I is not C-embedded.

We note that (as is mentioned in the Notes for Chapter 7) the class of Eberlein-compact spaces is contained in the class of Corson-compact spaces. The preservation of Eberlein-compact spaces under continuous functions is a basic result that has been extended to various other classes of subspaces of Σ-products. See in this connection Benyamini, Rudin & Wage (1977), Michael & Rudin (1977), Gul'ko (1977, 1978), Alster & Pol (1980), and Negrepontis & Tsarpalias (1981).

Appendix: preliminaries

We here discuss briefly the basic facts from set theory, topology and Boolean algebra, and measure theory that are used in this work.

A. Set theory and cardinal arithmetic

The set theory within which we work, denoted ZFC, is the Zermelo–Fraenkel set theory together with the axiom of choice.

A partially ordered set $\mathscr{P}$ is *inductive* if every *chain* (i.e., every linearly ordered subset) of $\mathscr{P}$ has upper bound in $\mathscr{P}$. *Zorn's lemma*, a form of the axiom of choice, is the statement that every non-empty inductive partially ordered set has a maximal element.

The ordinals are defined in such a way that an ordinal is the set of smaller ordinals. Every well-ordered set is order-isomorphic to a unique ordinal called its *order type*.

The ordinal successor of an ordinal ξ is denoted $\xi + 1$. An ordinal ξ is a *successor* or *non-limit* ordinal if there is ζ such that $\xi = \zeta + 1$; otherwise ξ is a *limit* ordinal. (In particular, 0 is a limit ordinal.) Ordinal variables are denoted by ξ, ζ, η, i, σ, τ and the like.

A *cardinal* (number) is an ordinal not in one-to-one correspondence with any smaller ordinal. The cardinal successor of a cardinal α is denoted α^+. A cardinal α is a *successor* or *non-limit* cardinal if there is a cardinal β such that $\alpha = \beta^+$; otherwise α is a *limit* cardinal. A cardinal α is a *strong limit* cardinal if $2^\beta < \alpha$ for all $\beta < \alpha$. Cardinal variables are denoted by α, β, γ, δ, κ, λ and the like; finite cardinal variables (natural numbers) are denoted by k, m, n. The least infinite cardinal is the set of finite cardinals and is denoted by ω.

Every infinite cardinal is a limit ordinal.

Definition. Let ξ be a limit ordinal. The *cofinality* of ξ, denoted $\mathrm{cf}(\xi)$, is the least ordinal ζ for which there is a function $f : \zeta \to \xi$

such that

f is *order-preserving* (i.e., $f(\eta') < f(\eta)$ for $\eta' < \eta < \zeta$), and f is *unbounded* (i.e., $\sup_{\eta<\zeta} f(\eta) = \xi$).

An infinite cardinal α is *regular* if $\alpha = \mathrm{cf}(\alpha)$, *singular* otherwise.

Clearly ω is a regular cardinal; and if α is an infinite cardinal then α^+ is a regular cardinal.

It is not difficult to verify that if ξ is a limit ordinal then $\mathrm{cf}(\xi) \leq \xi$ and $\mathrm{cf}(\xi)$ is a cardinal and $\mathrm{cf}(\mathrm{cf}(\xi)) = \mathrm{cf}(\xi)$. The following basic result is immediate.

A.1 *Theorem.* Let α be an (infinite) singular cardinal.

(a) There is a set $\{\alpha_\sigma : \sigma < \mathrm{cf}(\alpha)\}$ of cardinals such that $\alpha_{\sigma'} < \alpha_\sigma < \alpha$ for $\sigma' < \sigma < \mathrm{cf}(\alpha)$, and $\sum_{\sigma<\mathrm{cf}(\alpha)} \alpha_\sigma = \alpha$.

(b) If $\beta < \mathrm{cf}(\alpha)$ and $\{\alpha_\xi : \xi < \beta\}$ is a set of cardinals such that $\alpha_\xi < \alpha$ for $\xi < \beta$, then $\sum_{\xi<\beta} \alpha_\xi < \alpha$.

Notation. Let A be a set and let α and κ be cardinals. Then

$$
\begin{aligned}
&|A| \text{ denotes the cardinality of } A,\\
&\mathscr{P}(A) = \{B : B \subset A\},\\
&\mathscr{P}_\kappa(A) = \{B \in \mathscr{P}(A) : |B| < \kappa\},\\
&\mathscr{P}^*_\kappa(A) = \mathscr{P}_\kappa(A) \backslash \{\varnothing\},\\
&[A]^\kappa = \{B \in \mathscr{P}(A) : |B| = \kappa\}, \text{ and}\\
&\alpha^{\underset{\smile}{\kappa}} = \sum\{\alpha^\lambda : \lambda \text{ is a cardinal and } \lambda < \kappa\}.
\end{aligned}
$$

It is not difficult to show that

$$
\begin{aligned}
&|\mathscr{P}(A)| = 2^{|A|},\\
&|\mathscr{P}_\kappa(\alpha)| = \alpha^{\underset{\smile}{\kappa}} \quad \text{if } \omega + \kappa \leq \alpha^+,\\
&|[\alpha]^\kappa| = \alpha^\kappa \quad \text{if } \omega + \kappa \leq \alpha, \text{ and}\\
&\kappa \leq \alpha^{\underset{\smile}{\kappa}} \quad \text{if } 2 \leq \alpha.
\end{aligned}
$$

A.2 *Lemma.* Let $\alpha \geq 2$ and $\kappa \geq \omega$. Then $(\alpha^{\underset{\smile}{\kappa}})^{\underset{\smile}{\mathrm{cf}(\kappa)}} = \alpha^{\underset{\smile}{\kappa}}$.

Proof. It is sufficient to prove that

$$(\alpha^{\underset{\smile}{\kappa}})^\lambda = \alpha^{\underset{\smile}{\kappa}} \quad \text{for } \lambda < \mathrm{cf}(\kappa),$$

for then

$$(\alpha^{\underset{\smile}{\kappa}})^{\underset{\smile}{\mathrm{cf}(\kappa)}} = \sum_{\lambda<\mathrm{cf}(\kappa)} (\alpha^{\underset{\smile}{\kappa}})^\lambda = \sum_{\lambda<\mathrm{cf}(\kappa)} \alpha^{\underset{\smile}{\kappa}} = \alpha^{\underset{\smile}{\kappa}} \cdot \mathrm{cf}(\kappa) \leq \alpha^{\underset{\smile}{\kappa}} \cdot \kappa = \alpha^{\underset{\smile}{\kappa}}.$$

For $\mu' < \mu < \kappa$ and $f \in \alpha^{\mu'}$, we identify f with the function $\tilde{f} \in \alpha^\mu$

defined by

$$\begin{aligned} \tilde{f}(\xi) &= f(\xi) \quad \text{for } \xi \in \mu', \\ &= 0 \quad \text{for } \xi \in \mu \backslash \mu'; \end{aligned}$$

with this identification we have $\alpha^{\mu'} \subset \alpha^{\mu}$ for $\mu' < \mu < \kappa$.

Now let $\lambda < \mathrm{cf}(\kappa)$ and $f \in (\alpha^{\underline{\kappa}})^{\lambda}$. For $\eta < \lambda$ there is $\mu(\eta) < \kappa$ such that $f(\eta) \in \alpha^{\mu(\eta)}$. We set $\mu = \sup\{\mu(\eta) : \eta < \lambda\}$ and we note (since $\lambda < \mathrm{cf}(\kappa)$) that $\mu < \kappa$; it follows that $f[\lambda] \in \alpha^{\mu}$, i.e., that $f \subset (\alpha^{\mu})^{\lambda}$. From

$$(\alpha^{\underline{\kappa}})^{\lambda} \subset \bigcup_{\mu < \kappa} \alpha^{\mu \cdot \lambda}$$

we have

$$(\alpha^{\underline{\kappa}})^{\lambda} \leq \textstyle\sum_{\mu < \kappa} \alpha^{\mu \cdot \lambda} = \sum_{\mu < \kappa} \alpha^{\mu} = \alpha^{\underline{\kappa}},$$

as required.

A.3 Theorem. Let $\alpha \geq 2$ and $\kappa \geq \omega$.

(a) If κ is regular the $(\alpha^{\underline{\kappa}})^{\underline{\kappa}} = \alpha^{\underline{\kappa}}$;

(b) if κ is singular then $(\alpha^{\underline{\kappa}})^{\underline{\kappa}} = \alpha^{\kappa}$; and

(c) $((\alpha^{\underline{\kappa}})^{\underline{\kappa}})^{\underline{\kappa}} = (\alpha^{\underline{\kappa}})^{\underline{\kappa}}$.

Proof. (a) is immediate from Lemma A.2. To prove (b) we note first that if $\beta \leq \alpha^{\kappa}$ then $\beta^{\underline{\kappa}} \leq \alpha^{\kappa}$; it follows that $(\alpha^{\underline{\kappa}})^{\underline{\kappa}} \leq \alpha^{\kappa}$. For the reverse inequality let $\{\kappa_{\xi} : \xi < \mathrm{cf}(\kappa)\}$ be a set of cardinal numbers such that

$$\begin{aligned} &\mathrm{cf}(\kappa) < \kappa_{\xi} < \kappa \quad \text{for } \xi < \mathrm{cf}(\kappa), \text{ and} \\ &\kappa = \textstyle\sum_{\xi < \mathrm{cf}(\kappa)} \kappa_{\xi}; \end{aligned}$$

then

$$\alpha^{\kappa} = \alpha^{\sum_{\xi < \mathrm{cf}(\kappa)} \kappa_{\xi}} = \prod_{\xi < \mathrm{cf}(\kappa)} \alpha^{\kappa_{\xi}} \leq (\alpha^{\underline{\kappa}})^{\mathrm{cf}(\kappa)} \leq (\alpha^{\underline{\kappa}})^{\underline{\kappa}},$$

as required.

Statement (c) for regular κ follows from (a), while if κ is singular then from (b) we have

$$((\alpha^{\underline{\kappa}})^{\underline{\kappa}})^{\underline{\kappa}} = (\alpha^{\kappa})^{\underline{\kappa}} = \alpha^{\kappa} = (\alpha^{\underline{\kappa}})^{\underline{\kappa}}.$$

The following statement, known as Kőnig's lemma, was proved by Kőnig (1904) for countable index sets I and by Jourdain (1908) and Zermelo (1908) in general.

Theorem. Let I be a non-empty set and let $\alpha_i < \beta_i$ for $i \in I$. Then $\sum_{i\in I} \alpha_i < \prod_{i \in I} \beta_i$.

We omit the proof. We note two consequences of Kőnig's lemma.

A.4 Theorem. Let α be a cardinal. Then

(a) $\alpha < 2^\alpha$; and

(b) if α is infinite then $\alpha^{\mathrm{cf}(\alpha)} > \alpha$.

Proof. (a) Set $\alpha_i = 1$, $\beta_i = 2$, for $i < \alpha$.

(b) If α is regular we have

$$\alpha < 2^\alpha \leq \alpha^\alpha = \alpha^{\mathrm{cf}(\alpha)}$$

from (a). If α is singular we choose $\{\alpha_\sigma : \sigma < \mathrm{cf}(\alpha)\}$ as in Theorem A.1 and we set $\beta_\sigma = \alpha$ for $\sigma < \mathrm{cf}(\alpha)$.

Definition. An infinite cardinal α is *strongly inaccessible* if α is a regular cardinal such that $2^\beta < \alpha$ whenever $\beta < \alpha$.

Definition. Let α and κ be cardinals. Then α is *strongly κ-inaccessible* if

$$\kappa < \alpha, \text{ and}$$
$$\beta^\lambda < \alpha \text{ whenever } \beta < \alpha \text{ and } \lambda < \kappa.$$

If α is strongly κ-inaccessible we write $\kappa \ll \alpha$.

We note some consequences of this definition.

A.5 Theorem. Let α and κ be cardinals.

(a) If $\omega \leq \alpha, \kappa \leq \mathrm{cf}(\alpha)$ and $\kappa < \alpha$, then $\kappa \ll \alpha$ if and only if the following condition is satisfied:

If $\lambda < \kappa$ and $\beta_\xi < \alpha$ for $\xi < \lambda$ then $\prod_{\xi < \lambda} \beta_\lambda < \alpha$.

(b) If $\alpha \geq \omega$ then $\alpha^+ \ll (2^\alpha)^+$.

(c) If $\omega \leq \kappa \ll \alpha$ with κ singular and α regular, then $\kappa^+ \ll \alpha$.

Proof. (a) The necessity of the condition is clear. For the sufficiency set $\beta = \sum_{\xi < \lambda} \beta_\xi$; then $\beta < \alpha$ and hence $\prod_{\xi < \lambda} \beta_\xi \leq \beta^\lambda < \alpha$.

(b) is clear.

(c) Since $\kappa < \alpha$ and $\mathrm{cf}(\kappa) < \kappa$ we have $\kappa^+ \leq \kappa^{\mathrm{cf}(\kappa)} < \alpha$ (from Theorem A.4) and hence $\kappa^+ < \alpha$. For $\beta < \alpha$ and $\lambda < \kappa$ we have $\beta^\lambda < \alpha$; since α is regular and $\kappa < \alpha$ we have

$$\beta^{\underline{\kappa}} = \sum_{\lambda < \kappa} \beta^\lambda < \alpha.$$

The same argument now shows that $(\beta^{\underline{\kappa}})^{\underline{\kappa}} < \alpha$ for $\beta < \alpha$ and from Theorem A.3(b) we have

$$\beta^{\kappa} = (\beta^{\underline{\kappa}})^{\underline{\kappa}} < \alpha \quad \text{for } \beta < \alpha,$$

as required.

A.6 Lemma. Let α be an (infinite) singular cardinal, $\beta < \alpha$ and $\kappa \ll \alpha$. If $\kappa = \omega$, or κ is a successor cardinal, or $\mathrm{cf}(\kappa) \neq \mathrm{cf}(\alpha)$, or α is a strong limit cardinal, then there is a regular cardinal γ such that $\beta < \gamma < \alpha$ and $\kappa \ll \gamma$.

Proof. We assume without loss of generality that $\beta \geq \kappa$.

If $\kappa = \omega$ we set $\gamma = \beta^{+}$; if κ is the successor cardinal λ^{+} we set $\gamma = (\beta^{\lambda})^{+}$; and if α is a strong limit cardinal we set $\gamma = (2^{\beta})^{+}$. It is then clear that $\beta < \gamma < \alpha$ and $\kappa \ll \gamma$; the proof is complete in these cases.

We assume now that $\mathrm{cf}(\kappa) \neq \mathrm{cf}(\alpha)$ and we claim that if $\delta < \alpha$ then $\delta^{\underline{\kappa}} < \alpha$. If κ is the successor cardinal $\kappa = \lambda^{+}$ then $\delta^{\underline{\kappa}} = \delta^{\lambda} < \alpha$. If κ is a limit cardinal we choose sets $\{\alpha_{\sigma} : \sigma < \mathrm{cf}(\alpha)\}$ and $\{\kappa_{\eta} : \eta < \mathrm{cf}(\kappa)\}$ of cardinals such that

$$\alpha_{\sigma'} < \alpha_{\sigma} < \alpha, \kappa_{\eta'} < \kappa_{\eta} < \kappa \text{ for } \sigma' < \sigma < \mathrm{cf}(\alpha), \eta' < \eta < \mathrm{cf}(\kappa), \text{ and}$$
$$\alpha = \sup\{\alpha_{\sigma} : \sigma < \mathrm{cf}(\alpha)\}, \kappa = \sup\{\kappa_{\eta} : \eta < \mathrm{cf}(\kappa)\}$$

and we consider two cases.

Case 1. $\mathrm{cf}(\kappa) < \mathrm{cf}(\alpha)$. Then from $\delta^{\kappa_{\eta}} < \alpha$ we have

$$\delta^{\underline{\kappa}} = \sum_{\eta < \mathrm{cf}(\kappa)} \delta^{\kappa_{\eta}} < \alpha,$$

as required.

Case 2. $\mathrm{cf}(\alpha) < \mathrm{cf}(\kappa)$. For $\eta < \mathrm{cf}(\kappa)$ we have $\delta^{\kappa_{\eta}} < \alpha$ and hence there is $\sigma < \mathrm{cf}(\alpha)$ such that $\delta^{\kappa_{\eta}} < \alpha_{\sigma}$. Since $\mathrm{cf}(\alpha) < \mathrm{cf}(\kappa)$ and $\mathrm{cf}(\kappa)$ is a regular cardinal there is $\bar{\sigma} < \mathrm{cf}(\alpha)$ such that

$$|\{\eta < \mathrm{cf}(\kappa) : \delta^{\kappa_{\eta}} < \alpha_{\bar{\sigma}}\}| = \mathrm{cf}(\kappa);$$

hence $\delta^{\kappa_{\eta}} < \alpha_{\bar{\sigma}}$ for all $\eta < \mathrm{cf}(\kappa)$ and we have

$$\delta^{\underline{\kappa}} \leq \sum_{\eta < \mathrm{cf}(\kappa)} \delta^{\kappa_{\eta}} \leq \alpha_{\bar{\sigma}} \cdot \mathrm{cf}(\kappa) < \alpha.$$

The claim is established in all cases. We set $\gamma = ((\beta^{\underline{\kappa}})^{\underline{\kappa}})^{+}$. It follows from the claim (and the fact that α is a limit cardinal) that $\beta < \gamma < \alpha$;

and for $\delta < \gamma, \lambda < \kappa$ we have

$$\delta^\lambda \leq ((\beta^{\underset{\sim}{\kappa}})^{\underset{\sim}{\kappa}})^\lambda \leq ((\beta^{\underset{\sim}{\kappa}})^{\underset{\sim}{\kappa}})^{\underset{\sim}{\kappa}} = (\beta^{\underset{\sim}{\kappa}})^{\underset{\sim}{\kappa}} < \gamma$$

from Theorem A.3(c) and hence $\kappa \ll \gamma$.

A.7 Theorem. Let α be an (infinite) singular cardinal and $\kappa \ll \alpha$. If $\kappa = \omega$, or κ is a successor cardinal, or $\mathrm{cf}(\kappa) \neq \mathrm{cf}(\alpha)$, or α is a strong limit cardinal, then there is a set $\{\alpha_\sigma : \sigma < \mathrm{cf}(\alpha)\}$ such that

(i) α_σ is a regular cardinal for $\sigma < \mathrm{cf}(\alpha)$,

(ii) $\kappa \ll \alpha_\sigma$ for $\sigma < \mathrm{cf}(\alpha)$,

(iii) $\mathrm{cf}(\alpha) < \alpha_{\sigma'} < \alpha_\sigma < \alpha$ for $\sigma' < \sigma < \mathrm{cf}(\alpha)$, and

(iv) $\alpha = \sum_{\sigma < \mathrm{cf}(\alpha)} \alpha_\sigma$.

Proof. Since α is singular there is by Theorem A.1 a set $\{\gamma_\sigma : \sigma < \mathrm{cf}(\alpha)\}$ of cardinals with properties (analogous to) (iii) and (iv). We set $\beta_0 = \gamma_0$ and recursively, if $\sigma < \mathrm{cf}(\alpha)$ and $\alpha_{\sigma'}$ has been defined for all $\sigma' < \sigma$, we set

$$\beta_\sigma = \max\{(\sup\{\alpha_{\sigma'} : \sigma' < \sigma\})^+, \gamma_\sigma\}.$$

By Corollary A.6 there is α_σ satisfying (i) and (ii) and the condition $\beta_\sigma < \alpha_\sigma < \alpha$. It is clear that the family $\{\alpha_\sigma : \sigma < \mathrm{cf}(\alpha)\}$ is as required.

We note the following partial converse to Theorem A.7: if α is a singular cardinal and $\{\alpha_\sigma : \sigma < \mathrm{cf}(\alpha)\}$ a set of cardinals satisfying conditions (ii) and (iv) of Theorem A.7, then $\kappa \ll \alpha$. Indeed for $\beta < \alpha$ there is $\sigma < \mathrm{cf}(\alpha)$ such that $\beta < \alpha_\sigma$, and since $\kappa \ll \alpha_\sigma$ we have

$$\beta^\lambda < \alpha_\sigma < \alpha \text{ for } \lambda < \kappa.$$

Definitions. For a partially ordered set $\langle A, \leq \rangle$ and $a \in A$ we denote by $P(a)$ the set of (strict) predecessors of a; that is, $P(a) = \{b \in A : b < a\}$. A *tree* is a partially ordered set $\langle T, \preccurlyeq \rangle$ such that for $s \in T$ the set $P(s)$ is well-ordered by $\prec$. For $s \in T$ the unique ordinal number ξ such that $\langle \xi, \in \rangle$ is order-isomorphic to $\langle P(s), \prec \rangle$ is the *order* of s. We write

$$T_\xi = \{s \in T : \text{order of } s \text{ is } \xi\};$$

the elements of T_ξ are at *level* ξ. The *height* of T is $\sup\{\xi : T_\xi \neq \varnothing\}$. A *branch* of T is a well-ordered subset Σ of T such that if $s \in \Sigma$ then $P(s) \subset \Sigma$. Thus a branch is a (full, not necessarily maximal) well-ordered subset of T.

Definition. A cardinal α is a *weakly compact* cardinal if

(a) α is a strongly inaccessible (i.e., α is a regular strong limit cardinal) and

(b) for every tree $\langle T, \preccurlyeq \rangle$ of height α such that $|T_\xi| < \alpha$ for all $\xi < \alpha$, there is a branch Σ of T such that $|\Sigma| = \alpha$.

The following result, known as Tarski's recursion formula, is from Tarski (1925).

Theorem. Let α and κ be cardinals with α a limit cardinal and $\kappa > 0$, and let ξ be a limit ordinal such that $\kappa < \mathrm{cf}(\xi)$. Let $\{\alpha_\sigma : \sigma < \xi\}$ be a set of cardinals such that

$$\alpha_{\sigma'} < \alpha_\sigma < \alpha \text{ for } \sigma' < \sigma < \xi, \text{ and}$$

$$\sum_{\sigma < \xi} \alpha_\sigma = \alpha.$$

Then $\alpha^\kappa = \sum_{\sigma < \xi} \alpha_\sigma^\kappa$.

We omit the proof.

The *continuum hypothesis* (CH) is the statement $\omega^+ = 2^\omega$; the *generalized continuum hypothesis* (GCH) is the statement $\alpha^+ = 2^\alpha$ for every infinite cardinal α.

From the generalized continuum hypothesis it follows that every limit cardinal is a strong limit cardinal. We note some other consequences of GCH.

A.8 Theorem. Assume the generalized continuum hypothesis. Let α be an infinite cardinal.

(a) $\mathrm{cf}(\alpha)$ is the least cardinal κ such that $\alpha^\kappa > \alpha$;

(b) $\alpha^{\underset{\smile}{\mathrm{cf}(\alpha)}} = \alpha$; and

(c) if α is regular then $\alpha \ll \alpha^+$ (hence $\alpha^+ \ll \alpha^{++}$).

Proof. (a) Let $\kappa < \mathrm{cf}(\alpha)$. If α is the successor cardinal β^+ then

$$\alpha^\kappa = (2^\beta)^\kappa = 2^\beta = \alpha.$$

If α is a limit cardinal, there is a set $\{\alpha_\sigma : \sigma < \mathrm{cf}(\alpha)\}$ of cardinals such that

$$\kappa < \alpha_{\sigma'} < \alpha_\sigma < \alpha \text{ for } \sigma' < \sigma < \mathrm{cf}(\alpha), \text{ and}$$

$$\sum_{\sigma < \mathrm{cf}(\alpha)} \alpha_\sigma = \alpha,$$

and from Tarski's recursion formula we have

$$\alpha^\kappa = \sum_{\sigma < \mathrm{cf}(\alpha)} \alpha_\sigma^\kappa \le \sum_{\sigma < \mathrm{cf}(\alpha)} 2^{\alpha_\sigma} \le \alpha \cdot \mathrm{cf}(\alpha) = \alpha.$$

(b) From (a) we have

$$\alpha^{\underset{\smile}{\mathrm{cf}(\alpha)}} = \sum_{\kappa < \mathrm{cf}(\alpha)} \alpha^\kappa = \sum_{\kappa < \mathrm{cf}(\alpha)} \alpha = \alpha \cdot \mathrm{cf}(\alpha) = \alpha.$$

(c) For $\beta < \alpha^+$ and $\kappa < \alpha = \mathrm{cf}(\alpha)$ we have

$$\beta^\kappa \le \alpha^\kappa = \alpha < \alpha^+.$$

Most of the material of A.1–A.8 appears in extended form in Bachmann (1955) and in Comfort & Negrepontis (1974, section 1). The remainder of Appendix A, needed in 5.22–5.26, is unpublished work of Argyros.

Definition. Let α and κ be cardinals with $\alpha \ge \omega$ and $\kappa > 1$. Then

$$\sqrt[\kappa]{\alpha} = \min\{\beta : \beta^\kappa \ge \alpha\};$$

if in addition $\kappa > 2$, then

$$\sqrt[\underset{\smile}{\kappa}]{\alpha} = \min\{\sqrt[\lambda]{\alpha} : 1 < \lambda < \kappa\}.$$

We note that if $1 < \lambda \le \kappa$ then $2 \le \sqrt[\underset{\smile}{\kappa}]{\alpha} \le \sqrt[\lambda]{\alpha}$.

A.9 Lemma. Let $\alpha \ge \beta \ge \omega$ and $\sqrt[\kappa]{\alpha} \le \beta$ with $\kappa > 1$. Then $\sqrt[\underset{\smile}{\kappa}]{\beta} = \sqrt[\underset{\smile}{\kappa}]{\alpha}$.

Proof. It is clear that $\sqrt[\underset{\smile}{\kappa}]{\beta} \le \sqrt[\underset{\smile}{\kappa}]{\alpha}$. If $\sqrt[\underset{\smile}{\kappa}]{\beta} < \sqrt[\underset{\smile}{\kappa}]{\alpha}$ we have $\kappa \ge \omega$ and

$$\sqrt[\kappa]{\alpha} \le \beta \le (\sqrt[\underset{\smile}{\kappa}]{\beta})^\kappa < \alpha;$$

hence

$$(\sqrt[\underset{\smile}{\kappa}]{\alpha})^\kappa \le ((\sqrt[\underset{\smile}{\kappa}]{\beta})^\kappa)^\kappa = (\sqrt[\underset{\smile}{\kappa}]{\beta})^\kappa < \alpha,$$

a contradiction.

A.10 Theorem. Let $\alpha \ge \omega, \kappa > 1$ and $\sqrt[\underset{\smile}{\kappa}]{\alpha} < \alpha$. Then either

$$\sqrt[\kappa]{\alpha} = 2, \text{ or}$$

$$\mathrm{cf}(\sqrt[\underset{\smile}{\kappa}]{\alpha}) \le \kappa < \sqrt[\underset{\smile}{\kappa}]{\alpha}.$$

Proof. We suppose that $\sqrt[\underset{\smile}{\kappa}]{\alpha} > 2$. Then $2^\kappa < \alpha$. If $\kappa < \omega$ then $\kappa^\kappa < \omega \le \alpha$ and if $\kappa \ge \omega$ then $\kappa^\kappa = 2^\kappa < \alpha$; thus $\kappa < \sqrt[\underset{\smile}{\kappa}]{\alpha}$.

It remains to show that $\mathrm{cf}(\sqrt[\underset{\smile}{\kappa}]{\alpha}) \le \kappa$. We suppose that $\mathrm{cf}(\sqrt[\underset{\smile}{\kappa}]{\alpha}) > \kappa$, we choose a regular cardinal β such that

$$\sqrt[\kappa]{\alpha} < \beta \leq \alpha$$

(e.g., we set $\beta = (\sqrt[\kappa]{\alpha})^+$), we note from Lemma A.9 that $\sqrt[\kappa]{\alpha} = \sqrt[\kappa]{\beta}$, and we note that since

$$\beta \leq \alpha \leq (\sqrt[\kappa]{\alpha})^\kappa$$

there is a one-to-one function f from β into $(\sqrt[\kappa]{\alpha})^\kappa$. For $\xi < \beta$ the function $f(\xi)$ is a function from κ into $\sqrt[\kappa]{\alpha}$, and from the assumption $\mathrm{cf}(\sqrt[\kappa]{\alpha}) > \kappa$ we have

$$\sup(f(\xi)[\kappa]) < \sqrt[\kappa]{\alpha} \text{ for } \xi < \beta.$$

Since β is regular and $\sqrt[\kappa]{\alpha} < \beta$, there are $B \subset \alpha$ with $|B| = \beta$ and $\zeta < \sqrt[\kappa]{\alpha}$ such that

$$\sup(f(\xi)[\kappa]) < \zeta \quad \text{for } \xi \in B.$$

For $\xi \in B$ we have $f(\xi)[\kappa] \subset \zeta$ and hence $f(\xi) \in \zeta^\kappa$. Since $f|B$ is a one-to-one function, we have $|B| \leq |\zeta^\kappa|$; it follows that

$$\beta = |B| \leq |\zeta^\kappa| = |\zeta|^\kappa$$

and hence

$$\sqrt[\kappa]{\alpha} = \sqrt[\kappa]{\beta} \leq |\zeta| < \sqrt[\kappa]{\alpha},$$

a contradiction.

A.11. *Theorem.* Let α and κ be regular cardinals with $2 < \sqrt[\underline{\kappa}]{\alpha} < \alpha$ and set $\beta = (\kappa^{\underline{\kappa}})^+$. Then $\kappa \ll \beta$ and $\beta \leq \alpha$.

Proof. We have

$$(\kappa^{\underline{\kappa}})^\lambda \leq (\kappa^{\underline{\kappa}})^\kappa = \kappa^{\underline{\kappa}} < \beta \text{ for } \lambda < \kappa$$

by Theorem A.3(a), and hence $\kappa \ll \beta$.

We show next that $\kappa < \alpha$. From $2 < \sqrt[\underline{\kappa}]{\alpha}$ it follows that

$$2^\lambda < \alpha \text{ for } \lambda < \kappa.$$

We set $\gamma = \sqrt[\underline{\kappa}]{\alpha}$, we choose $\lambda < \kappa$ such that $\gamma = \sqrt[\lambda]{\alpha}$, and we note that if $\gamma < \kappa$ then

$$\gamma^\lambda \leq 2^\gamma \cdot 2^\lambda < \alpha,$$

a contradiction; it follows that

$$\kappa \leq \gamma < \alpha.$$

Now to show $\beta \le \alpha$ we consider two cases.

Case 1. κ is a successor cardinal. If $\kappa = \mu^+$ then

$$\kappa^{\underline{\kappa}} = (\mu^+)^\mu \le (2^\mu)^\mu = 2^\mu < \alpha$$

and hence $\beta \le \alpha$.

Case 2. κ is a limit cardinal. There is a set $\{\kappa_\sigma : \sigma < \kappa\}$ of cardinals such that

$$\kappa_{\sigma'} < \kappa_\sigma < \kappa \text{ for } \sigma' < \sigma < \kappa, \text{ and}$$
$$\kappa = \sum_{\sigma < \kappa} \kappa_\sigma .$$

For $\lambda < \kappa, \sigma < \kappa$ we have

$$(\kappa_\sigma)^\lambda \le 2^{\kappa_\sigma} \cdot 2^\lambda < \alpha$$

and from Tarski's recursion formula (with α, κ, ξ and α_σ replaced by κ, λ, κ and κ_σ, respectively) it follows that

$$\kappa^\lambda = \sum_{\sigma < \kappa} (\kappa_\sigma)^\lambda < \alpha,$$

the final inequality because α is regular and $\kappa < \alpha$.
Hence

$$\kappa^{\underline{\kappa}} = \sum_{\lambda < \kappa} \kappa^\lambda < \alpha$$

and $\beta \le \alpha$.

B. Topology and Boolean algebras

By a *space* in this work we mean a topological space satisfying the usual axioms. We state explicitly when a separation property is assumed or required; where none is mentioned, none is imposed. The topology of a space X, i.e., the set of open subsets of X, is denoted by $\mathscr{T}(X)$ or, when confusion is unlikely, by $\mathscr{T}$. We set

$$\mathscr{T}^*(X) = \{U \in \mathscr{T}(X) : U \ne \varnothing\}.$$

We do not distinguish notationally in this work between a cardinal number α and the discrete space whose underlying set of points is α.

For X a space and $A \subset X$, the closure of A in X is denoted $\bar{A}$ or $\mathrm{cl}_X A$ or $\mathrm{cl}\, A$, and the interior of A in X is denoted A° or int_X or $\mathrm{int}\, A$.

$\mathbb{R}$ denotes the ordered field of real numbers with its usual order topology. For $a, b \in \mathbb{R}$ with $a \leq b$ we set

$$[a, b] = \{x \in \mathbb{R} : a \leq x \leq b\}.$$

For spaces X and Y we denote by $C(X, Y)$ the set of continuous functions from X into Y; we often write $C(X)$ in place of $C(X, \mathbb{R})$.

Definition. Let X be a space.

(a) X is a *Hausdorff* space if for $x, y \in X$ with $x \neq y$ there are $U, V \in \mathscr{T}$ such that $x \in U$, $y \in V$ and $U \cap V = \emptyset$;

(b) X is a *regular* space if for $x \in U \in \mathscr{T}$ there is $V \in \mathscr{T}$ such that $x \in V \subset \bar{V} \subset U$; and

(c) X is *completely regular* if for $x \in U \in \mathscr{T}$ there is $f \in C(X)$ such that $f(x) = 0$ and $f(y) = 1$ for $y \in X \backslash U$.

The space $\mathbb{R}$ and the discrete spaces α are completely regular, Hausdorff spaces.

A non-empty family of sets has the *finite intersection property* if the intersection of every non-empty finite subfamily is non-empty.

A space X is *compact* if every non-empty family of closed subsets of X with the finite intersection property has non-empty intersection (equivalently, if every cover of X by open sets has a finite subcover).

Definitions. Let X be a space.

(a) A *base* for X is a set $\mathscr{B}$ of open subsets of X such that if $U \in \mathscr{T}$ there is $\mathscr{C} \subset \mathscr{B}$ such that $U = \cup \mathscr{C}$. The least cardinal number which is the cardinality of a base for X is the *weight* of X and is denoted $w(X)$.

(b) A *pseudobase* or *π-base* for X is a set $\mathscr{B}$ of non-empty open subsets of X such that if $U \in \mathscr{T}^*$ there is $B \in \mathscr{B}$ such that $B \subset U$. The least cardinal number which is the cardinality of a pseudobase for X is the *pseudoweight* or *π-weight* of X and is denoted $\pi w(X)$.

(c) A subset A of X is *dense* in X if $X = \mathrm{cl}_X A$. The least cardinal number which is the cardinality of a dense subset of X is the *density character* of X and is denoted $d(X)$.

B.1 *Theorem.* Let X be a space. Then

(a) $d(X) \leq \pi w(X) \leq w(X)$; and

(b) if X is a Hausdorff space then $|X| \leq 2^{2^{d(X)}}$.

Proof. (a) is clear.

(b) If A is dense in X then the function $X \to \mathscr{P}(\mathscr{P}(A))$ defined by

$$x \to \{U \cap A : x \in U \in \mathscr{T}\}$$

is a one-to-one function.

It has been proved by Pondiczery (1944) that the product of 2^α copies of the discrete space α has a dense subset of cardinality α; that is, $d(\alpha^{(2^\alpha)}) = \alpha$. The following statement, known often as the Hewitt–Marczewski–Pondiczery theorem, is an immediate consequence. (This was proved by Marczewski (1947) in the case $\alpha = \omega$ and by Hewitt (1946) in general.)

B.2 Theorem. Let $\alpha \geq \omega$ and let $\{X_i : \in I\}$ be a set of spaces with $|I| \leq 2^\alpha$ and with $d(X_i) \leq \alpha$ for $i \in I$. Then $d(\prod_{i \in I} X_i) \leq \alpha$.

We note a converse to Theorem B.2.

B.3 Theorem. Let $\alpha \geq \omega$ and let $\{X_i : i \in I\}$ be a set of spaces with $d(\prod_{i \in I} X_i) \leq \alpha$. If for each i there is a pair of non-empty, disjoint open subsets of X_i (e.g., if X_i is Hausdorff space such that $|X_i| > 1$), then $|I| \leq 2^\alpha$.

Proof. Let $U_i, V_i \in \mathscr{T}^*(X_i)$ with $U_i \cap V_i = \varnothing$ and let D be dense in $\prod_{i \in I} X_i$ with $|D| \leq \alpha$. The function

$$i \mapsto \{p \in D : p_i \in U_i\}$$

is one-to-one from I into $\mathscr{P}(D)$.

A proof of the Hewitt–Marczewski–Pondiczery theorem and of related results concerning κ-box products, using families (of functions) of large oscillation, is given by Comfort & Negrepontis (1972a, Theorem 3.1 ((a) $\Rightarrow$ (f)); 1974, Corollary 3.18 and Theorem 3.20).

Definitions. A subset A of a space X is *nowhere dense* (in X) if $\text{int}_X \, \text{cl}_X A = \varnothing$; A is of *Category* I if it is possible to write A in the form $A = \cup_{n<\omega} A_n$ with A_n nowhere dense for $n < \omega$, and of *Category* II otherwise.

A *Lusin set* of $\mathbb{R}$ is a subset A of $\mathbb{R}$ such that every uncountable subset of A is of Category II in $\mathbb{R}$.

The classical Baire category theorem is the statement that $\mathbb{R}$ is of

Category II (cf. Baire (1899) and Hausdorff (1914, Kapitel VIII, section 9. VIII)). The following consequence is due to Lusin (1914) (see also Oxtoby (1971, Proposition 20.1)).

B.4 Theorem. Assume the continuum hypothesis. There is a Lusin set $A \subset \mathbb{R}$ with $|A| = \omega^+$.

Proof. Let $\{A_\xi : \xi < \omega^+\}$ be an enumeration of the set of closed, nowhere dense subsets of $\mathbb{R}$. Let $x_0 \in \mathbb{R}$ and recursively, using the Baire category theorem at each stage, for $0 < \xi < \omega^+$ choose

$$x_\xi \in \mathbb{R} \backslash \cup_{\zeta < \xi} (A_\zeta \cup \{x_\zeta\})$$

and set $A = \{x_\xi : \xi < \omega^+\}$. It is clear that no uncountable subset of A is (contained in) the union of countably many of the sets A_ξ.

Definition. Let $\alpha \geq \omega$ and let X be a space.

(a) X is a *P_α-space* if the intersection of every family of less than α open subsets of X is open in X; and

(b) $X_{(\alpha)}$ is the space with underlying set equal to the underlying set of X and with the smallest topology $\mathscr{T}(X_{(\alpha)})$ such that

$$\mathscr{T}(X) \subset \mathscr{T}(X_{(\alpha)}) \text{ and}$$
$$X_{(\alpha)} \text{ is a } P_\alpha\text{-space.}$$

We note several simple consequences of this definition.

(1) $X_{(\omega)} = X$;

(2) $X = X_{(\alpha)}$ if and only if X is a P_α-space;

(3) $(X_{(\alpha)})_{(\alpha)} = X_{(\alpha)}$;

(4) if α is a regular cardinal and $\mathscr{B}$ is a base for (the topology of) X, then $\{\cap \mathscr{U} : \mathscr{U} \subset \mathscr{B}, |\mathscr{U}| < \alpha\}$ is a base for (the topology of) $X_{(\alpha)}$;

(5) if α is a singular cardinal then $X_{(\alpha^+)} = X_{(\alpha)}$;

(6) if X is a Hausdorff space then $X_{(\alpha)}$ is a Hausdorff space; and

(7) if X is a completely regular space then $X_{(\alpha)}$ is a completely regular space.

(We comment on the proof of (7). By (5), it is enough to consider the case in which α is regular. For $\alpha = \omega$ we have $X_{(\alpha)} = X$, and for $\alpha > \omega$ the space $X_{(\alpha)}$ has a base of open-and-closed sets. Indeed $\{f^{-1}(\{0\}) : f \in C(X)\}$ is such a base for $X_{(\omega^+)}$ and the statement for general (regular) $\alpha > \omega$ then follows from (4) and the relation $X_{(\alpha)} = (X_{(\omega^+)})_{(\alpha)}$.)

It follows from the Tychonoff product theorem and an embedding theorem of Tychonoff (1929) that a completely regular, Hausdorff space X is compact if and only if X is (homeomorphic to) a closed subspace of a power of the space $[0, 1]$. This result motivates the following definitions.

Definition. Let X be a completely regular, Hausdorff space.

(a) X is *realcompact* if X is (homeomorphic to) a closed subspace of a power of the space $\mathbb{R}$;

(b) X is *topologically complete* if X is homeomorphic to a closed subspace of a product of metrizable spaces.

It is clear that every compact space is realcompact and that every realcompact space is topologically complete.

Hewitt (1948) introduced and intensively studied realcompact spaces; his original definition concerned algebraic properties of the quotient fields $C(X)/M$ with M a maximal ideal of the ring $C(X)$. Much of the theory was developed independently (but not published) by L. Nachbin within the framework of the theory of uniform spaces; see Hewitt (1953), Nachbin (1952, 1954), and Gillman & Jerison (1960, Chapter 8; Notes to Chapter 8).

Responding to a question posed in the fundamental memoire of Weil (1937), Dieudonné (1939, p. 285) proved that a uniformizable space has a compatible complete uniformity if and only if it is (in our terminology) topologically complete.

Let $\mathbf{C}$ be a class of spaces and let X and Y be spaces with $X \subset Y$. Then X is said to be $\mathbf{C}$-*embedded* in Y if for every $f \in C(X, Z)$ with $Z \in \mathbf{C}$ there is $g \in C(Y, Z)$ such that $g | X = f$.

Definition. Let $\mathbf{C}$ be a class of spaces and let X and Y be spaces with X a $\mathbf{C}$-*embedded* subspace of Y.

(a) If $\mathbf{C} = \{[0, 1]\}$ then X is C^*-*embedded* in Y;

(b) if $\mathbf{C} = \{\mathbb{R}\}$ then X is *C-embedded in* Y; and

(c) if $\mathbf{C}$ is the class of metric spaces then X is *M-embedded* in Y.

Part (a) of the following theorem is due to Tychonoff (1929), Stone (1937) and Čech (1937); part (b) is due essentially to Hewitt (1948); part (c) (like (a) and (b)) is a special case of a result of van der Slot (1966) and Herrlich (1967) related to work of Freyd (1964, exercise 3J, 'The adjoint functor theorem'), Isbell (1964, I. 25–27), Kennison (1965), Herrlich & van der Slot (1967), Herrlich (1969) and Franklin (1971).

The Stone–Čech compactification and many of its subspaces are considered in detail by Gillman & Jerison (1960), Comfort & Negrepontis (1974) and Walker (1974). For a careful proof of Theorem B.5 and related topics see Comfort & Negrepontis (1975).

B.5 Theorem. Let X be a completely regular, Hausdorff space.

(a) There is a unique compact space, denoted βX and called the *Stone–Čech compactification* of X, in which X is dense and C^*-embedded; further, X is **C**-embedded in βX with **C** the class of compact, Hausdorff spaces.

(b) There is a unique realcompact space, denoted υX and called the *Hewitt realcompactification* of X, in which X is dense and C-embedded; further, X is **C**-embedded in υX with **C** the class of realcompact spaces.

(c) There is a unique topologically complete space, denoted γX and called the *Dieudonné topological completion* of X, in which X is dense and M-embedded; further, X is **C**-embedded in γX with **C** the class of topologically complete spaces.

The Stone–Čech compactification βX of X is unique in the following sense: if $B(X)$ is a compact space in which X is dense and C^*-embedded, then there is a homeomorphism h of βX onto $B(X)$ such that $h(x) = x$ for all $x \in X$. The spaces υX and γX of Theorem B.4(b) and (c) are analogously unique.

If $f \in C(X, Y)$ with X and Y completely regular, Hausdorff spaces, the unique $g \in C(\beta X, \beta Y)$ such that $g|X = f$ is called the *Stone extension* of f and is denoted $\bar{f}$.

The following result indicates that the spaces υX and γX are (naturally identified with) subspaces of βX.

B.6 Theorem. Let X be a completely regular, Hausdorff space. Then

$$\begin{aligned}\upsilon X &= \cap\{X' : X \subset X' \subset \beta X \text{ and } X' \text{ is realcompact}\}\\ &= \{p \in \beta X : X \text{ is } C\text{-embedded in } X \cup \{p\}\},\end{aligned}$$

and

$$\gamma X = \cap\{X' : X \subset X' \subset \beta X \text{ and } X' \text{ is topologically complete}\}$$
$$= \{p \in \beta X : X \text{ is } M\text{-embedded in } X \cup \{p\}\}.$$

We record several facts about the Stone–Čech compactification $\beta(\alpha)$ of the (discrete) space α. For less brief statements and for the proofs omitted here, see Gillman & Jerison (1960), Comfort & Negrepontis (1974), or Walker (1974).

(a) The set $\beta(\alpha)$ is (identified with) the set of ultrafilters on α;

(b) the inclusion $\alpha \subset \beta(\alpha)$ is achieved by identifying $\xi \in \alpha$ with the ultrafilter $\{A \subset \alpha : \xi \in A\}$;

(c) for $A \subset \alpha$ the set $\hat{A}$, defined by the rule

$$\hat{A} = \{p \in \beta(\alpha) : A \in p\},$$

is closed in $\beta(\alpha)$, and the topology of $\beta(\alpha)$ is the smallest in which each set $\hat{A}$ (with $A \subset \alpha$) is closed.

We note from (c) that $\mathrm{cl}_{\beta(\alpha)} A = \hat{A}$ for $A \subset \alpha$.

B.7 Lemma. Let $\alpha \geq \omega$.

(a) if $A, B \subset \alpha$ then $\hat{A} \cap \hat{B} = (A \cap B)\hat{}$;

(b) if $A \subset \alpha$ then $\hat{A}$ is open-and-closed in $\beta(\alpha)$; and

(c) $\{\hat{A} : A \subset \alpha\}$ is a base for $\mathscr{T}^*(\beta(\alpha))$.

Proof. (a) Since $\mathrm{cl}_{\beta(\alpha)} A = \hat{A}$ for $A \subset \alpha$, the inclusion $\hat{A} \cap \hat{B} \supset (A \cap B)\hat{}$ is clear. If there is $p \in \beta(\alpha)$ such that

$$p \in (\hat{A} \cap \hat{B}) \backslash (A \cap B)\hat{},$$

then we note that $p \in (A \backslash B)\hat{} \cap (B \backslash A)\hat{}$ and we choose $f : \alpha \to [0, 1]$ such that

$$f(\xi) = 0 \text{ if } \xi \in A \backslash B$$
$$= 1 \text{ if } \xi \in B \backslash A.$$

The (continuous) Stone extension $\bar{f}$ of f satisfies

$$\bar{f}[\hat{A}] = \{0\} \text{ and } \bar{f}[\hat{B}] = \{1\},$$

and we have $\bar{f}(p) = 0$ and $\bar{f}(p) = 1$, a contradiction.

(b) Set $B = \alpha \backslash A$. Since $\alpha = A \cup B$ we have

$$\beta(\alpha) = \hat{\alpha} = \hat{A} \cup \hat{B},$$

and since $\hat{A} \cap \hat{B} = \varnothing$ (from part (a)) we have

$$\hat{A} = \beta(\alpha) \backslash \hat{B}.$$

Since $\hat{B}$ is closed, the (closed) set $\hat{A}$ is open.

(c) If U is open in $\beta(\alpha)$ and $p \in U$, there is $B \subset \alpha$ such that $\hat{B} \supset \beta(\alpha) \backslash U$ and $p \notin \hat{B}$. Then with $A = \alpha \backslash B$ we have $\hat{A} \in \mathscr{T}^*(\beta(\alpha))$ from part (b), and

$$p \in \hat{A} \subset U,$$

as required.

We note that $\beta(A) = \hat{A}$ for $A \subset \alpha$.

Definition. Let $\alpha \geq \omega$, $A \subset \alpha$ and $p \in \hat{A}$. The *type of p relative to A*, denoted $T_A(p)$, is the set

$$T_A(p) = \{h(p) : h \text{ is a homeomorphism of } \hat{A} \text{ onto } \hat{A}\}.$$

We write $T(p)$ in place of $T_\alpha(p)$.

The following statements are clear for $\alpha \geq \omega$, $A \subset \alpha$, $\xi \in A$ and $p_0, p_1 \in \hat{A}$:

$T_A(\xi) = A$;

$T_A(p_0) = \{\bar{\tau}(p_0) : \tau \text{ is a permutation of } A\}$;

$|T_A(p_0)| \leq 2^{|A|}$ if $|A| \geq \omega$; and

$T_A(p_0) = T_A(p_1)$ if and only if $T_A(p_0) \cap T_A(p_1) \neq \varnothing$.

B.8 Lemma. Let $\alpha \geq \omega$ and $p, q \in \beta(\alpha)$. The following statements are equivalent.

(a) $q \in T(p)$;

(b) there are $f \in \alpha^\alpha$ and $A \in p$ such that $\bar{f}(p) = q$ and $f|A$ is one-to-one.

Proof. (a) $\Rightarrow$ (b). There is a permutation τ of α such that $\bar{\tau}(p) = q$. We set $f = \tau$ and $A = \alpha$.

(b) $\Rightarrow$ (a). If $p \in \alpha$ then $q = f(p) \in \alpha = T(p)$. If $p \in \beta(\alpha) \backslash \alpha$ then $|A| = |f[A]| \geq \omega$. There is $A_1 \in q$ such that $|\alpha \backslash A_1| = \alpha$ and there is $A_0 \in p$ such that $|\alpha \backslash A_0| = \alpha$, $A_0 \subset A$ and $f[A_0] \subset A_1$. There is then a permutation τ of α such that $\tau | A_0 = f | A_0$, and from $A_0 \in p$ we have

$$\bar{\tau}(p) = \bar{f}(p) = q$$

and hence $q \in T(p)$, as required.

B.9 Lemma. Let $\alpha \geq \omega$, $A \subset \alpha$, and $p, q \in \hat{A}$. If $T_A(p) \cap T_A(q) = \emptyset$ then $T(p) \cap T(q) = \emptyset$.

Proof. If the statement fails there is a permutation τ of α such that $\bar{\tau}(p) = q$. We set $B = A \cap \tau^{-1}(A)$ and we define a function f on A so that

$$\begin{aligned} f(\xi) &= \tau(\xi) \quad \text{for } \xi \in B, \\ &\in A \quad \text{for } \xi \in A \backslash B. \end{aligned}$$

Then $f \in A^A$, $B \in p$, $\bar{f}(p) = q$, and $f | B$ is one-to-one. It follows from Lemma B.8 (with A and B replacing α and A, respectively) that $q \in T_A(p)$. This contradiction completes the proof.

B.10 Corollary. Let $\alpha \geq \omega$, $A \subset \alpha$ and $p \in \hat{A}$. Then $T(p) \cap \hat{A} = T_A(p)$.

Proof. It is clear that $T_A(p) \subset T(p) \cap \hat{A}$. If $q \in T(p) \cap \hat{A}$ then from $q \in T(p) \cap T(q)$ and Lemma B.9 we have $T_A(p) \cap T_A(q) \neq \emptyset$ and hence $q \in T_A(p)$.

Definition. Let $\alpha \geq \omega$, $p \in \beta(\alpha)$ and $A \subset \alpha$.

(a) The *norm of* p, denoted $\|p\|$, is the cardinal number

$$\|p\| = \min\{|B| : B \in p\};$$

(b) p is *uniform on* A if $A \in p$ and $|A| = \|p\|$.

The set of ultrafilters uniform on A is denoted $U(A)$; in particular the set of uniform ultrafilters on α is denoted $U(\alpha)$.

The following fundamental theorem, due essentially to Hausdorff (1936), was first explicitly stated and proved by Pospíšil (1937).

B.11 Theorem. If $\alpha \geq \omega$ then $|U(\alpha)| = 2^{2^\alpha}$.

We omit the proof of this theorem.

B.12 Corollary. Let $\alpha \geq \beta \geq \omega$. There is $\{p_k : k < \omega\} \subset \beta(\alpha)$ such that

$$\begin{aligned} \|p_k\| &= \beta \quad \text{for } k < \omega, \text{ and} \\ T(p_{k'}) \cap T(p_k) &= \emptyset \quad \text{for } k' < k < \omega. \end{aligned}$$

Proof. Choose $A \in [\alpha]^\beta$. Since $|U(A)| = 2^{2^\beta}$ (by Theorem B.11) and $|T_A(p)| \leq 2^\beta$ for $p \in \bar{A}$, there is $\{p_k : k < \omega\} \subset U(A)$ such that

$$T_A(p_{k'}) \cap T_A(p_k) = \varnothing \quad \text{for } k' < k < \omega.$$

From Lemma B.9 we have

$$T(p_{k'}) \cap T(p_k) = \varnothing \quad \text{for } k' < k < \omega,$$

and from $p_k \in U(A)$ we have

$$\|p_k\| = |A| = \beta \quad \text{for } k < \omega,$$

as required.

Definition. Let $\alpha \geq \omega$ and $\kappa \geq \omega$. The space of *sub-κ-uniform ultrafilters* on α, denoted $N_\kappa(\alpha)$, is the space

$$N_\kappa(\alpha) = \{p \in \beta(\alpha) : \|p\| < \kappa\}.$$

We note that following the convention according to which the element η of α is identified with the ultrafilter $\{A \subset \alpha : \eta \in A\}$, we have $\alpha \subset N_\kappa(\alpha)$ for all $\alpha \geq \omega$, $\kappa \geq \omega$. We note further that

$$N_\kappa(\alpha) = \beta(\alpha) \quad \text{if } \kappa > \alpha \geq \omega.$$

It is clear that for infinite cardinal numbers α and κ we have

$$N_\kappa(\alpha) = \cup \{\mathrm{cl}_{\beta(\alpha)} A : A \in \mathscr{P}_\kappa(\alpha)\}.$$

B.13 Lemma. Let α and κ be infinite cardinals and K a compact subspace of $N_\kappa(\alpha)$. Then $|K \cap \alpha| < \kappa$.

Proof. For $p \in K$ there is $A(p) \in \mathscr{P}_\kappa(\alpha)$ such that $p \in \mathrm{cl}_{\beta(\alpha)} A(p)$. Since $\{\mathrm{cl}_{\beta(\alpha)} A(p) : p \in K\}$ is an open cover of K, there is finite $F \subset K$ such that $K \subset \cup_{p \in F} \mathrm{cl}_{\beta(\alpha)} A(p)$, and we have

$$K \cap \alpha \subset \bigcup_{p \in F} (\mathrm{cl}_{\beta(\alpha)} A(p)) \cap \alpha = \bigcup_{p \in F} A(p) \in \mathscr{P}_\kappa(\alpha).$$

Boolean Algebras

We assume that the reader is familiar with the definition and the basic properties of Boolean algebras. Here we list, often without proof, those elements of the theory that are essential to this work; detailed proofs and historical references can be found in Sikorski (1964).

Let B be a Boolean algebra. A subset F of B is a *filter* of B if

$$a \cap b \in F \quad \text{for } a, b \in F, \text{ and}$$
$$b \in F \quad \text{if } a \leq b \text{ and } a \in F.$$

A filter F is *proper* if $0 \notin F$; an *ultrafilter* of B is a proper filter not properly contained in any proper filter of B.

The *Stone space* of B, denoted $S(B)$, is the set of all ultrafilters of B with the (*Stone*) topology determined by the subbase

$$\{\{p \in S(B) : a \in p\} : a \in B\}.$$

This family is in fact a base for $S(B)$, and $S(B)$ is a compact, Hausdorff, totally disconnected space whose Boolean algebra of open-and-closed sets, denoted $\mathscr{B}(S(B))$, is isomorphic as a Boolean algebra to B via the correspondence (the *Stone isomorphism*)

$$\psi : B \rightarrow \mathscr{B}(S(B))$$

given by $\psi(a) = \{p \in S(B) : a \in p\}$.

If X is a space and $\mathscr{A}$ is a base for X then there is $\mathscr{A}' \subset \mathscr{A}$ such that $\mathscr{A}'$ is a base for X and $|\mathscr{A}'| = w(X)$. In particular there is $\mathscr{B}' \subset \mathscr{B}(S(B))$ such that $\mathscr{B}'$ is a base for $S(B)$ and $|\mathscr{B}'| = w(S(B))$. Every element of $\mathscr{B}(S(B))$, being open and compact, is the union of a finite subfamily of $\mathscr{B}'$. Thus we have the following simple consequence of Stone's isomorphism theorem.

B.14 *Theorem.* If B is a Boolean algebra then $w(S(B)) \leq |B|$. If in addition $|B| \geq \omega$ then $w(S(B)) = |B|$.

The fundamental duality of Stone is developed in Stone (1934, 1936, 1937).

A Boolean algebra B is *complete* if every subset of B has a supremum in B; and B is *σ-complete* if every countable subset of B has a supremum in B.

A space X is *extremally disconnected* if $\mathrm{cl}_X U$ is open for every open subset U of X; and X is *basically disconnected* if $\mathrm{cl}_X U$ is open for every cozero set U of X.

B.15 *Theorem.* A Boolean algebra B is complete if and only if $S(B)$ is extremally disconnected; and B is σ-complete if and only if $S(B)$ is basically disconnected.

B.16 Theorem. If X is a completely regular, basically disconnected c.c.c. space, then X is extremally disconnected.

Proof. Let U be a (non-empty) open subset of X. Let $\mathscr{V}$ be a family of cozero sets of X such that

(i) the elements of $\mathscr{V}$ are subsets of U,

(ii) the elements of $\mathscr{V}$ are pairwise disjoint, and

(iii) $\mathscr{V}$ is maximal with respect to (i) and (ii);

and set $V = \cup \mathscr{V}$. Since X is a c.c.c. space we have $|\mathscr{V}| \leq \omega$, and hence V is a cozero set of X. It follows from (i) and (iii) that V is dense in U; hence $\text{cl}_X U$ is equal to the (open) set $\text{cl}_X V$.

Definition. A subset U of a space X is *regular-open* if $U = \text{int}_X \text{cl}_X U$ (equivalently, if there is an open subset V of X such that $U = \text{int}_X \text{cl}_X V$). The set of regular-open subsets of X is denoted $\mathscr{R}(X)$, and we set

$$\mathscr{R}^*(X) = \{U \in \mathscr{R}(X) : U \neq \varnothing\}.$$

B.17 Theorem. Let X be a space. Relative to the operations $\wedge$, $\vee$, and $'$ defined by

$$U \wedge V = U \cap V,$$
$$U \vee V = \text{int}_X \text{cl}_X (U \cup V), \text{ and}$$
$$U' = \text{int}_X (X \setminus U),$$

the set $\mathscr{R}(X)$ is a complete Boolean algebra; for $\mathscr{A} \subset \mathscr{R}(X)$ the supremum of $\mathscr{A}$ in $\mathscr{R}(X)$, denoted $\vee \mathscr{A}$, is given by

$$\vee \mathscr{A} = \text{int}_X \text{cl}_X (\cup \mathscr{A}).$$

Definition. Let X be a space. The *Gleason space* of X, denoted $G(X)$, is the Stone space of the Boolean algebra of regular-open subsets of X; i.e., $G(X) = S(\mathscr{R}(X))$.

Thus for a space X, the Gleason space $G(X)$ of X is a compact, Hausdorff, extremally disconnected space; and the set $\mathscr{R}^*(X)$ is identified, via the Stone representation of $\mathscr{R}(X)$, with a base for $G(X)$.

If Y is a space and X is dense in Y, then the function defined on $\mathscr{R}(Y)$ by the rule

$$U \to U \cap X$$

is a one-to-one function (in fact, a Boolean algebra isomorphism) from $\mathscr{R}(Y)$ onto $\mathscr{R}(X)$. It follows that the Gleason spaces of X and Y are naturally homeomorphic; we have in particular this simple result.

B.18 *Theorem.* If X is a completely regular, Hausdorff space then the spaces $G(X)$ and $G(\beta X)$ are homeomorphic.

A continuous function f from a space Y onto a space X is said to be *irreducible* if there is no closed subset K of Y such that $K \neq Y$ and $f[K] = X$.

If X is a compact Hausdorff space and

$$p \in G(X) = S(\mathscr{R}(X)),$$

then it is easily seen that there is a unique element q of X such that

$$\{A \in \mathscr{R}(X) : q \in A\} \subset p.$$

The *canonical projection* $\pi : G(X) \to X$ is thus well-defined by the rule

$$\{\pi(p)\} = \cap \{\mathrm{cl}_X A : A \in p\}.$$

For $A \in \mathscr{R}(X)$ we have $\pi[\psi(A)] \subset \mathrm{cl}_X A$ and hence π is continuous. The following statement is now easily proved.

B.19 *Theorem.* Let X be a compact, Hausdorff space. The canonical projection $\pi : G(X) \to X$ is a continuous, irreducible function from $G(X)$ onto X.

B.20 *Corollary.* If X is a compact, Hausdorff space then $d(X) = d(G(X))$.

Proof. If D is dense in $G(X)$, then $\pi[D]$ is dense in X; hence $d(X) \leq d(G(X))$. For the reverse inequality let E be dense in X with $|E| = d(X)$, let $A \subset G(X)$ satisfy

$$|A \cap \pi^{-1}(\{x\})| = 1 \quad \text{for } x \in E,$$

and set $K = \mathrm{cl}_{G(X)} A$; since π is irreducible and $\pi[K] = X$, we have $K = G(X)$ and hence

$$d(G(X)) \leq |A| = d(X).$$

For compact Hausdorff spaces X the Gleason space $G(X)$ is defined (directly, by a method dual to the method used above) by

Gleason (1958). Its role as a projective space (in the category of compact Hausdorff spaces and continuous functions), based on work of Stone, MacNeille and Sikorski and described explicitly by Gleason (1958) and Rainwater (1959), is summarized and discussed by Comfort & Negrepontis (1974, section 2.33–2.41).

The following lemma is used in Theorem 2.20 to establish the relation $X \equiv G(X)$.

B.21 Lemma. Let X be a space, and define

$$\varphi^* : \mathcal{T}^*(X) \to \mathcal{R}^*(X) \subset \mathcal{T}^*(G(X))$$

by the rule $\varphi(U) = \text{int}_X \text{cl}_X U$.

Then φ^* is a function from $\mathcal{T}^*(X)$ onto $\mathcal{R}^*(X)$, and if $n < \omega$ and $\{U_k : k < n\} \subset \mathcal{T}^*(X)$, then $\cap_{k<n} U_k = \varnothing$ if and only if $\cap_{k<n} \varphi(U_k) = \varnothing$.

Proof. For $V \in \mathcal{R}^*(X) \subset \mathcal{T}^*(X)$, we have $\varphi(V) = V$ and hence φ is onto $\mathcal{R}^*(X)$.

For $\{U_k : k < n\} \subset \mathcal{T}^*(X)$, we have

$$\text{int}_X \text{cl}_X (\bigcap_{k<n} U_k) = \bigcap_{k<n} (\text{int}_X \text{cl}_X U_k);$$

it follows that if $\cap_{k<n} U_k = \varnothing$, then

$$\varnothing = \text{int}_X \text{cl}_X (\bigcap_{k<n} U_k) = \bigcap_{k<n} (\text{int}_X \text{cl}_X U_k) = \bigcap_{k<n} \varphi(U_k),$$

and conversely.

C. Measure theory

Definitions. Let X be a set. A non-empty family $\mathcal{S}$ of subsets of X is an *algebra* (*on* X) if $\mathcal{S}$ is closed under the operations of finite union, finite intersection, and complementation; $\mathcal{S}$ is a *σ-algebra* (*on* X) if $\mathcal{S}$ is an algebra closed under countable intersections.

If $\mathcal{A} \subset \mathcal{P}(X)$, then there is a smallest σ-algebra on X containing $\mathcal{A}$; this is called the σ-algebra *generated by* $\mathcal{A}$.

Let $\mathcal{S}$ be an algebra on X. A *finitely additive* (*positive*) *measure* μ on $\mathcal{S}$ is a set function

$$\mu : \mathcal{S} \to \mathbb{R} \cup \{\infty\}$$

such that

$\mu(\emptyset) = 0$,

$\mu(A) \geq 0$ for $A \in \mathscr{S}$, and

$\mu(A \cup B) = \mu(A) + \mu(B)$ for $A, B \in \mathscr{S}$ with $A \cap B = \emptyset$.

A finitely additive measure μ on $\mathscr{S}$ is called *σ-additive* (or, *countably additive*, or simply a *measure*) on $\mathscr{S}$ if $\mu(\cup_{n<\omega} A_n) = \sum_{n<\omega} \mu(A_n)$ for all sequences $\{A_n : n < \omega\}$ such that $A_n \in \mathscr{S}$ for $n < \omega$, $A_n \cap A_m = \emptyset$ for $n < m < \omega$, *and* $\cup_{n<\omega} A_n \in \mathscr{S}$.

The triple $(X, \mathscr{S}, \mu)$ is a *measure space* if μ is a measure on the σ-algebra $\mathscr{S}$ of subsets of X. The measure space $(X, \mathscr{S}, \mu)$ is *finite* if $\mu(X) < \infty$. If $\mu(X) = 1$ then $(X, \mathscr{S}, \mu)$ is a *probability measure space* (and μ is a *probability measure*).

C.1 *Theorem.* Let $(X, \mathscr{S}, \mu)$ be a measure space and let $\{S_n : n < \omega\} \subset \mathscr{S}$ with $\mathscr{S}_{n+1} \subset \mathscr{S}_n$ for $n < \omega$ and $\mu(\mathscr{S}_0) < \infty$. Then $\lim \mu(S_n)$ exists and is equal to $\mu(\cap_{n<\omega} S_n)$.

(For a proof see section 10.15 of Hewitt & Stromberg (1965).)

C.2 *Theorem.* Let $\{(X_i, \mathscr{S}_i, \mu_i) : i \in I\}$ be a non-empty family of probability measure spaces, and set $X = \prod_{i \in I} X_i$. Let $\mathscr{S}$ be the σ-algebra on X generated by the family

$$\{\pi_i^{-1}(S) : S \in \mathscr{S}_i, i \in I\}.$$

Then there is a unique probability measure μ on $\mathscr{S}$ such that

$$\mu(\pi_{i_1}^{-1}(S_{i_1}) \cap \ldots \cap \pi_{i_n}^{-1}(S_{i_n})) = \mu_{i_1}(S_{i_1}) \cdot \ldots \cdot \mu_{i_n}(S_{i_n})$$

for every faithfully indexed finite set $\{i_1, \ldots, i_n\} \subset I$ and $S_{i_k} \in \mathscr{S}_{i_k}$ $(1 \leq k \leq n)$. The measure μ is the *product measure* on X.

(A proof of this theorem can be found in section 22 of the Hewitt & Stromberg text (1965).)

C.3 *Theorem.* Let $(X, \mathscr{S}, \mu)$ be a measure space. Then there are a σ-algebra $\mathscr{M}_\mu$, the σ-algebra of *μ-measurable* sets, with $\mathscr{S} \subset \mathscr{M}_\mu$, and a measure $\bar{\mu}$ on $\mathscr{M}_\mu$, the *Carathéodory completion of* μ, such that

$\bar{\mu} | \mathscr{S} = \mu$, and

if $A \in \mathscr{M}_\mu$, $\bar{\mu}(A) = 0$ and $B \subset A$, then $B \in \mathscr{M}_\mu$.

(For a proof see Theorem 11.21 of Hewitt & Stromberg (1965).)

We denote by $\mathcal{N}_\mu$ the family of all $\bar{\mu}$-*null* sets, i.e., the sets A with $A \in \mathcal{M}_\mu$ and $\bar{\mu}(A) = 0$; then $\mathcal{N}_\mu$ is a σ-complete ideal of the (σ-complete) Boolean algebra $\mathcal{M}_\mu$.

Definitions. Let X be a space. A *Borel set of* X is an element of the σ-algebra generated by the family of open subsets of X; a *Baire set of* X is an element of the σ-algebra generated by the family of cozero sets of X.

A *Borel measure* (*on* X) is a measure defined on the Borel sets of X; a *Baire measure* (*on* X) is a measure defined on the Baire sets of X.

Let μ be a measure defined on a σ-algebra $\mathcal{S}$ containing the Borel sets of a space X. The measure μ is called *regular* if

$$\mu(A) = \sup\{\mu(K) : K \subset A, K \text{ compact}\}$$

for all $A \in \mathcal{S}$.

C.4 Theorem. Every Baire measure on a compact Hausdorff space X extends to a unique regular Borel measure on X, whose Carathéodory completion is also a regular measure.

(A proof of this theorem can be found in section 54 of the Halmos text (1950), or on p. 214 of the Berberian text (1965).)

C.5 Theorem. (The Hopf extension theorem). Let X be a set, $\mathcal{A}$ an algebra of subsets of X, and $\mu : \mathcal{A} \to \mathbb{R}$ a set function such that

$$0 \leq \mu(A) < \infty \quad \text{for } A \in \mathcal{A},$$

$$\mu(\emptyset) = 0,$$

$$\mu(A \cup B) = \mu(A) + \mu(B) \quad \text{if } A, B \in \mathcal{A} \text{ and } A \cap B = \emptyset, \text{ and}$$

$$\mu\Big(\bigcup_{n<\omega} A_n\Big) = \sum_{n<\omega} \mu(A_n) \quad \text{for all pairwise disjoint sequences}$$

$\{A_n : n < \omega\}$ such that $A_n \in \mathcal{A}$ for $n < \omega$ and $\bigcup_{n<\omega} A_n \in \mathcal{A}$.

Then there is a unique σ-additive measure ν defined on the σ-algebra on X generated by $\mathcal{A}$ such that

$$\nu(A) = \mu(A) \quad \text{for all } A \in \mathcal{A}.$$

(For a proof see sections 10.36 and 10.39 of Hewitt & Stromberg (1965).)

Definition. A *measure algebra* is an ordered pair $(\mathscr{B}, \lambda)$, where $\mathscr{B}$ is a complete Boolean algebra (with Stone space $S(\mathscr{B}) = \Omega$), and λ is a strictly positive regular Borel measure on Ω that is *normal* (i.e., $\lambda(U) = \lambda(\mathrm{cl}_\Omega U)$ for every open subset U of Ω).

Sometimes when there is no danger of confusion we call $\mathscr{B}$ itself a measure algebra, suppressing explicit mention of the measure λ.

Two measure algebras $(\mathscr{B}, \lambda)$ and $(\mathscr{C}, \nu)$ are *isomorphic* if there is a homeomorphism h of $S(\mathscr{B})$ onto $S(\mathscr{C})$ such that

$$\lambda(h^{-1}(A)) = \nu(A) \quad \text{for all Borel subsets } A \text{ of } S(\mathscr{C}).$$

Definition. A measure space $(X, \mathscr{S}, \mu)$ is *atomless* if for every $S \in \mathscr{S}$ with $\mu(S) > 0$ there is $T \in \mathscr{S}$ with $T \subset S$ such that $0 < \mu(T) < \mu(S)$.

C.6 *Theorem.* Let $(X, \mathscr{S}, \mu)$ be a probability measure space. Then the quotient $\mathscr{M}_\mu / \mathscr{N}_\mu$ is a complete Boolean algebra and there is a unique strictly positive, regular, normal, Borel measure λ on the Stone space Ω of $\mathscr{M}_\mu / \mathscr{N}_\mu$ such that $\lambda[A] = \bar{\mu}(A)$ for all $[A] \in \mathscr{M}_\mu / \mathscr{N}_\mu$.

If in addition $(X, \mathscr{S}, \mu)$ is atomless, then Ω is not a separable space.

(The pair $(\mathscr{M}_\mu / \mathscr{N}_\mu, \lambda)$ is called the *measure algebra of* the probability measure space $(X, \mathscr{S}, \mu)$.)

Proof. We show first, setting $\mathscr{A} = \mathscr{M}_\mu / \mathscr{N}_\mu$, that Ω has such a strictly positive regular Borel measure λ. For $A \in \mathscr{A}$ we choose an element S of the $\mathscr{N}_\mu$-equivalence class determined by A and we define $\nu(A) = \mu(S)$; it is clear then that $\nu: \mathscr{A} \to \mathbb{R}$ is a well-defined finitely additive measure on $\mathscr{A}$. Let $\varphi: \mathscr{A} \to \mathscr{B}(\Omega)$ be the isomorphism given by Stone's duality theorem between $\mathscr{A}$ and the Boolean algebra $\mathscr{B}(\Omega)$ of open-and-closed subsets of Ω, and define

$$\tilde{\nu} = \nu \circ \varphi^{-1} : \mathscr{B}(\Omega) \to \mathbb{R}.$$

Then $\tilde{\nu}$ is a well-defined finitely additive measure on $\mathscr{B}(\Omega)$, and if $\{U_n : n < \omega\}$ is a set of pairwise disjoint, non-empty elements of $\mathscr{B}(\Omega)$ then since Ω is compact we have $\cup_{n<\omega} U_n \notin \mathscr{B}(\Omega)$. It follows that the conditions of Hopf's extension theorem (Theorem C.5) are satisfied and there is a (unique) positive, σ-additive measure λ on the σ-algebra generated by $\mathscr{B}(\Omega)$ – i.e., on the set of Baire sets of

Ω– such that

$$\lambda(U) = \tilde{\nu}(U) = \nu(\varphi^{-1}(U)) \text{ for } U \in \mathscr{B}(\Omega).$$

The unique regular Borel extension of λ (cf. Theorem C.4), also denoted λ, is the required strictly positive measure on Ω. From the existence of λ it follows that Ω is a c.c.c. space.

Since the measure algebra $\mathscr{A}$ is a σ-complete Boolean algebra, the space Ω is basically disconnected. It then follows from Theorem B.15 that Ω is extremally disconnected. That λ is normal is immediate from the σ-additivity of μ.

To prove the last statement of the theorem we show first that $\lambda(\{p\}) = 0$ for all $p \in \Omega$. Indeed, suppose that there is $p \in \Omega$ such that $\lambda(\{p\}) = \delta > 0$. By the regularity of the Borel measure λ there is a sequence $\{B_n : n < \omega\}$ of open-and-closed subsets of Ω such that

$$p \in B_n \text{ and } B_{n+1} \subset B_n \text{ for } n < \omega, \text{ and}$$
$$\lambda(\cap_{n<\omega} B_n) = \delta.$$

With φ the Stone isomorphism as above there is $S_n \in \mathscr{M}_\mu$ such that

$$\varphi[S_n] = B_n \text{ for } n < \omega$$

(with $[S_n]$ the $\mathscr{N}_\mu$-equivalence class of S_n); we assume without loss of generality that $S_n \supset S_{n+1}$ for $n < \omega$. We have

$$\mu(S_n) = \lambda(B_n) \geq \lambda(\{p\}) = \delta \text{ for } n < \omega;$$

it then follows, setting $S = \cap_{n<\omega} S_n$, that $\mu(S) \geq \delta > 0$.

Since μ is atomless there is $T \in \mathscr{S}$ with $T \subset S$ such that $0 < \mu(T) < \delta$. We set $C = \varphi[T]$, so that

$$C \subset B_n \text{ for } n < \omega \text{ and}$$
$$0 < \lambda(C) < \delta.$$

We have either $p \in C$ or $p \in (\cap_{n<\omega} B_n) \backslash C$; in each case we conclude that $\lambda(\{p\}) < \delta$, a contradiction.

We show finally that Ω is not separable. Let $\{p_n : n < \omega\}$ be a countable subset of Ω. Since $\lambda(\{p_n\}) = 0$ for $n < \omega$ there is $\{B_n : n < \omega) \subset \mathscr{B}(\Omega)$ such that

$$p_n \in B_n \text{ and } \lambda(B_n) < 1/2^{n+1} \text{ for } n < \omega.$$

Let $S_n \in \mathscr{S}$ satisfy $\varphi[S_n] = B_n$ for $n < \omega$, and set

$$S = \cup_{n<\omega} S_n, \text{ and } T = X \backslash S, \text{ and } B = \varphi[T].$$

We have

$$\mu(S) \leq \sum_{n<\omega} \mu(S_n) = \sum_{n<\omega} \lambda(B_n) \leq \sum_{n<\omega} 1/2^{n+1} = 1/2$$

and hence $\mu(T) \geq 1/2$. Since $T \cap S_n = \emptyset$ for $n < \omega$ we have $B \cap B_n = \emptyset$ for $n < \omega$. Thus B is a non-empty, open-and-closed subset of Ω such that

$$B \cap \{p_n : n < \omega\} = \emptyset.$$

It follows that $\{p_n : n < \omega\}$ is not dense in Ω.

(That Ω is not a separable space also follows immediately from the fact that the probability measure λ on Ω is a normal measure such that $\lambda(\{p\}) = 0$ for all $p \in \Omega$.)

We note that a strictly positive, regular Borel measure need not be normal. Indeed, we have the following stronger result (cf. Sikorski 1964, section 21, Example F).

C.7 Theorem. There is an extremally disconnected, compact Hausdorff space with a strictly positive regular Borel measure but with no strictly positive normal Borel measure.

Proof. Let $X = G([0, 1])$ and denote by π the canonical projection of X onto $[0, 1]$ (cf. B.18 above). Let λ be a normal Borel measure on X and set $\mu(A) = \lambda(\pi^{-1}(A))$ for A Borel in $[0, 1]$. By a result of Marczewski (1934), there is a Borel set A of $[0, 1]$ such that $\mu(A) = 0$ and $[0, 1] \backslash A$ is of Category I in $[0, 1]$. Since the (continuous) function π is irreducible the Borel set $X \backslash \pi^{-1}(A)$ is of Category I in X, and since λ is a normal Borel measure we have $\lambda(X \backslash \pi^{-1}(A)) = 0$. It follows that

$$\lambda(X) = \lambda(\pi^{-1}(A)) + \lambda(X \backslash \pi^{-1}(A)) = \mu(A) + 0 = 0;$$

hence λ is not a strictly positive measure.

C.8 Definitions and Remarks. When the (discrete) Hausdorff space $\{0, 1\}$ is given the structure of a topological group (using addition mod 2 as the group operation), then for every non-empty index set I the set $\{0, 1\}^I$ is a compact, Abelian topological group. Let ν be the measure defined on the subsets of $\{0, 1\}$ determined by

$$\nu(\{0\}) = \nu(\{1\}) = 1/2$$

and μ the product measure on $\{0, 1\}^I$ (cf. C.2 above). It is then clear, since every open-and-closed subset of $\{0, 1\}^I$ depends on finitely many coordinates and each finite subproduct of $\{0, 1\}^I$ is a finite space, that μ is a strictly positive Baire measure on $\{0, 1\}^I$. We denote also by μ its unique regular Borel extension, and the (regular) Carathéodory completion of this latter (as given in C.4). It is not difficult to see that μ is translation-invariant in the sense that $\mu(A + x) = \mu(A)$ for all measurable subsets A of $\{0, 1\}^I$ and $x \in \{0, 1\}^I$ (where $A + x = \{a + x : a \in A\}$). Thus, μ is the unique *Haar probability measure* on the compact group $\{0, 1\}^I$.

(For details on the definition of Haar measure and its properties, see Hewitt & Ross (1963).)

C.9 Theorem (Maharam's Structure Theorem). Let $(\mathscr{B}, \lambda)$ be a measure algebra with Stone space $S(\mathscr{B}) = \Omega$. Then there are two unique cardinal numbers γ and δ, with $0 \leq \gamma \leq \omega$ and $0 \leq \delta \leq \omega$, a unique sequence $\{\alpha_n : n < \delta\}$ of cardinals such that

$$\omega \leq \alpha_n < \alpha_m \text{ for } n < m < \delta,$$

a unique sequence $\{p_i : i < \gamma\}$ of isolated elements of Ω, and a unique sequence $\{\Omega_n : n < \delta\}$ of non-empty open-and-closed subsets of Ω, such that

$p_i \neq p_j$ for $i < j < \gamma$,
$\{p_i : i < \gamma\} \cap \Omega_n = \varnothing$ for $n < \delta$,
$\Omega_n \cap \Omega_m = \varnothing$ for $n < m < \delta$,
$(\cup_{n<\delta} \Omega_n) \cup \{p_i : i < \gamma\}$ is dense in Ω, and
the measure algebra $(\mathscr{B}(\Omega_n), \lambda_n)$ is isomorphic to the measure algebra of $(\{0, 1\}^{\alpha_n}, \mathscr{B}_n, \mu_n)$ for $n < \delta$.

[Here $\mathscr{B}(\Omega_n)$ is the Boolean algebra of open-and-closed subsets of Ω_n, λ_n is the Borel probability measure on Ω_n defined by $\lambda_n(A) = \lambda(A)/\lambda(\Omega_n)$, $\mathscr{B}_n$ is the family of Borel subsets of $\{0, 1\}^{\alpha_n}$, and μ_n is the Haar probability measure on $\{0, 1\}^{\alpha_n}$.]

(For a proof of this fundamental theorem see the original work of Maharam (1942) or Semadeni (1971, section 26.5).)

Definition. Let X be a compact Hausdorff space. We set $C(X) = \{f : X \to \mathbb{R} : f \text{ is a continuous function}\}$, and

$$\|f\| = \sup\{|f(x)| : x \in X\} \text{ for } f \in C(X).$$

The linear space $C(X)$ with this norm $\|\cdot\|$ (called the *supremum* norm, or the norm of *uniform convergence*) is a Banach space.

A *linear functional* Λ on $C(X)$ is a linear function from $C(X)$ into $\mathbb{R}$; the linear functional Λ is *positive* if

$$\Lambda(f) \geq 0 \text{ for } f \geq 0, f \in C(X).$$

Every positive linear functional Λ on $C(X)$ is bounded (i.e., continuous).

C.10 *Theorem* (Riesz representation theorem). Let X be a compact Hausdorff space, and let Λ be a positive linear functional on $C(X)$ with $\Lambda(1) = 1$. Then there is a unique regular Borel probability measure μ on X such that

$$\Lambda(f) = \int f \, \mathrm{d}\mu \text{ for } f \in C(X).$$

(For a proof see Hewitt & Stromberg (1965, Theorem 12.36).)

C.11 *Theorem* (The Hahn–Banach Theorem). If K_0, K_1 are disjoint, closed convex subsets of a Banach space B with K_0 compact, then there are a bounded linear functional Λ on B and $\gamma_0, \gamma_1 \in \mathbb{R}$ such that

$$\Lambda(x) < \gamma_0 < \gamma_1 < \Lambda(y) \quad \text{for } x \in K_0, y \in K_1.$$

(For a proof of (this form of) the Hahn–Banach Theorem see Rudin (1973, Theorem 3.4) or Dunford & Schwartz (1958, Theorem 8 of V.2.7).)

References

Alexandroff, P. & Urysohn, P. (1929). *Mémoire sur les espaces topologiques compacts.* Verhandlingen Koninklijke Akademie van Wetenschappen, Afdeeling Natuurkunde, Sectie 1. Te Amsterdam. **14**, 1–96.

Alster, K. & Pol, R. (1980). On function spaces of compact subspaces of Σ-products of the real line. *Fundamenta Math.* **107**, 135–43.

Amir, D. & Lindenstrauss, J. (1968). The structure of weakly compact sets in Banach spaces. *Ann. Math.* (2) **88**, 35–46.

Anderson, R. D. (1966). Hilbert space is homeomorphic to the countable infinite product of lines, *Bull. Amer. Math. Soc.* **72**, 515–19.

Argyros, S. (1977a). Applications of isomorphic embeddings of $\ell^1(\Gamma)$. *Notices Amer. Math. Soc.* **24**, A-539. (Abstract.)

Argyros, S. (1977b). *Isomorphic embeddings of* $\ell^1(\Gamma)$ *in classical Banach spaces.* Doctoral Dissertation, Athens University. (In Greek.)

Argyros, S. (1978a). Isomorphic embedding of ℓ^1_m into m-dimensional subspaces of $C(\Omega)$: the regular cardinal case. *Notices Amer. Math. Soc.* **25**, A-234. (Abstract.)

Argyros, S. (1978b). On some isomorphic embeddings of ℓ^1_m and Z_m. *Notices Amer. Math Soc.* **25**, A-588. (Abstract.)

Argyros, S. (1980). A combinatorial theorem for families of functions. *Colloq. Math. Soc. János Bolyai* **23**, 301–22. (Proc. Budapest Colloquium on Topology, August, 1978 ed. A. Császár.) North-Holland, Amsterdam.

Argyros, S. (1981a). Weak compactness in $L^1(\lambda)$ and injective Banach spaces. Proc. Int. Conf. in Banach Spaces, Kent State University, August, 1979. *Israel J. Math.* To appear.

Argyros, S. (1981b). On non-separable Banach spaces. *Trans. Amer. Math. Soc.* To appear.

Argyros, S. (1981c). Boolean algebras without free families. *Algebra Universalis.* To appear.

Argyros, S. (1981d). On compact spaces without strictly positive measures. *Pacific J. Math.* To appear.

Argyros, S. & Haydon, R. (1981). Continuous functions on certain Gleason spaces. *Quarterly J. Math., Oxford* (2). To appear.

Argyros, S. & Kalamidas, N. (1981). The $K_{\alpha,n}$ property on spaces with strictly positive measures. *Canadian J. Math.* To appear.

Argyros, S., Mercourakis, S. & Negrepontis, S. (1983). Analytic properties of Corson-compact spaces. To appear.

Argyros, S. & Negrepontis, S. (1977). Isomorphic embedding of $\ell^1(\Gamma)$. *Notices Amer. Math. Soc.* **24**, A-538–A-539. (Abstract.)

Argyros, S. & Negrepontis, S. (1980). Universal embeddings of ℓ^1_α into $C(X)$

and $L^\infty(\mu)$. *Colloq. Math. Soc. János Bolyai* **23**, 75–128 (Proc. Budapest Colloquium on Topology, August 1978, ed. A. Császár). North-Holland, Amsterdam.

Argyros, S. & Negrepontis, S. (1982). Chain conditions for spaces of measures. To appear.

Argyros, S. & Tsarpalias, A. (1978a). Calibers of compact spaces. *Notices Amer. Math. Soc.* **25**, A-234. (Abstract.)

Argyros, S. & Tsarpalias, A. (1978b). Isomorphic embedding of $\ell^1(\Gamma)$ into subspaces of $C(\Omega)$, III. *Notices Amer. Math. Soc.* **25**, A-588. (Abstract.)

Argyros, S. & Tsarpalias, A. (1981). Calibers of compact spaces. *Trans. Amer. Math. Soc.* To appear.

Arhangel'skiĭ, A. V. (1976). Some topological spaces that arise in functional analysis. *Uspehi Mat. Nauk* **31:5**, 17–32. (In Russian.) (English translation: *Russian Math. Surveys* **31:5** (1976), 14–30.)

Arhangel'skiĭ, A. V. (1978). The structure and classification of topological spaces and cardinal invariants. *Uspehi Mat. Nauk* **33:6**, 29–84. (In Russian.) (English translation: *Russian Math. Surveys* **33:6** (1978), 33–96.)

Bachmann, Heinz (1955). *Transfinite Zahlen.* Ergebnisse der Mathematik und ihrer Grenzgebiete, vol. 1. Springer-Verlag, Berlin. (2nd edn 1967.)

Bagley, R. W., Connell, E. H. & McKnight, J. D., Jr. (1958). On properties characterizing pseudo-compact spaces. *Proc. Amer. Math. Soc.* **9**, 500–6.

Baire, R. (1899). Sur les fonctions de variables réelles. *Annali di Mat.* **3**, 1–123.

Balcar, Bohuslav & Franěk, František (1981). Independent families in complete Boolean algebras. *Trans. Amer. Math. Soc.* To appear.

Bell, Murray G. (1982a). On the combinatorial principle P_c. *Fundamenta Math.* To appear.

Bell, Murray G. (1982b). σ-n-linked remainders of ω. To appear.

Benyamini, Y. & Starbird, T. (1976). Embedding weakly compact sets into Hilbert space. *Israel J. Math.* **23**, 137–41.

Benyamini, Y., Rudin, M. E. & Wage, M. (1977). Continuous images of weakly compact subsets of Banach spaces. *Pacific J. Math.* **70**, 309–24.

Berberian, Sterling K. (1965). *Measure and Integration.* Macmillan, New York.

Bishop, Errett (1959). A minimal boundary for function algebras. *Pacific J. Math.* **9**, 629–42.

Bloch, Gérard (1953). Sur les ensembles stationnaires de nombres ordinaux et les suites distinguées de fonctions régressives. *Comptes Rendus Acad. Sci. (Paris)* **236**, 265–8.

Bockstein, M. (1948). Un théorème de séparabilité pour les produits topologiques. *Fundamenta Math.* **35**, 242–6.

Broverman, S., Ginsburg, J., Kunen, K, & Tall, F. D. (1978). Topologies determined by σ-ideals on ω_1. *Canadian J. Math.* **30**, 1306–12.

Bukovský, Lev (1968). ∇-model and distributivity in Boolean algebras. *Commentationes Math. Univ. Carolinae* **9**, 595–612.

Całczyńska-Karłowicz, M. (1964). Theorem on families of finite sets. *Bull. Acad. Polon. Sci. Sér. Sci. Math. Astronom. Phys.* **12**, 87–9.

Cantor, Georg (1882). Über unendliche lineare Punktmannigfaltigkeiten. *Math. Annalen* **20**, 113–21.

Čech, Eduard (1937). On bicompact spaces. *Ann. Math.* **38**, 823–44.

Cohen, Paul J. (1966). *Set Theory and the Continuum Hypothesis.* Benjamin, New York.
Comfort, W. W. (1967). A nonpseudocompact product space whose finite subproducts are pseudocompact. *Math. Annalen* **170**, 41–4.
Comfort, W. W. (1971). A survey of cardinal invariants. *General Topology and its Applications* **1**, 163–99.
Comfort, W. W. (1979). Products of spaces with properties of pseudo-compactness type. *Topology Proceedings* **4**, 51–66.
Comfort, W. W. (1980). Chain conditions in topological products and powers. *Colloq. Math. Soc. János Bolyai* **23**, 61–73 (Proc. Budapest Colloquium on topology, August 1978, ed. A. Császár.) North-Holland, Amsterdam.
Comfort, W. W. & Negrepontis, S. (1972a). On families of large oscillation. *Fundamenta Math.* **75**, 275–90.
Comfort, W. W. & Negrepontis, S. (1972b). Continuous functions on products with strong topologies. In *General Topology and Its Relations to Modern Analysis and Algebra III. Proc. Prague Topological Symp.*, Prague, 1971, ed. J. Novák, pp. 89–92. Academia, Prague.
Comfort, W. W. & Negrepontis, S. (1974). *The Theory of Ultrafilters.* Grundlehren der math. Wissenschaften, vol. 211. Springer-Verlag, Berlin.
Comfort, W. W. & Negrepontis, S. (1975). *Continuous Pseudometrics.* Lecture Notes in Pure and Applied Mathematics, vol. 14. Marcel Dekker, New York.
Comfort, W. W. & Negrepontis, S. (1978a). Chain conditions in topological products and powers. *Abstracts Budapest Colloq. on Topology.* (Abstract.)
Comfort, W. W. & Negrepontis, S. (1978b). Compact-calibres of topological spaces. *Proc. Int. Cong. of Mathematicians in Helsinki*, 1978. (Abstract.)
Corson, H. H. (1959). Normality in subsets of product spaces. *Amer. J. Math.* **81**, 785–796.
Corson, H. H. (1961). The weak topology of a Banach space. *Trans. Amer. Math. Soc.* **101**, 1–15.
Corson, H. H. & Isbell, J. R. (1960). Some properties of strong uniformities. *Quarterly J. Math. Oxford (2)* **11**, 17–33.
Dashiell, F. K. & Lindenstrauss, J. (1973). Some examples concerning strictly convex norms on $C(K)$ spaces. *Israel J. Math.* **16**, 329–42.
Davies, Roy O. (1967). An intersection theorem of Erdős and Rado. *Proc. Cambridge Philos. Soc.* **63**, 995–6.
Day, Mahlon M. (1955). Strict convexity and smoothness of normed linear spaces. *Trans. Amer. Math. Soc.* **78**, 516–28.
Devlin, Keith J. & Johnsbråten, Håvard (1974). *The Souslin Problem.* Lecture Notes in Mathematics, vol. 405. Springer-Verlag, Berlin.
Dieudonné, Jean (1937). Sur les fonctions continues numériques définies dans un produit de deux espaces compacts. *Comptes Rendus Acad. Sci. (Paris)* **205**, 593–5.
Dieudonné, Jean (1939). Sur les espaces uniformes complets. *Annales de l'école normale supérieure*, 3me série, **56**, 276–91.
van Douwen, Eric K. (1977a). Density of compactifications. In *Set-Theoretic Topology*, ed. George M. Reed, pp. 97–110. Academic Press, New York.
van Douwen, Eric K. (1977b). The Pixley–Roy topology on spaces of subsets. In *Set-Theoretic Topology*, ed. George M. Reed, pp. 111–34. Academic Press, New York.

Dunford, Nelson & Schwartz, Jacob T. (1958). *Linear Operators, Part I: General Theory*. Pure and applied mathematics series, vol. VII. Interscience, New York.

Dushnik, Ben (1931). A note on transfinite ordinals. *Bull. Amer. Math. Soc.* **37**, 860–2.

Efimov, B. (1968). Extremally disconnected bicompacta of continuum π-weight. *Doklady Akad. Nauk SSSR* **183**, 511–14. (In Russian.) (English Translation: *Soviet Math. Doklady* **9**, 1404–7.)

Efimov, B. & Engelking, R. (1965). Remarks on dyadic spaces, II. *Colloq. Math.* **13**, 181–97.

Engelking, R. (1965). Cartesian products and dyadic spaces. *Fundamenta Math.* **57**, 287–304.

Engelking, R. (1966). On functions defined on Cartesian products. *Fundamenta Math.* **59**, 221–31.

Engelking, R. (1968). *Outline of General Topology*. North-Holland, Amsterdam and PWN (Polish Scientific Publishers), Warszawa.

Engelking, R. & Karłowicz, M. (1965). Some theorems of set theory and their topological consequences. *Fundamenta Math.* **57**, 275–85.

Erdős, P. (1942). Some set-theoretical properties of graphs. *Univ. Nac. Tucumán. Revista A* **3**, 363–7.

Erdős, P. (1950). Some remarks on set theory. *Proc. Amer. Math. Soc.* **1**, 127–41.

Erdős, P. (1962) Review of Kurepa (1959b, 1939). *Math. Rev.* **23A, 444**.

Erdős, P. & Fodor, G. (1957). Some remarks on set theory. VI. *Acta Sci. Math. (Szeged)* **18**, 243–60.

Erdős, P., Galvin, F. & Hajnal, A. (1975). On set-systems having large chromatic number and not containing prescribed subsystems. In *Colloq. Math. Soc. János Bolyai*, **10**, 425–513. Proc. Keszthely, Hungary Colloquium. *Infinite and Finite Sets*, Vol. I, July, 1973, ed. A. Hajnal, R. Rado & Vera T. Sós. North-Holland, Amsterdam.

Erdős, P., Hajnal, A., Máté, A. & Rado, R. (1982). *Combinatorial Set Theory: Partition Relations for Cardinals*. Akadémiai Kiadó, Publishing House of the Hungarian National Academy of Sciences, Budapest.

Erdős, P., Hajnal, A., & Rado, R. (1965). Partition relations for cardinal numbers. *Acta Math. Acad. Sci. Hungary* **16**, 93–196.

Erdős, P. & Rado, R. (1956). A partition calculus in set theory. *Bull. Amer. Math. Soc.* **62**, 427–89.

Erdős, P. & Rado, R. (1960). Intersection theorems for systems of sets. *J. London Math. Soc.* **35**, 85–90.

Erdős, P. & Rado, R. (1969). Intersection theorems for systems of sets (II). *J. London Math. Soc.* **44**, 467–79.

Erdős, P. & Specker, E. (1961). On a theorem in the theory of relations and a solution of a problem of Knaster. *Colloq. Math.* **8**, 19–21.

Erdős, P. & Tarski, Alfred (1943). On families of mutually exclusive sets. *Ann. Math. (2)* **44**, 315–29.

Fakhoury, Hicham (1976). Étude du noyau d'un operateur défini sur l'espace de suites bornées et applications. *Bull. Sci. Math.* **100**, 44–55.

Fleissner, William G. (1978). Some spaces related to topological inequalities proven by the Erdős–Rado theorem. *Proc. Amer. Math. Soc.* **71**, 313–20.

Fleissner, W. & Negrepontis, S. (1979). Haydon's counterexample with not CH. *Notices Amer. Math. Soc.* **26**, A-381. (Abstract.)

Fodor, G. (1955). Generalization of a theorem of Alexandroff and Urysohn. *Acta Sci. Math. (Szeged)* **16**, 204–6.

Fodor, G. (1956). Eine Bemerkung zur Theorie der regressiven funktionen. *Acta Sci. Math. (Szeged)* **17**, 139–42.

Franklin, S. P. (1971). On epi-reflective hulls. *General Topology and its Applications* **1**, 29–31.

Fremlin, D. H. (1980). On Gaifman's example. Note privately circulated from University of Essex, Colchester, England.

Freyd, Peter (1964). *Abelian Categories.* Harper and Row, New York.

Frolík, Zdeněk (1960). The topological product of two pseudocompact spaces. *Czech. Math. J.* **10**, 339–49.

Frolík, Zdeněk (1967a). Sums of ultrafilters. *Bull. Amer. Math. Soc.* **73**, 87–91.

Frolík, Zdeněk (1967b). On two problems of W. W. Comfort. *Commentationes Math. Univ. Carolinae* **8**, 139–44.

Frolík, Zdeněk (1968). Fixed points of maps of βN. *Bull. Amer. Math. Soc.* **74**, 187–91.

Fuhrken, Gebhard (1965). Languages with added quantifier 'there exist at least $\aleph_\alpha$'. In *Proc. Int. Symp. on the Theory of Models*, Berkeley, 1963, ed. Addison *et al.*, pp. 121–31. North-Holland, Amsterdam.

Gaifman, H. (1964). Concerning measures on Boolean algebras. *Pacific J. Math.* **14**, 61–73.

Galvin, Fred (1980). Chain conditions and products. *Fundamenta Math.* **108**, 33–48.

Galvin, Fred & Hajnal, Andras (1981). On the relative strength of chain conditions. To appear.

Gerlits, J. (1980a). On subspaces of dyadic compacta. *Studia Math. Hungarica.* To appear.

Gerlits, J. (1980b). Continuous functions on products of topological spaces. *Fundamenta Math.* To appear.

Gillman, Leonard & Jerison, Meyer (1960). *Rings of Continuous Functions.* Van Nostrand, Princeton,

Ginsburg, John & Woods, R. Grant (1976). On the cellularity of $\beta X\ X$. *Proc. Amer. Math. Soc.* **57**, 151–4.

Gleason, Andrew M. (1958). Projective topological spaces. *Illinois J. Math.* **2**, 482–9.

Glicksberg, Irving (1952). The representation of functionals by integrals. *Duke Math. J.* **19**, 253–61.

Glicksberg, Irving (1959). Stone–Čech compactifications of products. *Trans. Amer. Math. Soc.* **90**, 369–82.

Gődel, Kurt (1938). *The consistency of the axiom of choice and the generalized continuum hypothesis. Proc. Nat. Acad. Sci. USA* **24**, 556–7.

Gődel, Kurt (1940). *The Consistency of the Axiom of Choice and the Generalized Continuum Hypothesis with the Axiom of Set Theory. Ann. Math. Studies*, no. 3. Princeton University Press, Princeton, N.J. 1940.

Goodner, Dwight B.(1950). Projections in normed linear spaces. *Trans. Amer. Math. Soc.* **69**, 89–108.

Gul'ko, S. P. (1977). On properties of subsets of Σ-products. *Soviet Math. Doklady* **18**, 1438–42.

Gul'ko, S. P. (1978). On the properties of some function spaces. *Doklady Akad. Nauk SSSR* **243**, 839–42. (In Russian.) (English translation : *Soviet Math. Doklady* **19** (1978), 1420–4.)

Gul'ko, S. P. (1979). On the structure of spaces of continuous functions and on the hereditary paracompactness of these spaces. *Uspehi Mat. Nauk* **34:6** 33–40. (In Russian.) (English translation: *Russian Math. Surveys* **34:6** (1979), 36–44.)

Hagler, James (1975). On the structure of S and $C(S)$ for S dyadic. *Trans. Amer. Math. Soc.* **214**, 415–28.

Hajnal, A. (1960). Some results and problems on set theory. *Acta. Math. Acad. Sci. Hungary* **11**, 277–98.

Hajnal, A. (1961). Proof of a conjecture of S. Ruziewicz. *Fundamenta Math.* **50**, 123–8.

Hajnal, A. & Juhász, I. (1967). Discrete subspaces of topological spaces. *Proc. Nederland Akad.*, Series A **70**, 343–356 [*Indag. Math.* **29**, 343–56].

Hajnal, A. & Juhász, I. (1969). Discrete subspaces of topological spaces, II. *Proc. Nederland Akad.*, Series A **72**, 18–30 [*Indag. Math.* **31**, 18–30].

Hajnal, A. & Máté, Attila (1975). Set mappings, partitions and chromatic numbers. In *Logic Colloquium '73, Proc. Logic Colloq.*, Bristol, 1973, ed. H. E. Rose & J. C. Shepherdson, pp. 347–79. North-Holland, Amsterdam.

Halmos, Paul R. (1950). *Measure Theory*. Van Nostrand, Princeton.

Halmos, Paul R. (1963). *Lectures on Boolean Algebras*. Van Nostrand, Princeton.

Haratomi, Keitaro (1931). Über höherstufige Separabilität und Kompaktheit (erster Teil). *Jap. J. Math.* **8**, 113–141.

Hausdorff, Felix (1914). *Grundzüge der Mengenlehre*. Veit, Leipzig. (Reprinted (1949) Chelsea Publishing Company, New York.)

Hausdorff, Felix (1936). Über zwei Sätze von G. Fichtenholz und L. Kantorovitch. *Studia Math.* **6**, 18–19.

Haydon, Richard (1977). On Banach spaces which contain $\ell^1(\tau)$ and types of measures on compact spaces. *Israel J. Math.* **28**, 313–24.

Haydon, Richard (1978). On dual L^1-spaces and injective bidual Banach spaces. *Israel J. Math.* **31**, 142–52.

Haydon, Richard (1980). Non-separable Banach spaces. In *Functional Analysis: Surveys and Recent Results II*, ed. K.-D. Bierstedt & B. Fuchsteiner, pp.19–30. North-Holland, Amsterdam.

Hebert, D. J. & Lacey, H. Elton (1968). On supports of regular Borel measures. *Pacific J. Math.* **27**, 101–18.

Hedrlin, Z. (1966). An application of Ramsey's theorem to the topological products. *Bull. Acad. Polon. Sci. Sér. Sci. Math. Astronom. Phys.* **14**, 25–6.

Herink, Curtis (1977). *Some applications of iterated forcing*. Doctoral Dissertation, University of Wisconsin.

Herrlich, Horst (1967). $\mathfrak{C}$-kompakte Räume. *Math. Z.* **96**, 228–55.

Herrlich, Horst (1969). On the concept of reflections in general topology. In *Contributions to Extension Theory of Topological Structures, Proc. Symp.* Berlin, 1967, pp. 105–14. VEB Deutscher Verlag der Wissenschaften, Berlin.

Herrlich, H. & van der Slot, J. (1967). Properties which are closely related to compactness. *Indag. Math.* **29**, 524–9.
Hewitt, Edwin (1946). A remark on density characters. *Bull. Amer. Math. Soc.* **52**, 641–3.
Hewitt, Edwin (1948). Rings of real-valued continuous functions I. *Trans. Amer. Math. Soc.* **64**, 45–99.
Hewitt, Edwin (1953). Review of a paper of T. Shirota. *Math. Rev.* **14**, 395.
Hewitt, Edwin & Ross, Kenneth A. (1963). *Abstract Harmonic Analysis I. Grundlehren der math. Wissenschaften* vol. 115. Springer-Verlag, Berlin.
Hewitt, Edwin & Stromberg, Karl (1965). *Real and Abstract Analysis*. Springer-Verlag, Berlin.
Horn, Alfred & Tarski, Alfred (1948). Measures in Boolean algebras. *Trans. Amer. Math. Soc.* **64**, 467–97.
Hung, H. H. & Negrepontis, S. (1973). Spaces homeomorphic to $(2^{\alpha})_{\alpha}$. *Bull. Amer. Math. Soc.* **79**, 143–6.
Hung, H. H. & Negrepontis, S. (1974). Spaces homeomorphic to $(2^{\alpha})_{\alpha}$, II. *Trans. Amer. Math. Soc.* **188**, 1–30.
Hušek, Miroslav (1972). Products as reflections. *Commentationes Math. Univ. Carolinae* **13**, 783–800.
Hušek, Miroslav (1976). Continuous mappings on subspaces of products. *Istituto Nazionale di Alta Matematica Symposia Math.* **17**, 25–41.
Isbell, J. R. (1964). *Uniform Spaces. Mathematical Surveys no. 12.* American Mathematical Society, Providence, Rhode Island.
Jech, Thomas J. (1967). Non-provability of Souslin's hypothesis. *Commentationes Math. Univ. Carolinae* **8**, 291–305.
Jech, Thomas J. (1971). *Lectures in Set Theory with Particular Emphasis on the Method of Forcing*. Lecture Notes in Mathematics, vol. 217. Springer-Verlag, Berlin.
Jech, Thomas J. (1978). *Set Theory*. Academic Press, New York.
Jensen, R. Björn (1968). Souslin's hypothesis is incompatible with V = L. *Notices Amer. Math. Soc.* **15** (1968), 935. [Abstract.]
Jensen, R. Björn (1972). The fine structure of the constructible universe. *Ann. Math. Logic* **4**, 229–308.
Jourdain, Philip E. B. (1908). On infinite sums and products of cardinal numbers. *Quarterly J. Pure and Applied Math.* **39**, 375–84.
Juhász, I. (1970). Martin's axiom solves Ponomarev's problem. *Bull. Acad. Polon. Sci. Sér. Sci. Math. Astronom. Phys.* **18**, 71–4.
Juhász, I. (1971). *Cardinal Functions in Topology*. Mathematical Centre Tracts 34. Mathematisch Centrum, Amsterdam.
Juhász, I. (1976). A generalization of nets and bases. *Periodica Math. Hungarica* **7**, 183–92.
Juhász, I. (1977). Consistency results in topology. In *Handbook of Mathematical Logic*, ed. Jon Barwise. *Studies in Logic and the Foundations of Mathematics*, vol. 90, pp. 503–22. North-Holland, Amsterdam.
Juhász, I. (1980). *Cardinal Functions in Topology–Ten Years Later*. Mathematical Centre Tracts No. 123. Mathematisch Centrum, Amsterdam.
Juhász, I. & Weiss, William (1978). Martin's axiom and normality. *General Topology and its Applications* **9**, 263–74.

Kalamidas, N. (1981). Chain conditions and their relation to Banach spaces of the form $C(X)$. Doctoral Dissertation, Athens University. (In Greek.)
Keisler, H. Jerome (1971). *Model Theory for Infinitary Logic.* North-Holland, Amsterdam.
Kelley, John L. (1954). Banach spaces with the extension property. *Trans. Amer. Math. Soc.* **72**, 323–6.
Kelley, John L. (1955). *General Topology*. Van Nostrand, Princeton.
Kelley, John L. (1959). Measures on Boolean algebras. *Pacific J. Math.* **9**, 1165–77.
Kemperman, J. H. B. & Maharam, Dorothy (1970). $\mathbb{R}^c$ is not almost Lindelof. *Proc. Amer. Math. Soc.* **24**, 772–4.
Kennison, J. F. (1965). Reflective functors in general topology and elsewhere. *Trans. Amer. Math. Soc.* **118**, 303–15.
Kister, J. M. (1962). Uniform continuity and compactness in topological groups. *Proc. Amer. Math. Soc.* **13**, 37–40.
Klee, Victor L., Jr. (1953). Convex bodies and periodic homeomorphisms in Hilbert space. *Trans. Amer. Math. Soc.* **74**, 10–43.
Klee, Victor L., Jr. (1956). A note on topological properties of normed linear spaces. *Proc. Amer. Math. Soc.* **7**, 673–4.
Knaster, B. (1945). Sur une propriété caractéristique de l'ensemble des nombres réels. *Mat. Sbornik* **16** (58), 281–8.
König, Denes (1927). Über eine Schlussweise aus dem Endlichen ins Unendliche. *Acta Litt. Sci. Szeged* **3**, 121–30.
König, J. (1904). Zum Kontinuumproblem. *Math. Annalen* **60**, 177–80.
Kunen, Kenneth (1968). *Inaccessibility properties of cardinals.* Doctoral Dissertation, Stanford University.
Kunen, Kenneth (1972). Ultrafilters and independent sets. *Trans. Amer. Math. Soc.* **172**, 299–306.
Kunen, Kenneth (1978). Saturated ideals. *J. Symbolic Logic* **43**, 65–76.
Kunen, Kenneth (1980). Weak P-points in N^*. *Colloq. Math. Soc. János Bolyai* **23**, 741–9. (Proc. Budapest Colloquium on Topology, August, 1978, ed. A. Császár.) North-Holland, Amsterdam.
Kunen, Kenneth (1981). A compact L-space under CH. *Topology and its Applications* **12**, 283–7.
Kunen, Kenneth & Tall, Franklin D. (1979). Between Martin's axiom and Souslin's hypothesis. *Fundamenta Math.* **52**, 173–81.
Kurepa, Georges (Gjuro Kurepa) (1935). *Ensembles ordonnès et ramifiès.* Doctoral Dissertation, University of Paris. (*Publ. Math. Univ. Beograd* **4**, 1–138.)
Kurepa, Georges (1939). Sur la puissance des ensembles partiellement ordonnés. *Comptes Rendus Soc. Sci. (Varsovie)* **32**, 62–7.
Kurepa, Georges (1950). La condition de Souslin et une propriété caractéristique des nombres réels. *Comptes Rendus Acad. Sci. (Paris)* **231**, 1113–14.
Kurepa, Georges (1952). Sur une propriété caractéristique du continu linéaire et le problème de Suslin. *Acad. Serbe Sci. Publ. Inst. Math.* **4**, 97–108.
Kurepa, Georges (1959a). On the cardinal number of ordered sets and of symmetrical structures in dependence on the cardinal numbers of its chains and antichains (Serbo-Croatian summary). *Glasnik Mat.-Fiz. Astronom. Društvo Mat. Fiz. Hrvatske Ser. II*, **14**, 183–203.
Kurepa, Georges (1959b). Sur la puissance des ensembles partiellement ordonnés

(Serbo-Croatian summary). *Glasnik Mat.-Fiz. Astronom. Drustvo Mat. Fiz. Hrvatske Ser. II*, **14**, 205–11.

Kurepa, Georges (1962). The cartesian multiplication and the cellularity numbers. *Publ. Inst. Math. (Beograd)* (N. S.) **2**, 121–39.

Lázár, D. (1936). On a problem in the theory of aggregates. *Compositio Math.* **3**, 304.

Losert, V. (1979). On the existence of uniformly distributed sequences in compact topological spaces II. *Monatshefte für Math.* **87**, 247–60.

Lusin, N. (1914). Sur une problème de M. Baire. *Comptes Rendus Acad. Sci. (Paris)* **158**, 1258–61.

Mägerl, G. & Namioka, I. (1980). Intersection numbers and weak* separability of spaces of measures. *Math. Annalen* **249**, 273–9.

Maharam, Dorothy (1942). On homogeneous measure algebras. *Proc. Nat. Acad. Sci. USA* **28**, 108–11.

Malyhin, V. I. & Šapirovskiĭ, B. È. (1973). Martin's axiom and properties of topological spaces. *Doklady Akad. Nauk SSSR* **213**, 532–5. (In Russian.) (English translation: *Soviet Math. Doklady* **14** (1973), 1746–51.)

Marczewski, Edward (1934). Remarques sur les fonctions complètement additives d'ensemble et sur les ensembles jouissant de la propriété de Baire. *Fundamenta Math.* **22**, 303–11.

Marczewski, Edward (1941). Remarque sur les products cartésiens d'espaces topologiques. *Comptes Rendus (Doklady) Acad. Sci. URSS* **31**, 525–7.

Marczewski, Edward (1945). Sur deux propriétés des classes d'ensembles. *Fundamenta Math.* **33**, 303–7.

Marczewski, Edward (1947). Séparabilité et multiplication cartésienne des espaces topologiques. *Fundamenta Math.* **34**, 127–43.

Mardesič, S. & Papič, P. (1955). Sur les espaces dont tout transformation réelle continue est bornée. *Hrvatsko Prirod. Društvo. Glasnik Mat. – Fiz. Astron.* ser. II, **10**, 225–232.

Marek, W. (1964). On families of sets. *Bull. Acad. Polon. Sci. Sér. Sci. Math. Astronom. Phys.* **12** (1964), 443–448.

Martin, D. A. & Solovay, R. M. (1970). Internal Cohen extensions. *Annals of Math. Logic* **2**, 143–78.

Marty, Roger Henry (1969). Mazur theorem and m-adic spaces. Doctoral Dissertation, Pennsylvania State University.

Mazur, S. (1952). On continuous mappings on Cartesian products. *Fundamenta Math.* **39**, 229–38.

Mibu, Y. (1944). On Baire functions on infinite product spaces. *Proc. Imperial Acad. Tokyo* **20**, 661–3.

Michael, E. (1962). A note on intersections. *Proc. Amer. Math. Soc.* **13**, 281–3; 1000.

Michael, E. & Rudin, M. E. (1977). A note on Eberlein compacts. *Pacific J. Math.* **72**, 487–95.

Miščenko, A. (A. Mishchenko), (1966a). Several theorems on products of topological spaces. *Fundamenta Math.* **58**, 259–84. (In Russian.)

Miščenko, A. (1966b). Uniformly closed mappings. *Fundamenta Math.* **58**, 285–308. (In Russian.)

Mostowski, A. (1969). *Constructible Sets with Applications. Studies in Logic and the Foundations of Mathematics*. PWN (Polish Scientific Publishers), Warwaw; and North-Holland, Amsterdam.

Mycielski, Jan (1964). α-incompactness of $\mathbb{N}^\alpha$. *Bull. Acad. Polon. Sci. Sér. Sci. Math. Astronom. Phys.* **12**, 437–8.

Nachbin, Leopoldo (1950). A theorem of the Hahn–Banach type for linear transformations. *Trans. Amer. Math. Soc.* **68**, 28–46.

Nachbin, Leopoldo (1952). On the continuity of positive linear transformations. In *Proc. 1950 Int. Congr. of Mathematicians*, pp. 464–5. American Mathematical Society, Providence, Rhode Island.

Nachbin, Leopoldo (1954). Topological vector spaces of continuous functions. *Proc. Nat. Acad. Sci. USA* **40**, 471–4.

Nagata, J. (1972). A survey of the theory of generalized metric spaces. In *General Topology and its Relations to Modern Analysis and Algebra III. Proc. Prague Topological Symp.*, pp. 321–31., ed. J. Novák. Prague, 1971.

Namioka, I. (1974). Separate continuity and joint continuity. *Pacific J. Math.* **51**, 515–31.

Negrepontis, S. (1969). The Stone space of the saturated Boolean algebras. *Trans. Amer. Math. Soc.* **141**, 515–27.

Negrepontis, S. (1978) Infinitary combinatorics in general topology and functional analysis. *Abstracts Budapest Colloq. on Topology*. (Abstract.)

Negrepontis, S. (1980). Combinatorial techniques in functional analysis. In *Surveys in General Topology*, ed. G. M. Reed, pp. 337–66. Academic Press, New York.

Negrepontis, S. (1981). Generalized Gaifman spaces. To appear.

Negrepontis, S. & Tsarpalias, A. (1981). A non-linear version of the Amir–Lindenstrauss method. *Israel J. Math.* **38**, 82–94.

Neumer, Walter (1951). Verallgemeinerung eines Satzes von Alexandroff and Urysohn. *Math. Z.* **54**, 254–61.

Noble, Norman (1967). A generalization of a theorem of A. H. Stone. *Archiv der Math. (Basel)* **18**, 394–5.

Noble, Norman (1969a). Products with closed projections. *Trans. Amer. Math. Soc.* **140**, 381–91.

Noble, Norman (1969b). Countably compact and pseudocompact products. *Czech. Math. J.* **19** (94), 390–7.

Noble, Norman (1972). *C*-embedded subsets of products. *Proc. Amer. Math. Soc.* **31**, 13–14.

Noble, N. & Ulmer, Milton (1972). Factoring functions on Cartesian products. *Trans. Amer. Math. Soc.* **163**, 329–40.

Novák, J. (1950). A paradoxical theorem. *Fundamenta Math.* **37**, 77–83.

O'Callaghan, Liam (1975). Topological endohomeomorphisms and compactness properties of products and generalized Σ-products. Doctoral Dissertation, Wesleyan University.

Oxtoby, John C. (1961). Cartesian products of Baire spaces. *Fundamenta Math.* **49**, 157–66.

Oxtoby, John C. (1971). *Measure and Category*. Graduate Texts in Mathematics. Springer-Verlag, Berlin.

Pelczynski, Aleksander (1958). On the isomorphism of the spaces *m* and *M*. *Studia Math.* **19**, 695–6.

Pełczynski, A. (1968). On Banach spaces containing $L^1(\mu)$. *Studia Math.* **30**, 238–46.

Pełczynski, Aleksander (1979) (in collaboration with Czesław Bessaga). Some aspects of the present theory of Banach spaces. In *Oeuvres de Stefan Banach*, vol. II, pp. 221–302. PWN–Editions Scientifiques de Pologne, Warszawa.

Piccard, Sophie (1937). Sur un problème de M. Ruziewicz de la théorie des relations. *Fundamenta Math.* **29**, 5–9.

Pixley, Carl & Roy, Prabir (1969). Uncompletable Moore spaces. *Proc. Auburn Topology Conf.*, pp. 75–85. Auburn University, Alabama.

Pol, Roman (1980). On a question of H. H. Corson and some related problems. *Fundamenta Math.* **109**, 143–54.

Pondiczery, E. S. (1944). Power problems in abstract spaces. *Duke Math. J.* **11**, 835–7.

Pospíšil, Bedrich (1937). Remark on bicompact spaces. *Ann. Math. (2)* **38**, 845–6.

Rainwater, John (1959). A note on projective resolutions. *Proc. Amer. Math. Soc.* **10**, 734–5.

Ramsey, F. P. (1930). On a problem of formal logic. *Proc. London Math. Soc.* (*2*) **30**, 264–86.

Rosenthal, Haskell P. (1970a). On injective Banach spaces and the spaces $L^\infty(\mu)$ for finite measures μ. *Acta Math.* **124**, 205–48.

Rosenthal, Haskell P. (1970b). On relatively disjoint families of measures, with some applications to Banach space theory. *Studia Math.* **37**, 13–36.

Rosenthal, Haskell P. (1974a). The heredity problem for weakly compactly generated Banach spaces. *Compositio Math.* **28**, 83–111.

Rosenthal, Haskell P. (1974b). A characterization of Banach spaces containing ℓ^1. *Proc. Nat. Acad. Sci. USA* **71**, 2411–13.

Rosenthal, Haskell P. (1975). The Banach space $C(K)$ and $L^p(\mu)$. *Bull. Amer. Math. Soc.* **81**, 763–81.

Rosenthal, Haskell P. (1978). Some recent discoveries in the isomorphic theory of Banach spaces. *Bull. Amer. Math. Soc.* **84**, 803–31.

Ross, K. A. & Stone, A. H. (1964). Products of separable spaces. *Amer. Math. Monthly* **71**, 398–403.

Rubin, M. & Shelah, S. (1981). On the expressibility of Magidor Malitz quantifiers. *J. Symbolic Logic*. To appear.

Rudin, Mary Ellen (1975). *Lectures on Set Theoretic Topology*. Regional Conference Series in Mathematics Number 23. American Mathematical Society, Providence, Rhode Island.

Rudin, Mary Ellen (1977). Martin's axiom. In *Handbook of Mathematical Logic*, ed. Jon Barwise. *Studies in Logic and the Foundations of Mathematics*, vol. 90, pp. 491–501. North-Holland, Amsterdam.

Rudin, Walter (1973). *Functional Analysis*. McGraw-Hill, New York.

Šapirovskiĭ, B. Ė. (1975). On embedding extremally disconnected spaces in compact Hausdorff spaces. b-points and weight of pointwise normal spaces. *Doklady Akad. Nauk SSSR* **223**, 799–800. (In Russian.) (English translation: *Soviet Math. Doklady* **16**, 999–1004.)

Šapirovskiĭ, B. Ė. (1977). On decomposition of a perfect mapping into an irreducible mapping and a retraction. *Abstracts 7th All-Union Topological Conference*, p. 205. Minsk. [Abstract.]

Šapirovskiĭ, B. È. (1979). Special types of embeddings in Tychonoff cubes. Subspaces of Σ-products and cardinal invariants. *Colloq. Math. Soc. János Bolyai* **23**, 1055–86. (Proc. Budapest Colloquium on Topology, August, 1978, ed. A. Császár). North-Holland, Amsterdam.

Semadeni, Zbigniew (1971). *Banach Spaces of Continuous Functions*. Monografie Matematyczne vol. 55. PWN (Polish Scientific Publishers), Warsaw.

Shanin, N. A. (1946a). A theorem from the general theory of sets. *Comptes Rendus (Doklady) Acad. Sci. URSS* **53**, 399–400.

Shanin, N. A. (1946b). On intersection of open subsets in the product of topological spaces. *Comptes Rendus (Doklady) Acad. Sci. URSS* **53**, 499–501.

Shanin, N. A. (1946c). On the product of topological spaces. *Comptes Rendus (Doklady) Acad. Sci. URSS* **53**, 591–3.

Shanin, N. A. (1948). On the product of topological spaces. *Trudy Mat. Inst. Akad. Nauk SSSR* **24**, 112 pages. (In Russian.)

Shelah, Saharon (1977). Remarks on cardinal invariants in topology. *General Topology and its Applications* **7**, 251–9.

Shelah, Saharon (1980). Remarks on Boolean algebras. *Algebra Universalis*. To appear.

Shelah, Saharon & Rudin, M. E. (1978). Unordered types of ultrafilters. *Topology Proceedings* **3**, 199–204.

Shoenfield, J. R. (1975). Martin's axion. *American Math. Monthly* **82**, 610–17.

Sikorski, Roman (1964). *Boolean Algebras*. Ergebnisse der Mathematik und ihrer Grenzgebiete, vol. 25, 2nd edn. Springer-Verlag, Berlin.

Simpson, Stephen G. (1970). Model-theoretic proof of a partition theorem. *Notices Amer. Math. Soc.* **17**, 964. (Abstract.)

van der Slot, J. (1966). *Universal topological properties*. ZW 1966–011. Math. Centrum, Amsterdam.

van der Slot, J. (1972). Compact sets in non-metrizable product spaces. *General Topology and its Applications* **2**, 61–5.

Solovay, R. M. & Tennenbaum, S. (1971). Iterated Cohen extensions and Souslin's problem. *Ann. Math. (2)* **94**, 201–45.

Sorgenfrey, R. H. (1947). On the topological product of paracompact spaces. *Bull. Amer. Math. Soc.* **53**, 631–2.

Souslin, M. (1920). Problème 3. *Fundamenta Math.* **1**, 223.

Spivak, Michael (1967). *Calculus*. W. A. Benjamin, New York.

Štěpánek, Petr & Vopěnka, Petr (1967). Decomposition of metric spaces into nowhere dense sets. *Commentationes Math. Univ. Carolinae* **8**, 387–404; 567–8.

Stone, A. H. (1948). Paracompactness and product spaces. *Bull. Amer. Math. Soc.* **54**, 977–82.

Stone, M. H. (1934). Boolean algebras and their applications to topology. *Proc. Nat. Acad. Sci. USA* **20**, 197–202.

Stone, M. H. (1936). The representation theorem for Boolean algebras. *Trans. Amer. Math. Soc.* **40**, 37–111.

Stone, M. H. (1937). Applications of the theory of Boolean rings to general topology. *Trans. Amer. Math. Soc.* **41**, 375–481.

Szpilrajn, Edward: see Marczewski, Edward.

Talagrand, Michel (1975). Sur une conjecture de H. H. Corson. *Bull. Sci. Math. (Série 2)* **99**, 211–2.

Talagrand, Michel (1977). Espaces de Banach faiblement $\mathcal{K}$-analytiques. *Comptes Rendus Acad. Sci. (Paris) Sér. A–B* **284**, A745–A748.

Talagrand, Michel (1978). En général il n'existe pas de relèvement linéaire borélien fort. *Comptes Rendus Acad. Sci. (Paris) Sér. A* **287**, 633–4.

Talagrand, Michel (1979a). Espaces de Banach faiblement $\mathcal{K}$-analytiques. *Annals of Math.* **110**, 407–38.

Talagrand, Michel (1979b). Deux generalizations d'un théorème de I. Namioka. *Pacific J. Math.* **81**, 239–51.

Talagrand, M. (1980a). Non-existence de certaines sections measurables et contre-examples en théorie du relèvement. In *Lecture Notes in Mathematics* vol. 794. *Measure Theory Oberwolfach 1979*, ed. D. Kölzow, pp. 166–75. (Proceedings of the Conference held at Oberwolfach, Germany, July 1–7, 1979.) Springer-Verlag, Berlin and Heidelberg.

Talagrand, M. (1980b). Séparabilité vague dans l'éspace des measures sur un compact. *Israel J. Math.* **37**, 171–80.

Tall, Franklin D. (1974). The countable chain condition versus separability–applications of Martin's axiom. *General Topology and its Applications* **4** (1974), 315–39.

Tall, Franklin D. (1976). The density topology. *Pacific J. Math.* **62**, 275–84.

Tall, Franklin D. (1977). First countable spaces with calibre $\aleph_1$ may or may not be separable. In *Set-Theoretic Topology*, ed. George M. Reed, pp. 353–8. Academic Press, New York.

Tall, Franklin D. (1979). Applications of a generalized Martin's axiom. *Comptes Rendus Math. Rep. Acad. Sci. (Canada)* **1**, 103–6.

Tarski, Alfred (1925). Quelques théorèmes sur les alephs. *Fundamenta Math.* **7**, 1–14.

Tennenbaum, S. (1968). Souslin's problem. *Proc. Nat. Acad. Sci. USA* **59**, 60–3.

Troyanski, S. L. (1971). On locally uniformly convex and differentiable norms in certain non-separable Banach spaces. *Studia Math.* **37**, 173–180.

Tsarpalias, A. (1977). A combinatorial theorem with applications in set-theory and functional analysis. *Notices Amer. Math. Soc.* **24**, A-538. (Abstract.)

Tsarpalias, A. (1978a). Infinitary combinatorial methods and weak compactness in the theory of Banach spaces. Doctoral Dissertation, Athens University. (In Greek.)

Tsarpalias, A. (1978b). Isomorphic embedding of ℓ^1_m into m-dimensional subspaces of $C(\Omega)$: the singular cardinal case. *Notices Amer. Math. Soc.* **25**, A-234. (Abstract.)

Tsarpalias, A. (1980). A combinatorial theorem. *J. Combinatorial Theory*, series A. To appear.

Tulcea, A. Ionescu (1974). On pointwise convergence, compactness, and equicontinuity. II. *Advances in Math.* **12**, 171–7.

Tychonoff, A. (1929). Über die topologische Erweiterung von Räumen. *Math. Annalen* **102**, 544–61.

Ulmer, Milton Don (1970). Continuous Functions on Product Spaces. Doctoral Dissertation, Wesleyan University.

Vašak, L. (1980). On one generalization of weakly compactly generated Banach spaces. *Studia Math.* **70**. To appear.

Verbeek, A. (1970). *A note on intersections*. Math. Centrum Amsterdam (Afd. Zuivere Wisk.) ZW 05 70. (In Dutch.)

Vopěnka, Petr (1965). Properties of ∇-model. *Bull. Acad. Polon. Sci. Sér. Sci. Math. Astronom. Phys.* **13**, 441–4.

Wage, Michael L. (1979). Almost disjoint sets and Martin's axiom. *J. Symbolic Logic* **44**, 313–18.

Wage, Michael. L. (1980). Weakly compact subsets of Banach spaces. In *Surveys in General Topology*, ed. George M. Reed, pp. 479–94. Academic Press, New York.

Walker, Russell C. (1974). *The Stone–Čech Compactification*. Ergebnisse der Math. und ihrer Grenzgebiete, vol. 83. Springer-Verlag, Berlin.

Weil, André (1937). *Sur les espaces à structure uniforme et sur la topologie générale*. Actualités Scientifiques et Industrielle No. 551. Hermann & Cie., Paris.

Williams, Neil H. (1977). *Combinatorial Set Theory*. Studies in Logic and the Foundations of Mathematics, vol. 91. North-Holland, Amsterdam.

Zachariades, Th. (1981). Applications of combinatorics to the embeddings of l_m^1 Doctoral Dissertation, Athens University. (In Greek.)

Zermelo, E. (1908). Untersuchungen über die Grundlagen der Mengenlehre I. *Math. Annalen* **65**, 261–81.

Subject Index

Index of Symbols